From Additive Manufacturing to 3D/4D Printing 3

Series Editor
Jean-Charles Pomerol

From Additive Manufacturing to 3D/4D Printing 3

Breakthrough Innovations: Programmable Material, 4D Printing and Bio-printing

Jean-Claude André

iSTE

WILEY

First published 2018 in Great Britain and the United States by ISTE Ltd and John Wiley & Sons, Inc.

ISTE Ltd
27-37 St George's Road
London SW19 4EU
UK

www.iste.co.uk

John Wiley & Sons, Inc.
111 River Street
Hoboken, NJ 07030
USA

www.wiley.com

Library of Congress Control Number: 2017954665

British Library Cataloguing-in-Publication Data
A CIP record for this book is available from the British Library
ISBN 978-1-78630-232-8

Contents

Acknowledgments

Sincere thanks to 3A, BeAm, 3DCeram, 3D Systems, Fives, Nanoscribe, Poietis, Prodways, XtreeE, Alex (alias Alexandre Martel, co-founder of 3D Natives.com) and Laurent Dupont, head of the Lorraine Fab Living Lab® for their effective cooperation, and particularly for graciously providing certain images.

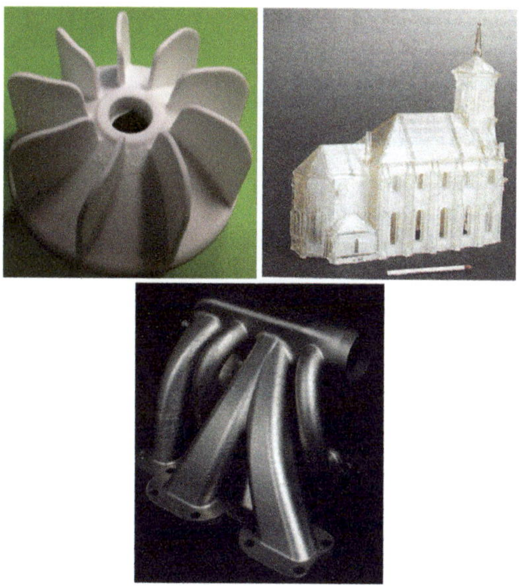

Prototype part (3DCeram, 2017), reproduction of the Église de Bonsecours in Nancy (LRGP, 1994) and metallic part (3A – Applications Additives Avancées, 2017)

Foreword

The evocative expression "3D printing" has been overtaken in everyday speech by the expression generally preferred by scientists and engineers, "additive manufacturing". In both cases, it is a matter of manufacturing objects in successive layers, and soon every workshop and every school will have a 3D printer and engage in additive manufacturing. Self-service workshops known as fab-labs already offer users the possibility to create their own objects. However, the adventure is not over, as "4D" is coming up over the horizon with materials that evolve over time, not to mention "bio-printing", which aims to create organs to be used to repair the living. Furthermore, the 3D printing of tomorrow, which will be performed without layers, threatens to make the term "additive manufacturing" obsolete, thereby making it possible to return to the initial concept of 3D printing. Whatever the case may be, we are faced with not only a very active and booming world, but also a complex world that calls on numerous skills in physics, engineering, chemistry of materials and mechanics with a resolutely multidisciplinary and convergent approach.

To understand the origin of the ideas in additive manufacturing/3D printing, learn about the current state of what is known and explore the developments to come, what could be better than to ask one of the inventors of the technology and one of the first French patent holders in the field, Jean-Claude André, to share his knowledge with us? This led to the idea of this 3-volume edition that I am pleased to present; a work that is both erudite and prospective, as its intention is to start at the genesis of the ideas that led to additive manufacturing to anticipate the impact and future of still "additive" technologies and, beyond this, to encourage reflection on the interactions between science and society of today and tomorrow.

If the first patents date back to 1984, an era where lasers, photo-materials and computer-aided design had already been mastered, was the idea of additive manufacturing completely disruptive as would be said today. What was creative was to put all of this knowledge together to come to something entirely new. Nevertheless, approval for the concept of additive manufacturing came rather quickly. It is on this basis that other additive methodologies, currently many of them with very specific niches, could be developed. These range from prototype and industrial parts to art, variable spatial scales – from the decameter to the nanometer –, from the inert to the living, from industrial organizations to very delocalized forms of manufacturing, etc.

On the basis of these works with varied applicative and societal spectrums, some of which are in the process of becoming stabilized, others to be invented, the principles of additive manufacturing can serve as an example, even as a "laboratory" to better understand how the interactions between research and society can (and must) develop, whether this is through new scientific concepts and the associated concepts of creativity, interdisciplinary scientific and technological operations, the popularization of public research, links with society in terms of the creation of new markets and jobs, and also forms of responsibility and ethics.

Throughout these three volumes, the author would like to invite you to reflect on the circuits between the applications that pose new scientific questions and prior research which opens the door to new applications or new products. The more we progress in the field of new niches, the more previously unasked scientific questions are considered, questions whose answers (if they exist) are supported and encouraged by public authorities and industry, which are gaining awareness of an immense industrial and/or medical market, as is the case for bio-printing. From dream to reality, scientists are often in the position to anticipate the length of the path; however, a dynamic is created. This leads to cultural changes and changes in practices, particularly concerning the importance of creativity, sharing enthusiasm for research, openness with others, the multiplying (and sometimes inhibiting) effect of public actors, on the one hand, and the economic world, on the other, as this work illustrates wonderfully.

This saga of additive manufacturing, told by one of its inventors, teaches us that creativity alone does not suffice; it is necessary to have a good dose of perseverance as well, and it is, of course, necessary to keep moving after the first failures. In addition, this shows us that sometimes the research structures and the environment are not entirely receptive to innovation, even when success comes relatively quickly.

Jean-Claude André also explains with great enthusiasm how we give shape to an idea to feed our intuition, which in turn increases creativity. On the whole, these three volumes provide a wealth of information on additive manufacturing, and additionally, they illustrate and encourage veritable reflection on the task of a researcher and research structures, as well as the role of creativity in research, and finally, they invite us to rethink and reinforce the relations between science and society.

Jean-Charles POMEROL
President of the Incubateur AGORANOV
and the ISTE Editions Scientific Committee

Preface

"We have too often forgotten that specialists are created from amateurs, just as soldiers are made from civilians". [LAT 07]

"In France, strangely enough, it is not those used to sailing the seas, the specialists of the real and tangible, who are asked for advice guiding the flagship, but the members of a caste who stay at port and who, for the most part, have only purely theoretic knowledge of the sea". [BEI 12]

"Technology has taken on a new breadth and organization. Here, I am searching for its specific structure, and I have noticed that it exists as a system, in other words, as an organized whole". [ELL 04]

"Those in the organization who have ideas to do things otherwise or better are divided into two categories: those who do not dare and those who dare. Those who do not dare understand very well the risks and the importance of new ideas, but they are paralyzed by risk taking and the fear of displeasing. Having never tried anything, they have not known failure and are thus unharmed by reproach […], they are quitters. Those who dare, the innovators, move forward by challenging conventional ideas, organizations, and sometimes procedures. They stir up fears and a lack of understanding and are truly criticized…". [PHI 12]

"Science has largely renounced an interdisciplinary vision allowing the merits of different results to be faced". [THO 83]

"Theory is when everything is known and nothing works. Practice is when everything works and no one knows why. Here, we have united theory and practices: nothing works... and no one knows why!". [EIN 07]

"These creatures of man [machines] are exacting. They are now reacting on their creators, making them like themselves. They want well-trained humans; they are gradually wiping out the differences between men, fitting them into their own orderly functioning, into the uniformity of their own regimes". [VAL 57]

"Speaking of discipline is designating the scientific activity as a particular form of the division of labor in the social world". [FAB 06]

"The imagination is brilliant in that it produces images that enlarge reality and really invent it". [GUÉ 15]

"In cultural terms, no enterprise is built with dreams alone and none without. Action, if it is to be successful, is by necessity guided by practical circumstances. But the goal of any action is defined, implicitly or explicitly, by the deep nature of the human being, his dreams, his vision of life, his culture. The dynamics of life, the challenge of risk and uncertainty, today require from us a new creative effort leading to the reconstruction and to the re-conquest of the notion of progress, which the philosophies and the ideologies of certainty have shuttered almost to the point of destruction". [GIA 90]

"Researching is inventing the world, it is setting new rules of functioning for an ephemeral world. Not like the tyrants who also invent a new world for themselves, but impose it upon others. The researcher does not recreate the world, but rather unravels it to make it. He/She imagines one, then compares it with the real world to clarify it and not to exhaust it. Researching is an endless quest". [ROS 01]

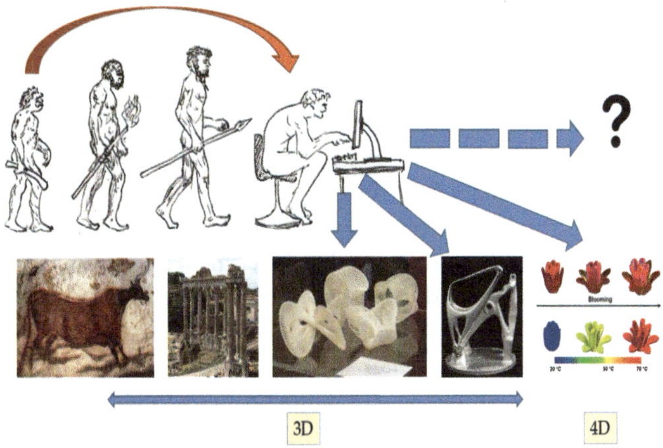

Figure 1. *From additive manufacturing to 3D/4D printing*

This book (in three volumes) is the result of a demand that has been repeated countless times for different reasons, notably among these, of the oversight and the reminder of the oversight to cite a French school that in 1984 succeeded to patent the first additive manufacturing process, stereolithography, several weeks before the Americans (who were working on the same subject, without either party knowing it). However, at the same time, thirty or so years later, it is a history lesson that can be told about a process concept, tossed out in France, without any malice of course, by "clairvoyant hierarchists", the explosion of the research team who felt their future was blocked and an American technical-economic development which has today led to several books and more than 50,000 scientific publications on additive manufacturing, because consequent applicative markets exist with profitable enterprises (and also because there is an immense attraction field around this subject that conditions the actions of a great number of researchers).

So why have we entitled these three volumes "From Additive Manufacturing to 3D/4D Printing"? First, it was about locally bringing material and/or energy to perform a transformation (e.g. from a powder to a solid or from a liquid to a solid). The expression "additive" then takes on its true meaning. But for a short time now, researchers have been developing (or working on) new processes that allow this change to be avoided through the additions mentioned at the start of this paragraph. It thus becomes possible to create an object in one go. Moreover, the use of so-called "smart" materials authorizes the introduction of a complementary parameter, i.e. time or functionality. The 4D aspect is thereby introduced.

The first volume on additive manufacturing is strongly linked to the existence of an effective economic market, one that is already significant, stemming from technological research in the engineering sciences connected to an essential component, that of materials (and of manipulating them to prepare them for manufacturing). It will take several decades for 3D technology to emerge and find its place as a robust technology for manufacturing objects in quite diverse domains. This situation, linked from the start to a strong attractiveness on the part of industrial R and R&D services, has allowed for "field" experimentation with competent users who are more and more demanding in terms of manufacturing qualities (without seeking in this preface to define what this quality, a true portmanteau, represents). Mastery by users, on the one hand, and competition between the bearers of knowledge pertaining to different 3D printing knowledge, on the other, are translated into new demands to be satisfied. In this framework, this demand has in fact made up one of the driving forces of incremental research, a "technology pull" described in Volume 2 (at least as much as is known (or published)).

A solution is good if and only if the concept, its demonstration with the right people, a culture of industrial innovation, and time and finances effectively come together. Maybe at that time, in 1984, there was a closed system of opinion and self-centered management that had not even thought of a possible debate on futuristic technological openings. This conformity to a manufacturing follower style of thinking was more and more often considered to be obsolete. But there was also, beyond socio-economic milieus, an incredible viscosity with many scientists: the most common attitude was not openness to other explicative schemata, but in the majority of cases, the ignorance and/or refusal to accept their existence. Tricks that only imperfectly fit into our ethics as researchers (at the time) must be made and likely developed.

According to estimation methods, the revenue from additive manufacturing lies somewhere between 5 and 40 billion euros (we could think that this is an estimation of the number of protesters in a claim by the police or trade unions!). Some speak of a revolution and others imagine senseless promises (which, according to Audétat [AUD 15], could put every emerging sector in danger); in short, things are booming at present with seven main stabilized technologies and a new kind of governance (Jeremy Rifkin's "makers"). This appreciative placement of the normalizers into categories is indeed rather artificial. Beyond a recent manufacturing technique that associates computer science and matter, 3D printing, with cheaper and cheaper home machines (down to a few hundred euros), constitutes a paradigm shift that impacts product design (which can even be defined, thanks to "open-source" systems), creation (from heavy industry to one's "garage"), consumption and the business models that result from them (from market activity, a new handicraft and DIY (Do-It-Yourself) to counterfeiting).

In fact, the progression rates are always in the double figures (between 20 and 40% per year), which leads some to believe that the additive manufacturing processes will continue to evolve for a long time to become a widespread technology, as they increasingly occupy ever-new applicative niches, quashing the other manufacturing methods that made up the skeleton of 20th Century industrial manufacturing. But what do tens of billions of euros per year represent for the world relative to France's "small" debt amounting to 2 trillion euros? It is therefore difficult to project a future which leads to a possible hegemony of additive manufacturing; besides, it would be more interesting to explore how intelligent synergies can be implemented with technology that emerged long before 1984. Yet, as is resurfaced in Volume 2, there are spaces, still relatively empty, where an attempt is made to challenge the very concept of adding material to processes.

The early 21st Century is marked by the "hegemonic" presence of the digital transition with the technological and practical complements of additive manufacturing processes likely to affect Western society in a quick and profound way. "In the face of radical innovation markets, where the first arrivers can acquire decisive, dominant positions and make the passage of other markets and the economic actors in place disappear, keeping a distance and watching things happen can lead to considerable social and economic costs" [FRA 17]. To go beyond this already uncertain space and become involved in disruptive innovations implies taking risks, thus accepting potential failure, facing their possible negative consequences, and being capable of learning all the lessons this teaches. "If we do not proactively incorporate innovation, this will end up being imposed all the same, in an even more disruptive manner" [FRA 17]. In short, it may be useful to anticipate.

In roughly a century, the number of researchers in Europe has gone from a few thousand to a few million, and despite some disturbances, this trend is continuing. Research activities have been the subject of reassuring discourse on the researcher's independence, on the one hand, and on the other, of a certain programming of research with the aim of achieving goals: security (before the fall of the "Iron Curtain", for example) and economic developments (from mass production with ECSC projects to information and communication sciences and technologies) participating in different forms of competition from France and the European Union.

On this basis, the stereotypical image of the scientist, responsible for the truth and good, is still part of the idealized image, which often positions him/her very highly in relation to a social reality of which he/she only has an imperfect mastery. The will to achieve the best "research efficiency" has led to the promotion of rather mono-disciplinary processes that are easier to manage from "peers", referents of a discipline. On the one hand, in-depth scientific study is maintained by actors from the same field provided that the guarantee of excellence is defined and respected; on the other, for the State, it is easier to realize international comparisons discipline after discipline. Indeed, and this is necessary to remember, without really noticing it, we have gone from a limited worldwide scientific elite to mass research (with tens of thousands of scientific journals) which represents a characteristic that is not discussed by developed nations: research must indeed allow society to respond to the great challenges that loom today: employment, progress, security, global warming, health and quality of life, sustainable development, etc.

Without seeking to speak of two worlds exploring different paradigms, one of in-depth study, the other of responding to social demand (even its anticipation), for this aim would be too limited, rather we look at evolutions translated by a research program that takes account of the different and sometimes antagonistic imperatives (see Volumes 2 and 3). This situation actually shows, at least in part, that the researcher is an element of society who is not independent, even if forms of "grand isolation" have long protected him. But, in the European Charter for Researchers signed by France at the CNRS (National Center for Scientific Research) in 2005, a reminder is given that "Researchers should focus their research for the good of mankind and to expand the frontiers of scientific knowledge, while enjoying the freedom of thought and expression, and the freedom to identify methods by which problems are solved, according to recognised ethical principles and practices."

Without this having been noticed by most of the research actors financed by the State, even if the notion of good is not easily defined (in any case, it does not simply mean the absence of evil), this sentence is a reminder of the role of research centers as a social (or societal) actor, implying new approaches like functioning through interdisciplinary projects and strategic reflections negotiated by stakeholders, stemming from a new prospective work. Considering their importance for the development of citizens' quality of life, research associated with technology is an element that is really starting to be discussed. Indeed, it has participated in the "natural" evolution of things and technological progress has long allowed man to be free from a number of material constraints. In this framework, the rhythm of implementing research results has been greatly modified and complicated, thanks to a more and more frequent coming-and-going between "manufacturing" and research, and thanks to a hybridization of technologies as well as, to a lesser extent, modes of research action (added value from the ability to interact). To work in the new economy of knowledge with greater partnership, there is a need for better reflection on creativity, innovation, and the societal impact of scientific and technological activities. So then, in today's context of growing co-constructed and contractual research actions, must/can we break away from the "researcher's temptation of innocence", of the consoling illusion of "neutral" science, or of the simple transfer of responsibility to the deciders/financers? It will be understood that these are somewhat the stakes of the current evolutions/revolutions applicable to additive manufacturing, and particularly to its future.

With the concept of informed matter, there must be a possibility to modify the shape of objects in time (4D printing), to print living matter (bio-printing), etc. It is thus conceivable to come closer to life by flirting with its possible prolongation! This questioning, like 3D printing pushed to its limits (nanomanufacturing, micro-fluidics, electronics and robotics) associated with other domains, does not correspond to an economic market present today, but instead, if researchers, breaking with the traditions of incremental innovation, succeed (thanks to a bit of creativity and epistemic exploration), immense markets (relative to the "modest" market today amounting to 10 billion euros per year) should open up. The illustrative example of bio-printing which could correspond to a market worth several hundred billion euros per year is a great demonstration of the stakes linked to research concerning initial findings, presented in Volume 3.

If it is necessary to put some of this enthusiasm into perspective, the "classic" additive manufacturing technologies, which have already successfully demonstrated their numerous capacities of industrial development, offer application fields, some of which are very recent and possible, thanks, in particular, to disciplinary research, enabling existing manufacturing processes to be improved. This concentration on a clearly identified objective, process–material optimization, has limited more creative research leading to weaker programming and support for "divergent" researchers, whose numbers, for various reasons, are rather limited in the world of research. Nevertheless, these new applications called 4D printing, bio-printing, 5D printing, etc. result from more complex interdisciplinary activities that, if they succeed, could open markets, no longer in the 4-digit range (billions of euros around the world), but in all likelihood in the 5- or 6-digit range!

There are thus (at least) two types of challenge in additive manufacturing, one is the realization of 3D pieces which contribute a (the most) crucial input relative to the more traditional manufacturing techniques (prototyping, foundry, soldering, etc.) and the other is more prospective on openings in new fields with renewed approaches (and with the associated difficulties). For this reason, with the publisher (ISTE), there was a wish to present the 3D domain in three parts, one with validated scientific and technological bases (certainly with potential redundancies relative to other works on this subject) and the others based on a field of possibilities that offers new epistemological questions, terrible risk-taking, but considerable stakes.

In the first three volumes, it was actually about writing two open "scenarios" that were slowly constructed within a framework, but without a very strict preliminary plan, the scenarios in which the elements were to be introduced and discussed would be spread in an *a priori* graded manner. Each chapter has some degree of autonomy, which can be translated by possible repetitions (as few as possible, however), with a "history" that is progressively fed thanks to the in-depth reading of hundreds (thousands?) of publications, numerous times meandering through and delving into beautiful ideas and scientific meetings for debates, sometimes with success. The gray literature has been a vital source for what is happening in the field at times, which explains the numerous references to the websites in some chapters.

In Volumes 1 and 2, the reader is sensitively placed within the "summary table of disciplines" published in 1829 by Auguste Comte with an "institutional" organization for scientific disciplines, enabling incremental research and development in additive manufacturing. In Volume 3, the idea is to place the reader in a less programmable mode of functioning, with a recursive, systematic and self-organizing character of knowledge, a better willfulness in processes, which sets it apart from the first two, yet it is nevertheless complementary (because it is still constructed using what is known). However, a bit of naivety and/or ignorance may allow for progress to be made in the domain by tackling new paths of creation from a small amount of scientific and technical knowledge in a less "professional" manner, but full of enthusiasm towards a new world to be explored.

An intentional artifact (linked to the engineer and/or designer's work) may be considered a means of connecting an "internal" environment, the substance, the functioning, and the organization of the artifact itself and an external environment, the surroundings in which it is implemented. If the two environments are compatible, the artifact responds to the specifications. As underlined by H.A. Simon [SIM 04] in another framework, the knowledge of an artifact as an additive manufacturing machine "benefits from an advantage on the knowledge of nature, for it is based on valid, previous foundations whose ends will be perverted with a certain dose of new willingness to give projects intelligibility and openings on society." This notion can also be found within the facts in the three works, but with different interdisciplinary openings.

In Volume 3, for the researcher who studies the behaviors associated with the intrusion of temporal aspects and functionality in additive manufacturing, the systems operate for sufficiently long, entirely determined times. But, like "self-organization" phenomena, they can become very off-balance and sensitive to factors considered to be negligible near equilibrium. This is the intrinsic activity of the increasingly complex system, with an increasingly nonlinear behavior, which determines how it is possible to describe its relationship to the environment, which thus generates the type of intelligibility that will be pertinent to understand its possible stories. It is thus not only a matter of an applicative field with its constraints, but also of a theoretical domain to be approached and interrogated in order to resolve the end/means equation in a robust way so as to achieve it.

It will be understood that the epistemological foundations of the reflections in Volume 3 are based on the complexity paradigm, where interdisciplinarity is projected as one of the means of study. The disciplinary approach is too often divided, fragmentary and linear, hence a master idea aiming to know how to percolate through disciplinary borders so that the complexity paradigm can truly spread, notably because the recomposition of thought categories can no longer be based on borders and disciplinary subjects, but on boundary subjects based on the creative, the divergent, who, having no fear of recursiveness, hope to legitimately respond to the great risks society must face.

This change in delivering research for a more systematic approach does not hope to be the indicator of a field of scientific disciplines that, hoping to keep its power, loses its authority, even if current societal issues still cannot handle constructive forms of subordination well. It aims for a real, responsible integration of activities open towards society, bearers of meaning, allowing new research in additive manufacturing to be made to emerge as credible scientific evidence of movements that are materializing.

The evocation of different attractors of disruptive innovation in 3D manufacturing is the focus of Volume 3, in addition to its scientific and technical aspects. The author uses his experiences in this volume to recreate a bit of the history of new additive manufacturing processes, which could, in case of success, invade our daily lives in some years. It is in the spirit of creating a history, and interiorizing it by trying with the time and means available to re-establish them with a personal vision, with the risk of committing mistakes, of having failed with a promising idea. But this is the price to pay.

In the three volumes on the subject of additive manufacturing, it is shown that in relation to almost every problem, there is in fact a creative avant-garde with low inertia: this is carried out by groups of divergent researchers working in practice on the problem at hand. Then there are all the followers, who will structure the "paradigm" and engage it only in forms of conservatism authorizing research to improve processes or materials ("programmable" research). It will take years, even decades, for this paradigm to change positions – often with shoves (linked to the work of the creative by following information provided by the avant-garde). "Paths must be transformed into roads, the ground leveled, etc., so that the landscape will transform significantly until it becomes the main group's parking place" was written by L. Fleck in 1935. Could this context, in terms of research, be adapted to economic development? These characteristics of considering time, and its management, are the elements to be taken into consideration in a process of spatial and temporal transformation of matter that displays significant advantages.

Thus, beyond scientific aspects, indispensible techniques will be discussed to examine how the edifice of additive manufacturing was and is being built through its cultural filters and filters of understanding and interpretation. Anticipating the future of the field of additive manufacturing in the larger sense, to be in a position to prepare ourselves, is considered one of the keys for the long-term durability and competition of companies. This imperative to think of the future, to add to this divergent thought to create new devices for creating objects with the adapted material, devices that are functional, adaptive, "smart", etc., today seems even more significant considering the instability of the environment, the speed of evolution and the generalization of uncertainty. In such a context, research locations must be "offerers" of concepts, of their demonstration to anticipate the productive industrial future, not to mention the technological, economic and governance systems in which, on shorter and shorter reference times, companies evolve (undergoing nonlinear dynamics, splits and breaks). This mission is not only meant for individual researchers, but also for everything around them: research units, their administration and also (and above all else) the proactivity of economic milieus.

In terms of tomorrow and the future, can we not foresee new means of creating objects? At present, we have mastered synthesis, the way in which the objects are constructed. But we could also ask ourselves if it wouldn't be possible to develop systems in which we could give objects an intentionality, thus giving it the choice to look for itself for what changes it needs to make, thus moving onto self-organization with the selection of necessary elements that it would extract from a "bank" for the edification of the final object. This would go beyond the 4D printing that tackles the functional and evolutionary assembly of materials that should be able to come together to create an upgradeable object and that could be made easier through "programmable matter": "Programmable materials and objects that are themselves created would thus make assembly factors and heavy installation procedures superfluous… Robotization, the heart of progress in 20th Century productivity, could thus be integrated into the products themselves, with, as can be imagined, some ethical problems to be taken into consideration" [FRA 17]. Let us thus dream together of this future. The process attempted in these volumes therefore aims to try to question a present (it is impossible to know if this present will likely be able to achieve all its goals) and to determine the conceptual elements that could lead to an original future with access to new applicative niches by exploiting revisited paradigms.

Beyond the exhaustion of the reserves and consequences, it is also the way in which we understand scientific policy to be carried out by taking into consideration different world actors that should evolve to stimulate this nascent domain. In the reflection these books are aiming to create in its readers, it will likely be a matter of proposing changes to be undergone, which correspond to the conceptual displacement of the economy allowed by technology towards a new economy of creativity making a better effort to consider social, economic, organizational, geopolitical, even emerging environmental constraints. It is a form of "design thinking" that is thus to be considered. A reflection on the processes to help the integration of societal data, far from its disciplinary culture, would probably also be projectable (if only on the organizational aspects). In the end, it would be a matter of demonstration, through changes negotiated with the responsible authorities (some of whom are mute), leading to better exploration of the complexity, which can be done well, if not better, maybe with less equipment, but otherwise in a context of social and/or socio-economic demand that it would be advantageous to anticipate, if not follow. The paradigm shift would then take place thanks to scientific initiatives, which are marginal today, which remain aporetic in the paradigm in crisis, and which should be muted in a new scientific era, less framed, applied to 3D printing.

These three volumes can serve to think about the future in the domain that remains exciting for the author after more than 30 years since his 1984 patent, so that we can again find its place concerning its abilities of industrial creation and development in an ever more competitive environment. 3D, 4D, even 5D technologies constitute a path of promotion (among others that stem from the author's competence) of this desire for renewal.

NOTES.–

– For these three volumes, the search for the greatest possible number of specific or general visions concerning the subject of additive manufacturing, which can help the reader, has led to the presentation of the bibliography chapter by chapter and in alphabetical order. In fact, it was almost impossible to classify the bibliography through the numbering of entries.

– Some repetitions in the chapters of these three volumes may exist in an attempt to give them certain coherence and to provide them some degree of autonomy.

Jean-Claude ANDRÉ
Research Director at CNRS
October 2017

Bibliography

[AUD 15] AUDÉTAT M., *Sciences et technologies émergentes: pourquoi tant de promesses*?, Herrmann, Paris, 2015.

[BEI 12] BEIGBEDER C., *Puisque c'est impossible, faisons-le*, J.C. Lattès, Paris, 2012.

[EIN 07] EINSTEIN A., quoted by Debonneuil M., *L'espoir économique: vers la révolution du quaternaire*, Bourin, Paris, 2007.

[ELL 04] ELLUL J., *Le système technicien*, Le cherche midi, Paris, 2004.

[FAB 06] FABIANI J.L., "A quoi sert la notion de discipline?" in BOUTIER J., PASSERON J.C., REVEL J. (eds), *Qu'est-ce qu'une discipline?*, EHESS, Paris, 2006.

[FRA 17] FRANCE STRATÉGIE, "2017/2027 – Répondre à l'innovation disruptive – Actions critiques", available at: http://www.strategie.gouv.fr/publications/20172027-repondre-linnovation-disruptive-actions-critiques, 2017.

[GIA 90] GIARINI O., STAHEL W.R., *Les limites du certain: affronter les risques dans une nouvelle économie de service*, Presses Polytechniques et Universitaires Romandes, Lausanne, 1990.

[GUÉ 15] GUÉRIN M., *La croyance de A à Z; un des plus grands mystères de la philosophie*, Encre marine, Paris, 2015.

[LAT 07] LATOUR B., *L'espoir de Pandore; pour une version réaliste de l'activité scientifique*, La Découverte, Paris, 2007.

[PHI 12] PHILIPPE J., "L'innovation managériale, comment innover dans l'univers bancaire?", in EUROGROUP CONSULTING, L'art du management de L'innovation dans le service public, Eurogroup, Paris, 2012.

[ROS 01] ROSE J., *Profession quasi-chercheur*, L'Harmattan, Paris, 2001.

[SIM 04] SIMON H.A., *Les sciences de l'artificiel*, Folio-Essais, Paris, 2004.

[THO 83] THOM R., *Paraboles et catastrophes*, Champs Science, Paris, 1983.

[VAL 57] VALÉRY P., *Œuvres complètes*, La Pléiade, Paris, 1957.

Introduction

"Fashion always conjugates the taste for imitation and the taste for change, conformism and individualism, aspiration to merge into the given social group and the desire to differentiate ourselves from it, even by small details". [SIM 03]

"Technologies...do not only produce instruments which transform our life, they alter the reality which surrounds us and reorganize our social lives. Such movement has got carried away since the industrial revolution". [KLE 11]

"We then count upon technological progress to later get ourselves out of the problems that we come up against and which we know to be real – habits and attitudes we have not got out of". [LER 11]

"Does the risk not seem in the end both the impoverished and simplistic manner in which Man from technoscientific societies, who does not succeed in making sense of his/her unhappiness, realizes what is happening to him/her". [DUP 02]

"The definition of a given problem tends to freeze in the position defended by bureaucratic agencies and to thus resist all forms of transformation. Cyert and March [CYE 63], who were organizational theorists, wrote that organizations seek to overcome uncertainty by following routinized procedures. Such organizations do not anticipate

problems, but rather respond to the feedback effects generated by their own behavior. Hence, they tend to 'reel from crisis to crisis' by relying upon standardized procedures for decision-making. This mode of operation, in private as much as in public institutions, enables us to explain how when a new problem arises, it can be misunderstood and be subject to inappropriate treatment". [REI 10]

"All desire originates from a given need, that is to say a form of deprivation, indeed that is to say a form of suffering. Satisfaction puts an end to it. However... the satisfying of any particular wish does not cause sustainable and unshakable consent. It is like alms that they throw to the beggar. They save him/her today, so as to prolong his/her misery until tomorrow". [SCH 14]

"The main role of the state may be to somewhat promote original projects, by taking the risk of investing in research which leads either to nowhere, or to something other than what we anticipate. In short, this is high-risk research in which the point of a given discipline seems to call for a form of transdisciplinary exchange". [LEC 97]

"All forms of opinion are simply assessment. Such assessment leads to a given reputation, and it is in demonstrating allegiance to reputations born of conversations which are apparently trivial, that we adopt both in market places and in polling stations". [KAT 09]

"The disruptive innovations occurred so intermittently that no company had a routinized process for handling them. Furthermore, because the disruptive products promised lower profit margins per unit sold and could not be used by their best customers, these innovators were inconsistent with the leading companies' values". [CHR 03]

"In Mandarin, innovation means the concepts of both learning and copying". [GOD 11]

I.1. Introduction

NOTE.– The beginning of this introduction partly resumes that suggested at the beginning of Volume 2 [AND 17b], to the extent that there are a certain number of common features.

Additive manufacturing is made up of seven standardized processes, according to the French standards NF: E 67-001 and international standards ISO 17296-2:2015E. It involves the following processes [MAD 17]:

– VAT photopolymerization or stereolithography;

– material jetting;

– binder jetting;

– powder bed fusion;

– material extrusion;

– direct energy deposition (DED);

– sheet lamination.

The themes in Volume 1 [AND 17a] have already shown all of the benefits of additive manufacturing within numerous spheres of application with the technological specifications, which were able to be stipulated. The emergence of a small revolution translates through various inclusion difficulties within a manufacturing culture, which already has its own constraints. This leads to the promise of accomplishments while everything within the manufacturing environment is far from being stable. The performance is sometimes insufficient, as is the time taken for manufacturing [DGE 16]. This is the case even if increasingly new materials, software and processes reach a mature phase of their product lifecycle. Three-dimensional (3D) printing has been introduced in a slightly surreptitious manner by proposing local production of objects, which would otherwise be practically impossible. Even though there still remain numerous outstanding issues in this sphere, transition to the growth phase still occurs with double-digit gains.

> "Additive manufacturing is often presented as being likely to radically alter how objects are designed, as well as produced. However, the production system organization and the mass distribution of the given product still remain largely dependent on a number of significant technical developments. The pace of such advances is difficult to anticipate. At the current time, machines and processes do not allow us to respond to all industrial production constraints. Progress may, in particular, have to be achieved to increase the production rate". [COE 17]

At the same time, the emergence of a given technology removes numerous design phases by supplying a tool that (in principle) sits between the computer and the object concerned. This re-empowers designers and "makers" who can produce innovative objects "in their own garage workshops". The enlargement of this new relationship between conception and the given device may lead to cultural developments in the production process. This can be accompanied by desired aspects of reintegration of some production nationally, and possible "uberization" of given activities.

These components will still increase or develop from scientific, technological and social perspectives, in equal measure. These have comprised a unique sphere in which the author has "floundered" in a sense. In Volume 1, he has tried to share with you a little of his hopes, with suggestions and opinions that you may choose to ignore. From the author's perspective, the important aspect is to fuel reflections around the emergence of ideas for "informed" discussions.

In this book, volume 3 of the set, the spirit of the text is not entirely the same. Indeed, it is envisaged that the sphere should be open to unexpected forms. This is because the scales, constraints and other factors introduced into additive manufacturing will change over time. In this setting, which remains vague, indeed yet to be constructed, there are still limited numbers of published knowledge sources. In these circumstances, it is difficult to "explore" the sphere. This leads the author to simply mention the elements of information which he has retained from his own reading. It will be interesting, in a few years, to examine how these spheres will develop a structure, so as to move towards preferred directions, which will then appear more clearly. This analysis will be all the more interesting for France (but perhaps not France alone). This is because it will be necessary not only to establish a link between scientific and technological sets of themes, but also individuals, who are currently greatly "diluted" within the national research mechanism [FCM 17].

The German psychologist, Wolfgang Köhler (cited by [CAS 17]) explains in this regard that "the most delightful moments of the history of knowledge take place when facts which have been up to that point specific data, are suddenly related to other facts which were apparently remote, and thus appear in a new light. Or, as Steve Jobs said, more simply, creativity simply consists of linking things together". In these spheres, the author leaves it to the reader to discover upon reading Volume 3 that there are areas where we have to jump forward to practice serendipity, with a necessity of taking risks. It is at this cost that risks taken by designers allow us, without doubt, to structure new fields stemming from additive manufacturing and to speak like Steve Jobs.

Similarly, Arthur Koestler [KOE 65] confirmed that creative originality does not consist of creating *ex nihilo* ideas, but rather by combining diagrams and well-established structures, through, as it were, hybridization. "The act of creation is not creation in the sense of the Old Testament. It is not creation from nothing. It discovers, mixes up, and synthesizes facts, ideas, competencies, and technologies existing already. The entire invention will be all the more surprising as the given parts become more familiar".

The technologies, which will be described here, are characterized by their major interdependence with other spheres: nanotechnologies, biology, medicine, chemistry and energy (and its storage). Of equal importance is the digital sphere, which must never be forgotten. The breakthroughs, which will be made within the spheres set out in the following chapters, are conditioned to a large extent by the progress, which is both made (and expected), in their cutting-edge research. "This interdependence of technologies makes so-called foresight exercises, regarding technological development, fairly complicated endeavors. This is all the more true since innovations, in particular breakthrough innovations, spring up at interfaces. Yet, these interfaces increase in figures by reason of the possibilities of the information society, as well as the major interdependence of the most recent technological advances. These favor interactions between technical and professional spheres, which were previously less connected. Furthermore, the given technology's lifespan is uncertain, and may be long. Sometimes many years pass between the first positive laboratory results and deployment of a given technology" [COE 17].

In 2016, the DGE (Direction Générale des Entreprises – the French Directorate General for Enterprise), a division of the Ministry of Industry, published a report entitled "Technologies clés, 2020" (Key technologies, 2020), with 47 targeted technologies. Within Volume 2 of this series, the attentive reader is able to see that 10 of these 47 technologies will be directly or indirectly impacted by the breakthrough technology activities set out in the present work. These are the following technologies:

– cell and tissue engineering;

– health imaging;

– collaborative robotics and human technologies;

– microfluidics;

– advanced and active (smart) materials;

– autonomous robots;

– new hardware–software integrations;

– modeling, simulation and digital engineering;

– Internet of Things;

– not forgetting, of course, the subject of this work – additive manufacturing.

Lin *et al.'s* view [LIN 14] upon the development of additive manufacturing is shown in Figure I.1. This passes from the phase of producing monofunctional parts to the use of nanotechnologies in additive manufacturing. In this volume, concerning the emerging activities with an essentially incremental origin, some of the recommendations of authors of other works will be included. These recommendations cover areas where additive manufacturing is pushed to its limits. In this figure, the fields placed on the left essentially correspond to the subject matter set out in Volume 2, and those on the right to that in Volume 3.

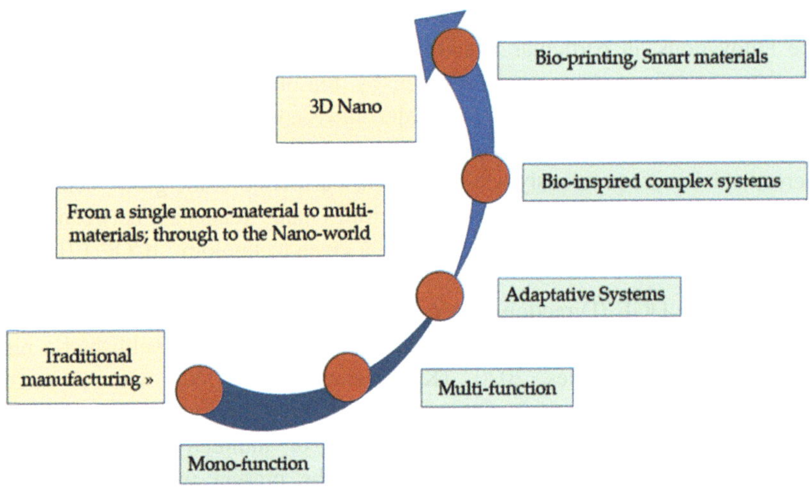

Figure I.1. *A medium- to long-term view of additive manufacturing. For a color version of the figure, see www.iste.co.uk/andre/printing3.zip*

However, who has not dreamed of controlling matter simply by thinking about it? Kaku [KAK 14], in his forecasting study, placed "morphing" or programmable matter within development trends, but with an actual emergence beyond the 2030s. Within the rationale of Kaku, where scientific and technological constraints may be excluded, why should we not evoke the spatial transformation of a given object by the specific contribution of energy, whatever its initial form takes?

Moreover, the chemist Joseph De Simone has presented a machine capable of 3D printing designs to a TED Conference in Vancouver, in the West of Canada. These designs were printed as if they emerged from a liquid metal, reminiscent of the dreaded robot T-1000 of "Terminator 2" arising from a given pool of silver [CUL 15]. However, alas, this only related to the stereolithography CLIP process, which is highly clever but extremely far from this view of man's genuine control over matter.

To return to a more scientific and technological viewpoint, indeed slightly philosophical in additive manufacturing, an observation which is even superficial in nature shows that the latter tends to organize itself into ordered structures. Moreover, who has not been filled with wonder by the view of a "Romanesco" cauliflower with its fractal shapes? At that point, in theory, two options may be at work. The first consists of controlling a set of physicochemical and energetic systems so that the desired object appears without returning to traditional principles of rational construction, voxel after voxel, of the "additive manufacturing" type. The other, which is more reasonable, aims to produce, as has been shown in Volume 1, a given object by the addition of matter. From the manufacturing point of view this is all the same somewhat original, as normally machining methods by the elimination of matter are practiced. Attempting, at this point, to leave the "simple" accumulation of additive manufacturing voxels, it appears useful to instigate a short reflection on this particular course. This may, perhaps, reinvigorate the field, in any case by replacing it within a given potential technological dimension.

Alvaro [ALV 04] said, "the self-organizing entity is made up of parts whose mutual interactions together determine dynamically the entirety that they form, which occurs as a stable spatial-temporal pattern. The self-organization 'identity' is thus the structure constructed, but we can also say that this identity is created through this form of organization. This involves the idea that self-organization is both a process and a system".

Two viewpoints have been developed to elaborate a mathematical theory of the concept of morphogenesis:

– one favors the model (the equations that describe the system studied) and uses analytical tools, particularly the bifurcation theory, to determine which structures develop and in what way;

– the other applies to classify scenarios by which given forms appear, without attaching to a particular design.

These two complementary approaches have enabled considerable progress in the understanding of morphogenesis for around 30 years, a common thread in some of the chapters of this volume. In these nonlinear systems exploiting the potentialities of morphogenetic engineering there will appear variable phenomena: these include bistability, saturation or, on the contrary, hyper-amplification, oscillations and other effects [JOS 12].

"The engineer's role is to use the technical knowledge available to him/her to devise innovations which are an acceptable compromise for all stakeholders. It is said that technical perfection poisons the economy! It is necessary to be ready to 'tinker' to satisfy the greatest number of people. Technological resources are such that you cannot believe an engineer who assures you, without batting an eyelid, that there is only one efficient profitable solution and that you must choose it" [CAL 11]. On reading these sentences, the non-mastery of the inverse problems in the morphogenesis aspects which have just been introduced could (should) have closed the scientific debate. Nowadays, there is not yet a realistic solution!

Within the cultural approach of engineering sciences, the quest for the understanding of reality relies largely upon "detour strategies, in other words, the construction of given designs or the production of virtual occurrences" [GRA 95]. However, these detours exploit the knowledge of what is possible and, as far as is feasible, of what is simple, by imposing causalities.

What this introduction tries to show is that it is necessary, without doubt, to make progress in finding a compromise between perfect causalist systems with the possibility of moving voxels on deliberate courses, and a global morphogenetic view, exploring complex systems. This is moving largely towards these difficult situations through which we are "navigating" in this text. However, previously we thought it appeared interesting to see if we could start by considering additive manufacturing applications within the spheres where it is pushed to its limits.

As de Brabandere [DE 16] stated, "It is very important to realize that we construct our thinking, our ideas, our judgments and our stereotypes more on the basis of ignorance than on knowledge. In order to design, we are, on the one hand, required to forget, and, on the other hand, to take a form of distance". Let us therefore endeavor to move forward in this direction.

I.2. Bibliography

[ALV 04] ALVARO M., "Auto-organisation, autonomie et identité", *Revue internationale de philosophie*, vol. 228, pp. 135–150, 2004.

[AND 17a] ANDRÉ J.-C., *From Additive Manufacturing to 3D/4D Printing 1*, ISTE Ltd, London and John Wiley & Sons, New York, 2017.

[AND 17b] ANDRÉ J.-C., *From Additive Manufacturing to 3D/4D Printing 2*, ISTE Ltd, London and John Wiley & Sons, New York, 2017.

[CAL 11] CALLON M., LACOSTE A., "Défendre l'innovation responsable", *Debating Innovation*, vol. 1, pp. 5–18, 2011.

[CAS 17] CASTILLO A., "La créativité ou l'art d'exhumer des analogies cachées", available at: https://www.letemps.ch/economie/2017/02/16/creativite-lart-dexhumer-analogies-cachees, 2017.

[CHR 03] CHRISTENSEN C.M., *The Innovator's Dilemna*, Harper Business, New York, 2003.

[COE 17] COE (Conseil d'Orientation pour l'emploi), "Automatisation, numérisation et emploi – Tome 1 – Les impacts sur le volume, la structure et la localisation de l'emploi", available at: http://www.coe.gouv.fr/IMG/pdf/COE_170110_Rapport_Automatisation_numerisation_et_emploi_Tome_1.pdf, 2017.

[CUL 15] CULTUREBOX, "Le robot liquide de " Terminator 2 " source d'inspiration d'une imprimante 3D", available at: http://culturebox.francetvinfo.fr/tendances/numerique/le-robot-liquide-de-terminator-2-source-dinspiration-dune-imprimante-3d-214501, 2015.

[CYE 63] CYERT R.M., MARCH J.G., *A Behavioral Theory of the Firm*, Prentice-Hall, Englewood Cliffs, 1963.

[DE 16] DE BRABANDERE L., "Les Big Data pourraient-ils nous rendre plus créatifs?", available at: http://www.ladn.eu/innovation/datas-datas/luc-de-brabandere-plus-dinformations-moins-de-certitudes/, 2016.

[DGE 16] DGE - DIRECTION GÉNÉRALE DES ENTREPRISES, "Technologies clés – Préparer l'industrie du futur 2020", available at: http://www.entreprises.gouv.fr/files/files/directions_services/politique-et-enjeux/innovation/technologies-cles-2020/technologies-cles-2020.pdf, 2016.

[DIR 16] DIRECTION GÉNÉRALE DES ENTREPRISES, "Technologies Clés 2020 – Préparer l'industrie du futur", available at: http://www.entreprises.gouv.fr/politique-et-enjeux/technologies-cles-2020, 2016.

[DUP 02] DUPUY J.P., *Pour un catastrophisme éclairé – Quand l'impossible est certain*, Le Seuil, Paris, 2002.

[FCM 17] FCM (FILIÈRE CARNOT MANUFACTURING), "Cartographie des acteurs clés de la R&D en Fabrication Additive en France", available at: http://www.instituts-carnot.eu/sites/default/files/images/CartoFabAdd_CETIM-filiere-Manufacturing-juil2017.pdf, forthcoming, 2017.

[GOD 11] GODET M., *Bonnes nouvelles des conspirateurs du futur*, Odile Jacob, Paris, 2011.

[GRA 95] GRANGER G.G., *Le probable, le possible et le virtuel*, Odile Jacob, Paris, 1995.

[JOS 12] JOSHI S.D., DAVIDSON L.A., "Epithelial machines of morphogenesis and their potential application in organ assembly and tissue engineering", *Biomechanics and Modeling in Mechanobiology*, vol. 11, pp. 1109–1121, 2012.

[KAK 14] KAKU M., *Une brève histoire du futur; comment la science va changer le monde*, Flammarion, Paris, 2014.

[KAT 09] KATZ E., "Influence et réception chez Gabriel Tarde; un modèle pour la recherche sur l'opinion et la communication", in *La réception*, Revue Hermès, Paris, 2009.

[KLE 11] KLEIN E., *Le small bang des nanotechnologies*, Odile Jacob, Paris, 2011.

[KOE 65] KOESTLER A., *Le Cri d'Archimède: l'art de la découverte et la découverte de l'art*, Calmann-Lévy, Paris, 1965.

[LEC 97] LECOURT D., *L'avenir du progrès*, Textuel, Paris, 1997.

[LER 11] LEROUX T., *Le laboratoire des pollutions industrielles: Paris 1770–1830*, Albin Michel, Paris, 2011.

[LIN 14] LIN D., NIAN Q., DENG B. *et al.*, "3D printing of complex structures: man made or toward nature?", *ACS Nano*, vol. 8, pp. 9710–9715, 2014.

[MAD 17] MADEELI, "Fabrication Additive: Enjeux et Impacts en Occitanie Pyrénées-Méditerranée", available at: http://www.madeeli.fr/wp-content/uploads/2017/01/Fabrication-Additive-en-Occitanie-Etude-2016.pdf, 2017.

[REI 10] REICH M.R., "Le cas de la pollution du Michigan par le PBB", in AKRICH M., BARTH Y., RÉMY C. (eds), *Sciences, politique et environnement*, Mines Paristech, Paris, 2010.

[SCH 14] SCHOPENHAUER A., *L'art d'avoir toujours raison*, Librio, Paris, 2014.

[SIM 03] SIMMEL G., cited by Lipovetsky G. "Luxe éternel, luxe émotionnel", in LIPOVETSKY G., ROUX E. (eds), *Le luxe éternel ; de l'âge du sacré au temps des marques*, Gallimard, Paris, 2003.

PART 1

Programmable Smart/Intelligent Matter and 4D Printing

Introduction to Part 1

"Although we are positioned to see things from the perspective of intelligent objects, we should first observe that the significant is still distinctive whilst the intelligible is always universal". (Aristotle, cited by Nadeau [NAD 16])

"The development of 3D printing technology falls within the post-industrial framework looking towards the functionality economy. The digital economy amplifies this trend". [GHI 16]

"Two ways to reach the 'truth' are:

1) Non-explicability consisting in saying that the object in question is too complex and based on what we know today, it will never be possible for us to master;

2) Intelligibility by creating a given 'relay' concept. This pushes the question further back (e.g. the existence of God). This is the occurrence of vitalism introduced by Woodger in 1929". [BUN 03]

"Paradigm shifts make scientists, within their particular sphere of research, see everything differently. To the extent that they only have access to the world through which they see, we may be led to say that after a revolution, scientists are reacting to a different world". [KUH 62]

"Far from equilibrium, the irreversible processes are a source for consistency. The appearance of this coherent activity around matter – the 'dissipative structures' – impose upon us a fresh look, and a new way of positioning ourselves in relation to the system which we define and manipulate. Whilst equilibrium and quasi-equilibrium exist, the behavior of the system is, for a sufficiently long time, entirely determined by boundary conditions; we will henceforth acknowledge a certain level of autonomy. This enables us to speak of structures which are far from equilibrium such as 'self-organization' phenomena". [PRI 09]

"The 'principle of scale relativity' postulates that the fundamental laws of nature should be valid, whatever the 'scale status' for the system of reference. It is relationships of scale alone which have a given direction, never an absolute scale. Such a concept is also imposed as a fundamental principle which enables 'compelling the description of a non-differentiable space–time relationship'". [CHA 12]

"The pessimist sees the difficulty in each opportunity; the optimist sees the opportunity in every difficulty". [LEM 17]

"There is nothing more dangerous than an idea when it is simply an idea". [ALA 69]

"Those who tell you that the only alternative is to slow down are impostors. To slow down, it is indeed necessary to use a given level of energy which counteracts the given acceleration of time". [SMA 11]

"Without talent, ingeniousness or curiosity, neither chance nor necessity suffices to produce innovation". [FOR 02]

"Behavior is motivation when passed through the filter of opportunity". [SHI 11]

"Transgressing the boundaries between disciplines [is] a subversive undertaking as it has every chance of ignoring the so-called havens of commonly accepted ways of thinking". (Greenberg cited by Staune [STA 15])

"Imagine that an excess of scientific and technological innovation is the solution: this amounts to a desire to run ever faster so as not to fall over". [CÉR 17]

> *"The present metamorphosis marks the beginning of a new era, and it is not just law that it almost reformulates. Beyond a production technique, 3D printing is a paradigm shift which impacts upon product design, product creation and product consumption and the business models which flow from them" [CHA 16] ... It is envisaged that we will expand upon this dynamic here by widening the dimension of the field covered: going from 3D to reach the fourth dimension! The discussion is in fact important because it evokes a possible transformation of how we produce objects without needing to assemble components that are manufactured separately. "Each product might ultimately be capable of self-assembly and implementation within extreme environments may be facilitated by the use of "Programmable Matter". Programmable materials which build themselves thus make the use of assembly plants and heavy installation procedures superfluous. Robotization, at the crux of twentieth century productivity gains, would thus be integrated into the products themselves". [GUI 14]*

Kupiec and Sonigo [KUP 00] pose the issue of producing simple forms from clay. They introduce a Lamarckian principle in which the material is placed in *ad hoc* molds. Any foundry worker might produce these (whether from metal or polymers). They evoke Darwinian thinking with a paste that would change shape continuously, spontaneously, and by a process of natural selection or production by the bench scientist. The point is that we would have to wait for the appearance of the desired form (we do not know when this would happen). "Here the result does not pre-exist [unlike the mold]. These transformations are decided on by a given cause… which does not relate to the particular result and may stabilize another given form". In this section, we are in an in-between phase. This involves the creation of a given object and spatial, temporal and operational evolutions, to varying degrees.

The idea of self-organization was introduced succinctly in the first volume concerning additive manufacturing research pushed to its limits. By the process of making matter intelligent (see the concept introduced 20–30 years ago by Alain Le Méhauté), it is possible to create 3D artifacts which may be considered to be interfaces between an "internal" environment, the smart substance and the organization of the artifact itself, and an external environment. In principle, the aim of this is controlled by the engineer. "Although the internal environment is adapted to the external environment or *vice versa*, the artifact will follow the assigned goals" [LE 04]. This concept of smart/intelligent matter is not entirely new, everyone knows about it, not least the media: smart clothes, photochromic lenses which become darker depending upon the intensity of outside lighting, concrete or carpets including optical fibers, and other related technology [GAI 16]; also to be considered is the perceptual aspect of plants being supported in the vertical position [INR 16]. In this volume, it is a matter of associating this information principle with additive manufacturing.

At the same time, for some, programmable matter may be "to some extent the link between the Internet of Things and the 'science-fiction' universe of Eric Drexler's nanotechnology. This consists of his universal assemblers capable of creating everything and anything – and potentially destroying the world simply with a stroke of grey goo. In a nutshell, we may say that programmable matter is situated at the micro-level and not the nano-level. Projects within this sphere essentially rest upon the creation of small components, like grains of sand, which are likely to fit together to adopt any shape: an object, such as a car… even the image of a human being or a kind of avatar which might occur in the real world" [SUS 14]. If the general idea is close to this aspect of temporal (and functional) evolution, this promise, which is slightly apocalyptic, is far from the intention of these chapters.

Ramirez-Ferrero [RAM 15] showed a positioning of additive manufacturing technologies using a Hype diagram (see Figure I.1). The Hype cycle is a representation of emerging technologies developing at a given historical moment in time. We may implement this concept for a given technology (each innovative technology going through the various stages of the cycle) or produce, on a regular basis, a diagram representing all emerging technologies, and their position on the cycle. This latter process is what the Gartner group [GAR 15] did. The principle of their diagram is a registered trademark owned by the group. Each cycle breaks down into five key phases, shown in Table I.1.

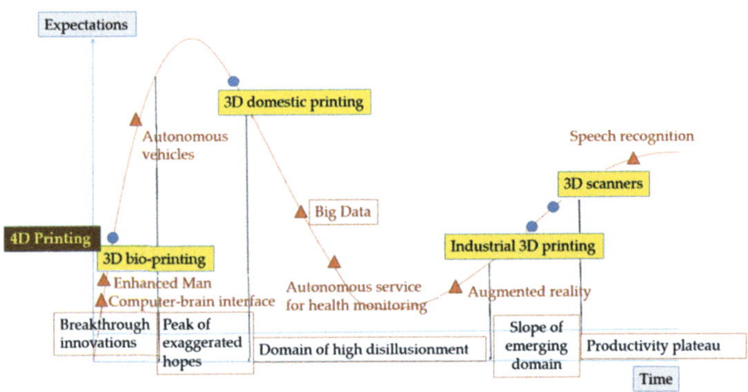

Figure I.1. *Hype diagram and 3D technologies. For a color version of the figure, see www.iste.co.uk/andre/printing3.zip*

Phase number	Designation	Commentary and description
1	Launching of technology	The proof of concept has been demonstrated. The disruptive arrival on the market of a new technology: these are not products which are easily usable, but more prototypes of a technology considered to have potential
2	Peak of exaggerated expectations	Media hype results in exaggerated and unrealistic expectations. Startups are being created to develop and market products based on the new technology
3	Abyss of despair	The products available do not succeed in responding to the exaggerated hopes which had previously been expressed. As a result, the media sources thriving on sensationalism reject that which they previously loved, especially if the technology takes time to emerge. We might witness a stock market crash: this is the so-called "anti-Hype"
4	Illumination "slope"	Some companies persist/survive and develop the so-called second-generation products. We start to understand genuine advantages and practical applications of technology. We witness a progressive and robust development of the market
5	Production or productivity plateau	Within this last phase, technology works well, is acknowledged as legitimate and enables the development of third-generation products. The scope of applications is variable according to whether the technology is widely applicable or only serves a niche market

Table I.1. *Various phases of the Hype cycle*

Figure I.1, traditionally used for innovation, enables the positioning of current 3D market evolutions. Even further (and without doubt, still further, for a given length of time in some cases) biological applications are connected by fundamental research. This is the case even if high promises are attached to this biotechnology (see Part 2 of this book). On the contrary, it may be that a market is already mature with industrial applications which are already economically profitable (mechanical engineering, aviation and construction, or another sphere). Some applications are already (almost) industrialized within highly varied spheres. Examples include the pharmaceutical sector, surgery, dentistry, food, design, construction and other sectors. These are generally predictable operations, which are developing and which have been mainly mentioned or shown in Volume 1.

In that chapter, that constituted a sphere which is rarely tackled in additive manufacturing, it seemed important to the author to develop the notion of smart/programmable matter (this will further serve a purpose in this third volume) before tackling a number of recent studies linked to the exploration of the use of the fourth dimension to open up development avenues for 3D printing. This provides a complementary element to the so-called "cold" or inert structure of functional material, by providing scalable functionalities according to an external energy supply with diverse origins. On this basis, the shape of the object can change and become like that which has just been mentioned, that of actuators.

In 2014, Frost and Sullivan [FRO 14] produced an analysis of 4D technologies, introduced into the Bibliography section of this chapter. They reflected upon the various possible spheres of application (see [MAN 14, CAR 14]). These are shown in Figures I.2 and I.3 with a five-year view (see Figure I.3). A considerable potential market then opens up. This would be necessary (because it is necessary to anticipate this and provide technological proposals), no doubt first investigating the relevant academic research.

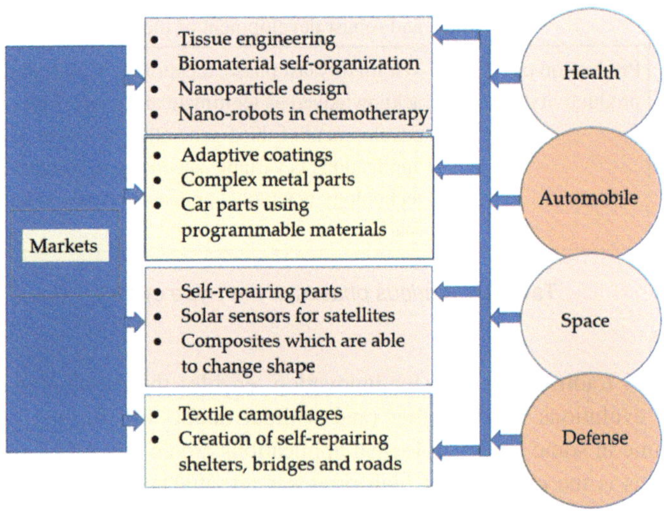

Figure I.2. *Possible applications for 4D printing methods. For a color version of the figure, see www.iste.co.uk/andre/printing3.zip*

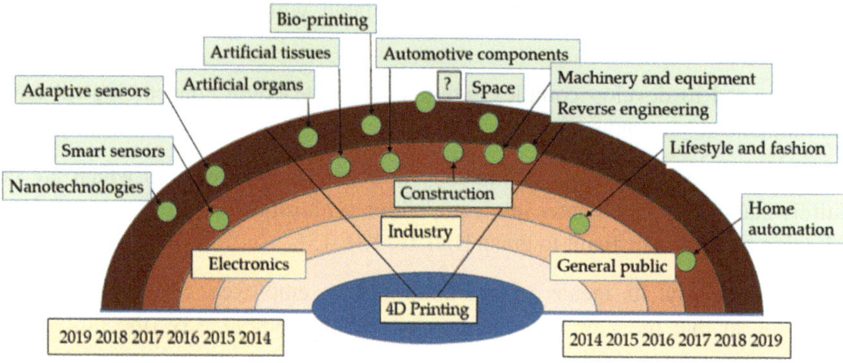

Figure I.3. *Possible applications (which are potentially positive) for 4D printing methods. For a color version of the figure, see www.iste.co.uk/andre/printing3.zip*

It may be the same for applications (including the medical aspect targeted by 4D Printing [4DP 14] with the possibility of having adaptive 3D structures available; the military aspect with possibilities for camouflage and active armor and other equipment [4DP 14a]). Other applications may be envisaged to produce chemical reactors, smart filtration processes (the active filtration function to process pollutants and other devices) and active valves [BAK 15, TSA 13]. The list may be indeed much longer.

Although additive manufacturing and its basic concept is around 30 years old, these two coupled domains are much younger. However, it is possible to base numerous aspirations within a quasi-virgin domain, with barriers that will have to be overcome for it to flourish, for various reasons:

– the development of additive manufacturing leads to consequent investments for the implementation of one (or more) solution(s) for additive manufacturing production. DGE [DGE 16] state "it takes a long time to achieve a return on investment as the fixed costs are high and the given entity has to reach a critical mass to absorb them";

– scientific and technological exploration of the "traditional" process–material couplings to be met, with an emerging community and a high external demand [GRE 15, WEL 15];

– funding orientated towards incremental innovations having a high potential for useful concrete applications and economic benefits [LES 16];

– inertial approach to research;

– low “stock” of intellectual capital available within academic or industrial research milieu;

– the need to develop complex intelligence by teaching epistemological bases and fundamental concepts of complex thought [IES 16];

– the complex emergence of radically new and disruptive ideas which are not entirely stabilized, which enter into competition with robust technologies, involving risk-taking which is difficult to grasp [INI 16, KIE 15];

– the company, its means of operation and its employees should adapt, even transform so that the new concept develops in an optimal way (the problem of managing change for Guillard [GUI 16]);

– difficulty of interdisciplinary approaches and support for creativity;

– the approach of new forms of complexity;

– other factors.

It will be understood that if the previous chapters of Volumes 1 and 2 had already illustrated some unstructured aspects in terms of original research proposals, this descriptive and open nature will be maintained in this volume because the “paradigmatic nucleation” of the field is still required. The innovation cycle is made up of three main phases, the phase with R&D investment to demonstrate the “revolutionary/ground-breaking” invention, the phase of testing when the company transforms the given innovation into a saleable product, where it attracts a proportion of the public for whom awareness has been raised by the given novelty, and the phase of introducing the product to the market [DOC 15].

Figure I.4 from Potstada and Zybura [POT 14] concerns creativity and economic forecasting applied to additive manufacturing. It highlights two aspects. The lower part of the figure shows incremental improvement and the top part shows a given disruption. It is this axis which is targeted in this part.

Obviously, we are positioned at the beginnings of a genuine breakthrough which will moreover be of use in discussions on bio-printing. The 3D sphere is subject to a significant estimated demand by citizens of many parts of the world [ZAR 16, LEI 16]. The achievement of these technologies by opening up new horizons towards 4D aspects and intelligent matter make up the following part.

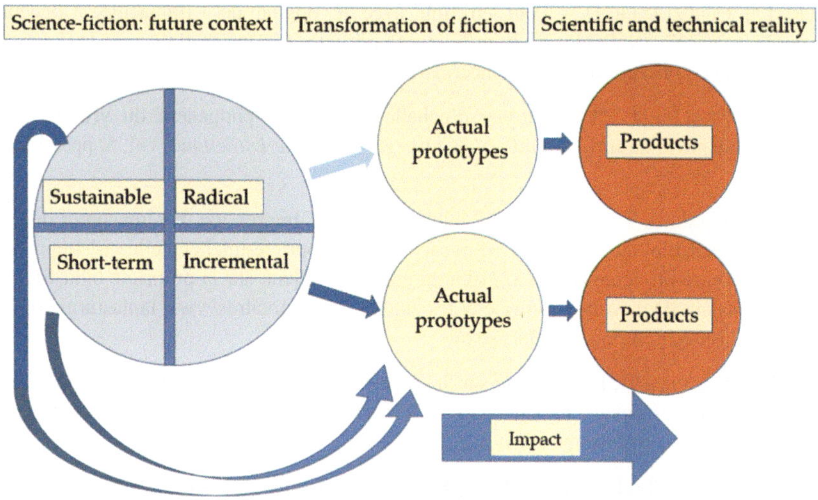

Figure I.4. *Transition from science fiction to innovation. For a color version of the figure, see www.iste.co.uk/andre/printing3.zip*

Ray Kurzwell has said "by 2020 there will be a whole host of products available immediately to buy for pennies on the dollar and to print straight away. It will become the norm for people to have printers in their homes" (quoted by Jackson [JAC 16]). It remains, unlikely however, that these will be 4D printers.

I.1. Bibliography

[4DP 14a] 4D PRINTING, "Three medical uses for 4D printing", available at: http://www.4dprinter.net/4d-printing-medical-applications/, 2014.

[4DP 14b] 4D PRINTING, "Three reasons why the military is interested in 4D printing", available at: http://www.4dprinter.net/three-reasons-why-military-interested-4d-printing/, 2014.

[ALA 69] ALAIN E.C., *Propos sur la religion*, Presses Universitaires de France, Paris, 1969.

[AND 16] ANDRÉ J.C., "Impression 3D d'organes", *Colloque National MECAMAT AUSSOIS* available at: http://aussois2016.event.univ-lorraine.fr/DOCUMENT/expose_du_soir_JC_Andre.pdf, 2016.

[BAK 15] BAKARICH S.E., ILL R.G., IN HET PANHUIS M. *et al.*, "4D printing with mechanically robust, thermally actuating hydrogels", *Macromolecular Rapid Communications*, vol. 36, pp. 1211–1217, 2015.

[BUN 03] BUNGE M., *Emergence and Convergence: Qualitative Novelty and the Unity of Knowledge*, University Toronto Press, Toronto, 2003.

[CER 17] CÉRÉZUELLE D., "Pour en finir avec le dogme de l'Immaculée Conception de la science", available at: http://sciences-critiques.fr/pour-en-finir-avec-le-dogme-de-limmaculee-conception-de-la-science/, 2017.

[CHA 12] CHALINE J., "La relativité d'échelle dans la morphogenèse du vivant: fractal, déterminisme et hasard", *Annales de Chirurgie Plastique Esthétique*, vol. 5, pp. 517–526, 2012.

[CHA 16a] CHARLES T., "La Propriété intellectuelle et l'Impression 3D: rencontre du 3ème type", available at: http://www.finyear.com/La-Propriete-intellectuelle-et-l-Impression-3D-rencontre-du-3eme-type_a37216.html and Livre Blanc sur la propriété industrielle de l'Impression 3D dans la plasturgie et les enjeux réglementaires/www.laplasturgie.fr/livre-blanc-plasturgie-impression-3d/, 2016.

[CHA 16b] CHARLES T., Livre Blanc sur la propriété industrielle de l'impression 3D dans la plasturgie et les enjeux réglementaires, White Paper, available at: www.laplasturgie.fr/livre-blanc-plasturgie-impression-3d, 2016.

[DGE 16] DGE, "Technologies clés – Préparer l'industrie du futur 2020", available at: http://www.entreprises.gouv.fr/files/files/directions_services/politique-et-enjeux/innovation/technologies-cles-2020/technologies-cles-2020.pdf, 2016.

[DOC 15] DOCKNER E.J, SIYAHHAN B., "Value and risk dynamics over the innovation cycle", *Journal of Economic Dynamics and Control*, vol. 61, pp. 1–16, 2015.

[FOR 02] FORAY D., "Ce que l'économie néglige ou ignore en matière d'analyse de l'innovation", in ALTER N. (ed.), *Les logiques de l'innovation*, La Découverte, Paris, 2002.

[FRO 14] FROST AND SULLIVAN, "Advances in 4D printing", available at: http://www.frost.com/sublib/display-report.do?id=D545-01-00-00-00&bdata=aHR0cDovL3d3dy5nb29nbGUuZnIvdXJsP3NhPXQmcmN0PWomcT0mZXNyYz1zJnNvdXJjZT13ZWImY2Q9MiZ2ZWQ9MGFoVUtFd2kwNWRPbmxKTFNBaFdDVVJvS0hhcklDRlFRRmdnc01BRSZ1cmw9aHR0cCUzQSUyRiUyRnd3dy5mcm9zdC5jb20lMkZwcm9kJTJGc2VydmxldCUyRnJlc29sdmVob3N0JTJGZDU0NSZ1c2c9QUZRakNORS01NWFKN0FFZ2ZXUEVKRU16N01nNFJyd0lzQUB%2BQEJhY2tAfkAxNDg3MTYzNzU2Mjg3, 2014.

[GAI 16] GAIGNAIRE A., "Ingénieurs du futur recherchés", available at: https://www.lesaffaires.com/dossier/profession-ingenieur/ingenieurs-du-futur-recherches/591065, 2016.

[GHI 16] GHILASSENE F., "L'impression 3D et la question de la redevance pour copie privée, Report, Observatoire de la Propriété Intellectuelle", available at: https://www.inpi.fr/sites/default/files/impression_3d_redevance_pour_copie_privee.pdf, 2016.

[GAR 15] GARTNER, "Gartner hype cycle", available at: http://www.gartner.com/technology/research/methodologies/hype-cycle.jsp, 2015.

[GRE 15] GRESS D.R., KALAFSKY R.V., "Geographies of production in 3D: theoretical and research implications stemming from additive manufacturing", *Geoforum*, vol. 60, pp. 43–52, 2015.

[GUI 14] GUIGUE S., "L'impression 4D rend la matière programmable", available at: http://www.atelier.net/trends/articles/impression-4d-rend-matiere-programmable_430671, 2014.

[GUI 16] GUILLARD J.C., "L'innovation appliquée, l'antidote contre le Darwinism Digital", Press release available at: http://www.journaldunet.com/solutions/expert/65558/l-innovation-appliquee--l-antidote-contre-le-darwinism-digital.shtml, 2016.

[INI 16] INIGO E.A., ALBAREDA L., "Understanding sustainable innovation as a complex adaptive system: a systemic approach to the firm", *Journal of Cleaner Production*, vol. 126, pp. 1–20, 2016.

[INR 16] INRA, "Le tour de force perceptif des plantes pour se maintenir à la vertical", Press release, available at: http://www2.cnrs.fr/sites/communique/fichier/cp_moulia.pdf, 2016.

[IES 16] IESF, Relever les défis d'une économie prospère et responsable, White paper, available at: http://home.iesf.fr/752_p_43968/parution-des-propositions-des-iesf.html, 2016.

[JAC 16] JACKSON B., "Futurist Ray Kurzweil makes 3D printing predictions", available at: https://3dprintingindustry.com/news/futurist-ray-kurzweil-reiterates-3d-printing-predictions-100504/, 2016.

[KIE 15] KIETZMANN J., PITT L., BERTHON P., "Disruptions, decisions, and destinations: enter the age of 3-D printing and additive manufacturing", *Business Horizons*, vol. 58, pp. 209–215, 2015.

[KUH 62] KUHN T., *The Structure of Scientific Revolutions*, University of Chicago Press, Chicago, 1962.

[KUP 00] KUPIEC J.J., SONIGO P. (eds.), *Ni Dieu, ni gène; pour une autre théorie de l'hérédité*, Points Sciences, Paris, 2000.

[LE 04] LE MOIGNE J.L., "Quoi de plus naturel que les sciences de l'artificiel?", in SIMON H.A. (ed.), *Les sciences de l'artificiel*, Gallimard, Paris, 2004.

[LEI 16] LEIST S.K., ZHOU J., "Current status of 4D printing technology and the potential of light-reactive smart materials as 4D printable materials", *Virtual and Physical Prototyping*, vol. 11, pp. 249–262, 2016.

[LEM 17] LE MONDE, "Le dictionnaire des citations", attributed to Winston Churchill, available at: http://dicocitations.lemonde.fr/citations/citation-44664.php, 2017.

[LES 16] LESAGE S.E., HURTEAU P., L'innovation en recherche universitaire: Austérité et arrimage aux intérêts du secteur privé, Report, available at: http://iris-recherche.s3.amazonaws.com/uploads/publication/file/web_NouveauTexteEtude_RechercheUniversitaire_vli_02.pdf, 2016.

[MAJ 13] MAJIDI C., "Soft robotics: a perspective – Current trends and prospects for the future", *Soft Robotics*, vol. 1, pp. 5–11. 2013.

[MAN 14] MANGIOLA S, "4D printing on the rise", available at: http://machinedesign.com/3d-printing/4d-printing-rise, 2014.

[NAD 16] NADEAU E., *Philosophies de la connaissance*, Presses Universitaires de Montréal, Montreal, 2016.

[POT 14] POTSTADA M., ZYBURA J., "The role of context in science fiction prototyping: the digital inustrial revolution", *Technological Forecasting & Social Change*, vol. 84, pp. 101–114, 2014.

[PRI 09] PRIGOGINE I., STENGERS I., *Le temps et l'éternité*, Champs Sciences, Paris, 2009.

[RAM 15] RAMIREZ-FERRERO M., "Market impact and perspectives of 3D printing technologies", available at: http://www.dima3d.com/en/market-impact-perspectives-3d-printing-technologies/, 2015.

[SHI 11] SHIRKY C., *Cognitive Surplus: How Technology Makes Consumers into Collaborators*, Penguin, New York, 2011.

[SMA 11] SMATI R., *Eloge de la vitesse*, Eyrolles, Paris, 2011.

[STA 15] STAUNE J., "*Les clés du futur: réinventer ensemble la société, l'économie et la science*", Plon, Paris, 2015.

[SUS 14] SUSAN R., "Retour sur la matière programmable", available at: http://www.internetactu.net/2014/07/11/retour-sur-la-matiere-programmable/, 2014.

[TSA 13] TSAI E.Y., 4D Printing; Toward Biomimetic Additive Manufacturing, Thesis, MIT, Boston, 2013.

[WEL 15] WELLER C., KLEER R., PILLER F.T., "Economic implications of 3D printing: market structure models in light of additive manufacturing revisited", *International Journal of Production Economics*, vol. 164, pp. 43–56, 2015.

[WOO 29] WOODGER J.H., *Biological Principles: A Critical Study*, Keagan Paul, Trench, Trubaer & Co., London, 1929.

[ZAR 16] ZAREK M., LAYANI M., ELIAZAR S. *et al.*, "4D printing shape memory polymers for dynamic jewelry and fashionwear", *Virtual and Physical Prototyping*, vol. 11, no. 4, available at: http://dx.doi.org/10.1080/17452759.2016.1244085, 2016.

1

Programmable Matter or Smart Matter, Stimulated Organization and 4D Printing

"It is not reason which speaks within technical objects, it is desire, and it is concrete existence which will bear the brunt of these dreams". [CÉR 11]

"The most successful entrepreneurs don't begin with brilliant ideas; they discover them. The great advantage of working in this way is that when trying to do something new and uncertain, we rarely know what we don't know". [SIM 11]

"The answer reassures us. The question merely serves to worry us. In the literal sense of the term, it compels us to react, indeed to invent. Raising the question, out of necessity, forces us to compel us to change our emphasis, to reconsider the hypotheses, to review the rules and to unearth the obscure". [LAG 12]

"Something that it has distractedly enabled [Man] into the present which would open up the future to him/her". [ZAM 89]

"Speed both demands and creates insensitivity to anything that can delay, to all forms of friction and hesitation which make us feel that we are not alone in the world; by slowing down, we again become capable of understanding, of getting acquainted with, of acknowledging what we have and what is transmitted to us, to think and imagine and, in the same process, to create with others relationships are not themselves captured". [STE 13]

"The concept of engineering is not realizing the given use until we discover its existence". [FRE 94]

"There are therefore no 'rules of induction' of general application, thanks to which we may derive the solution mechanically or infer hypotheses or theories from empirical data. To go from the stage of data to theory, it is necessary to undertake a creative work of imagination". [HAM 14]

"Initially, we learn that thinking too quickly may be a very bad idea. Infobesity surrounds us and it is a serious threat to our creativity". [LEW 16]

"The undertaking will be difficult, as it goes against the dogma of the Science's so-called 'Immaculate Conception'. According to this, scientific truths being designed without sin, are all beneficial. So that the sanctity of Research will not be diminished by the uses that the World puts it to and that it is not for those who are ignorant to discuss the use of funds – indeed always insufficient – which are allocated to them". [CÉR 17]

"'Festina Lente' (Hurry up slowly) was the motto of Augustus". [LEW 16]

1.1. Introduction

In the introduction to this book, DNA origami was mentioned as an example of a self-organizing system. Indeed, this was already chemically stimulated organization as it is necessary for us to have attached links in the DNA fixing onto the macromolecule to define a given form anticipated by the designer. The notion of stimulated self-organization used by the scientific community is perhaps not adapted to reality, or nature doing what it can. It is this issue which we will investigate in this chapter and indeed those that follow.

This is the case for molecular motors whose movement is enabled by external energy sources. Richard Feynman had predicted in the 1950s the development of molecular nanomachines such as those procured by Jean-Pierre Sauvage in 1983 when he invented a chemical method that enables the intertwining of two molecules taking the form of rings, thus forming a chain known as catenane [CNR 16, COL 01]. Thanks to an electric or light stimulation, it was widely shown that it is possible to cause molecular rotation in a controlled manner. This is similar to the given operation of a motor. We may therefore consider these examples as already being representative of what might be programmable matter. This has been of no interest to researchers for several years without an ability to detect a trend (whether linear or exponential) in the growing number of publications associated with the key words "programmable matter" (see Figure 1.1). The total number supplied by university libraries demonstrates a figure of the order of 1000 publications which should be compared to the 50,000–100,000 concerning additive manufacturing.

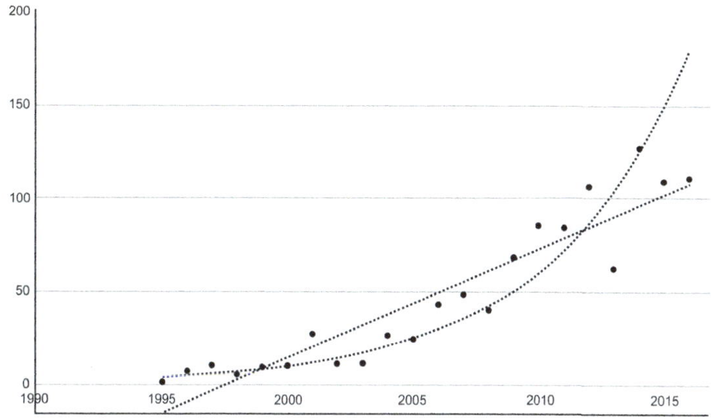

Figure 1.1. *Development of the number of publications on the subject "programmable matter" (University of Lorraine, France)*

As the focus of the subject is 3D printing, it appeared necessary to leave this sphere to approach the concept of additive manufacturing, which effectively involves the notion of additivity. We need at least two molecules to begin discussing additive manufacturing.

The aim of this chapter is to examine how we can exploit the overall principle of gathering separate components. We can thus make up a "self-organized" shape (to keep the term we used previously). Thus having reviewed this principle, it will be possible to envisage examples of self-organization which will partly serve as examples of 4D printing.

1.2. Natural (spontaneous) self-organization

Stating that matter can, in certain situations, be ordered means that the mold can, as a first approximation, perhaps be described geometrically in a simple way, by using for ease the concept of symmetry (for example, the case of a crystal). This is the case even if there are sub-structures which are more or less complicated, which can even take on a fractal aspect. However, the general form remains simple. We may think that structures, which are more or less self-organized, may be combined with some types of physical or physicochemical conditions [CHO 15, CHO 96].

The term "self-organization" designates a spontaneous or stimulated formation of a dynamic order leading to forms of organization stemming from a set of homogeneous or non-homogeneous units, which are interacting. Having placed the energy in the correct place and at the right time, in a given environment, the initial system, whether or not it is random, is organized to achieve a given form, if possible that sought by the given experimenter [MOR 04]. From a thermodynamic point of view, this "shaping" assumes that the transport of matter and energy within the system which is created should be subject to functional constraints in the absence of an external energy supply. The object is constructed, having the appearance of causal relationship (the laws of universal physics), which may be shown by a type of selection between several potential alternatives [CAM 74]: "Thus, its role is the construction of higher-level entities, formed from the (re)structuring of units which are intrinsically made up of material causality" [MOR 06].

1.2.1. *Nonlinearities*

The dissipative structures are processes in which fluctuation is stabilized to make up a dynamic macroscopic configuration. This is enabled by the flow of energy through the system under construction. These macroscopic structures necessitate, so as to increase in size, a continuous supply of energy in a range adapted to its cohesion. "The term 'self-organization' instead implies spontaneous creation of a given order or organization that refers to an agent-based identity, which is able to self-determine or form itself in a direction which involving a functional plasticity" [ALV 04]. Ultimately, we are concerned with respecting Bruter's concept [BRU 76]. He wrote, "Events shape the object with time. It is, above all, piercing the secrets of its history, of the line which it comes from, both the culmination and the projection".

A classic example is that of living systems which manifest behaviors and self-supporting capabilities. These include growth, differentiation, self-repair (at least partially) and other features. Such systems are capable of altering their behavior and their structure according to exterior parameters (nutrients, for example, temperature, gravity and other factors).

Quoted by Thuillier [THU 80], Stéphane Leduc wrote: "The elementary act of life is diffusion and osmosis". It was this professor of medical physics from Nantes who manufactured shapes and structures "resembling those of living beings" (in the early 20th Century). In 1910, he published a work entitled: "Physicochemical theory of life and spontaneous generations". Of course, it naturally conflicted with the stance of the French Academy of Sciences. Figure 1.2 represents one of the osmotic productions of Leduc [THU 80]. He wrote in 1912: "When we come to know the physical mechanism for the production of a given object or phenomenon... it becomes possible... to reproduce the object or the phenomenon. At that point, science becomes synthetic. Biology is a science like all others… it must sequentially be descriptive, analytical and synthetic" [LED 12].

Figure 1.2. *Osmotic productions by Leduc (copy from [LED 10]; see also [THU 80, EAS 09])*

This causalist theory and the methods flowing from it are to be applied to these constructs that aim to obtain finished and optimized products after "they have been initiated", in particular as regards matter and energy consumption, the notion of the voxel which may then reach the size of a molecule, indeed a cation or an atom then becomes relevant. The idea is that, as for all of the (or the majority of) engineering sciences, these methods are still often thought to be determinist: the various phenomena are controlled by laws that we must discover and apply as faithfully as possible. There might therefore not be (too much) space for the random.

Although this was the case, based upon an energy spatially positioned within a known manner, with cooperative components able to self-assemble, it would suffice to wait a "sufficient length of time" to reach an equilibrium leading to an object taking a given shape. Moreover, resuming the same experiment, all things being equal, should provide the same object (taking the same shape and size). Thus, by using this causalist hypothesis, it is not possible to predict the evolution of shapes by inferring them uniquely from the application of simple laws which may be able to be identified. This result illustrates an evolution that develops in a non-determinist way by the interaction of natural systems both with each other and their environment. However, by way of example, Prigogine [PRI 94] recalls this experiment. If we cool sodium chlorate at rest, the crystals produced are optically inactive. If, during the same time period, we shake the solution, we obtain either levorotatory crystals or dextrorotatory crystals. This is a bifurcation. The first provides either a skewed shape or a straight form and the initial germs which will participate in the crystallization process are either skewed or straight. There is therefore a break in the symmetry of the reactive system. This example serves to show that irreversible phenomena do not necessarily reduce down to an increase in chaos. However, there can exist constructive roles for this irreversibility of time as is the case with life. Hawking [HAW 89] recalls on this subject: "A strong thermodynamic arrow is needed for intelligent life forms to act". These aspects will be reconsidered in Part 2 on bio-printing.

In these conditions, Turing [TUR 52] showed, using other examples, how reactions between chemical substances that he called morphogens, coupled with a process of diffusion of these substances through living tissues, could give rise to the appearance of specific structures or regularities (spots on a coat, the shape of a crystal, a shell or other examples). In his founding article, he introduced a mathematical model known as "reaction-diffusion equations" as well as the study of this model in certain relatively simple cases.

The dynamics of this basic model which he explains results from the coupling between chemical reactions and reagent diffusion, within a system which is continuously fed. The Turing structures correspond to periodic spatial variations of concentrations of chemical species [LES 13, ZEN 13, SHA 15, JOS 12] which compete with the mechanisms for molecular transport. To observe spatial variations in concentrations, it is therefore necessary for other mechanisms to come into play. It introduces a scenario involving two species A and B, known as "morphogens" and the combination of the following properties:

– species A is capable of inducing its own production;

– species A can produce the second species B;

– species B inhibits the production of A;

– B diffuses quicker than A.

It is thus sufficient that a small local fluctuation at a given point in space induces a slight excess of A so that the production of both A and B is higher than the average at this point. B, diffusing quicker than the excess of A, creates an inhibitory crown around the initial point, isolating the peak of A by an area which is richer in B. These random evolutions can happen in other places in the given area with other peaks of species A. "These activator peaks arise from selective ranges of fluctuations, which in very small quantities yet numerous, which spontaneously modifies the state of the homogeneous mixture. The distribution which is dynamically the most stable of these peaks forms a periodic pattern, such as spots that we may note on the cheetah" [CHO 15].

James Clerk Maxwell wrote in 1876 (cited by Thom, [THO 83]: "When the state of affairs is such that an infinitely small variation in the present state will only alter the future state by an infinitely small quantity, the state of the given system, whether at rest or moving is known as stable. However, when an infinitely small variation in the present state may cause a finite difference in a finite time, the condition of the system is known as unstable. It is obvious that the existence of unstable conditions makes the prediction of future events impossible, if our knowledge of the present state is only approximate and inaccurate". The question of the control of the "correct" main parameters of influence on stability or instability is therefore paramount, but risks not being sufficient.

1.2.2. *Achieving the desired form?*

Once science starts to understand the way that nature has selected certain forms and not others, as much in the physical as the biological world, it can seek to apply/transpose corresponding mechanisms with a view to resolving engineering problems. This is advantageous for the manufacturing of given artifacts, tools or final articles. The large variety of given applications in a generic form such as the manufacture of a functional object is not inspired from nature. It is worthwhile investigating this if we seek to use knowledge of "morphogenesis" to achieve the desired form through self-organization [BAQ 04].

This is the inverse problem applicable to the direct manufacturing of objects (and not by additive manufacturing, on a voxel-after-voxel basis) and to create (in a way) three-dimensional structures. These are of varying sizes from a pseudo-homogeneous environment. This is a traditional problem "normally" not respecting entropy principles. It forces us to know from what energy and in what form it should be available. This is because the process is anti-dissipative. Indeed, there is nothing in the demonstration of the second principle, which enables us to assert that there exists a causal relationship between entropy variation and evolution towards a chaotic state. However, as soon as several entities interact, or the effects are not proportional to causes, there may be the possibility to make nonlinear characteristics available, with specific effects (for example, snowflakes within systems outside the equilibrium environment with the provision of continuous matter or "photochemical muscles", which introduce tensions of the manner of mammalian muscles). A multi-scale approach could be an asset to dissipate a given energy surplus. In these recursiveness conditions, the structure is as it were both a cause and consequence of this dissipation (self-similarity, see Figure 1.3). Self-organization thus gives the impression that the structure emerges from the combination of interactions between constitutive entities (including energy).

Thom [THO 83] analyzed problems connected with the control of morphogenesis by relying upon the notion of the model "black box" with its inputs and outputs (some of which revert to the recursiveness of inputs). When choosing Ui inputs, it is possible to measure Sj outputs. This is provided that their existence has been identified and, thereafter, to define a relationship between the Ui and Sj respective inputs and outputs, as Turing proposed in [TUR 52]. The central question, well known in process engineering, is from the experiments completed to reconstruct the inner workings, which happens within the black box. Within complex systems, there may exist "exceptional" situations for which the relationships {Ui, Sj} come out what we expect as "normal". The research into these "divergent", but real, situations may lead to new and necessary information becoming more apparent as to what happens in the "box". At such times, if we do not know where to look (and what to find) in these situations from an experimental point of view, it indeed falls to modeling to help search for peculiarities. This takes place at different scales.

Figure 1.3. *Self-similarity – fractal pyramid produced in the 1990s [DIO 92/93]. For a color version of the figure, see www.iste.co.uk/andre/printing3.zip*

Laurent Larger [LAR 15] states that nonlinear dynamic systems apply to all practical situations or movements appearing in space and/or time and for which there is a determinist description of development, taking the form of a differential equation (solely the temporal dynamic) or a partial derivative equation (spatio-temporal dynamics). Our world is of course 100% dynamic and nonlinear, but the current technological control relating to its operation is very often limited to situations with linear approximations. The example of morphogenesis illustrates not only the relevance of the nonlinear approach (a simple, deterministic, local dynamic model explains how nature constructs globally stable, reproducible, and robust forms, at the macroscopic level), but also the gap of scientific progress existing in the understanding, and then in the mastery and exploitation, of the associated mechanisms for concrete applications, with enormous technological potentials.

Finally, we must underline here the significance of the local and global notions within nonlinear dynamic phenomena. It is indeed necessary to understand the fact that a highly complex solution, at the global level, can effectively emerge in a determinist way from a uniquely local property (the dynamic determinist equation and local nonlinear coupling, which apply to all of the environments studied). It is relatively simple within its formulation.

From a point of view of modeling within the sphere, we may consider a system that evolves in time stemming from a given form to achieve another form (a dynamic approach) or impose upon stable form constraints of different forms (chemical, energy or other spheres) which will lead to a distortion (a thermodynamic approach). It may be useful to introduce, in the first instance, temporal elements which enable the definition of development at a given time. In the second instance, thermodynamic and chemical potential components are introduced. That being so, starting from a given Ui set, it is possible to define the Sj outputs in theory, provided that all determining factors are known. The converse problem aims to define the Sj outputs defining instructions and researching the Ui which meet the objective. This is a traditional problem that is always difficult in the case of linear systems, which are generally poorly conditioned (see deconvolution not least owing to noise and measurement errors; see [MUG 08]). This will be all the more the case since we will be positioned within nonlinear approaches with loud signals. That being said, the issue is the size as the control of the inverse problem can constitute an integral means to produce an object from a matrix which is easy to produce. The latter aspect and the distortion function in terms of time and energy supplied may enable the production of an object taking a complex form.

Chauvet [CHA 06] considered that when two units A and B are subject to certain conditions, they can do nothing other than join together (see the works of Turing mentioned in the introduction). For example, the dynamics of two biochemical pathways carrying a reagent from one to the other would be, per this principle, more stable than the dynamics of each entity when separated. This is the principle of stabilizing self-association per Chauvet, [CHA 06], Staetzler, Sonnenschein and Soto, [STA 11] and, Nykter *et al.* [NYK 08]. It can operate the same for enzymes (see [KEK 15]) with divisions between species which necessitate limited transport by molecular diffusion (solution of the Von Smoluchowski equation within a potential field; Kekenes-Huskey *et al.* [KEK 12]). Self-organized systems can be very stable relative to the range of a number of parameters, but they can diverge suddenly according to small variations of one or several influencing variables to reach a new stable state [NYK 08]. "Turing" evolutions with fractal approaches can also be put forward (see [BIZ 11]).

Thus, between the possible illusion of perceptible self-organization and the inverse problem, constituting a significant challenge, allowing the production of the researcher's or engineer's desired organized structure, there is a considerable gap to get over. Would it perhaps be necessary to consider nonlinear, out of balance and multi-scale phenomena? Perhaps it would be necessary to view the biological aspect (as was introduced in Volume 1 with DNA and DNA origamis or "smart" matter. Other biomimetic solutions are also envisaged using nanoparticles: Sarikaya *et al.* [SAR 03] Nikitin *et al.* [NIK 10]. Slocik and Naik, [SLO 10], Tamerler and Sarikaya, [TAM 07], Chiu *et al.* [CHI 11], Maye *et al.* [MAY 09], Laxminarayana, [LAX 12]). These involved original experimental pathways from the traditional physicochemical sphere to produce customized three-dimensional structures (although going beyond the nanometric sizes of origamis). Perhaps it would also be necessary to move away from traditional reductionist methods to involve holistic thinking in order to explore this small element of complexity. Nowadays, this point partially moves away from the topic explored, but this work is an enormous task which, at least from this author's point of view, must be looked into more deeply. This will be shown with the example of bio-printing (see Part 2).

Put simply, what the engineer seeks is to place matter "in the right place" and "at the right time" so that the organization leads to the object sought (and not something to which to get closer or nearer). As when dealing with an insufficient control of systems, he/she is obliged to attempt to find situations where he/she resumes control of the overall complexity, through using "smart" matter. The results of these efforts are shown below.

1.3. "Smart" matter

The underlying concept is highly different. It starts from a given geometry so that it becomes, through the supply of energy, a mold corresponding to a given application (see Figure 1.4). By playing with spatial division with a range of form of this energy (from these energies), there may be also the possibility of having an adaptive system, such as an actuator, a structure evolving in time and perhaps in its functionalities [ZAR 17, LEH 02]. In passing, ideas of this sort were proposed in comics ([TUR 15]; Franquin published in Beaux-Arts, [BEU 15]). Only science-fiction films approach this type of subject thoroughly.

Figure 1.4. *The principle of using "smart" matter*

This context does not mean spontaneous complexity discussed above has been abandoned, but rather that the engineer and/or the designer attempts to take control of automated phenomena which are difficult to control, by accepting the manipulation of non-living matter and energy, so that initial shapes are tailored by their will. This section thus explores stimulated conversion of matter pathways. Lehn [LEH 02], said, backed up by Figure 1.5 in this paper, the "smart" matter referred to her falls within the exploration of adaptive supramolecular systems.

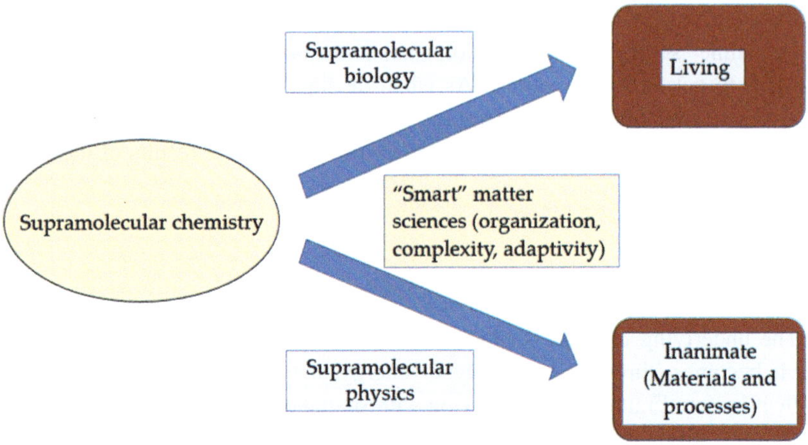

Figure 1.5. *Smart matter and supramolecular systems. For a color version of the figure, see www.iste.co.uk/andre/printing3.zip*

1.3.1. *Active polymers: photochemical muscles*

One particular approach concerns the effect of light on polymer structure (the now famous expression "photochemical muscles"; Nakano, [NAK 10], Yoshino *et al.* [YOS 10]; Yu *et al.* [YU 03], Yu *et al.* [YU 15], Khoo *et al.* [KHO 15], André, [AND 17]) as Figure 1.6 indicates. This is extracted from Kuksenok and Balzs [KUK 13]. The concept involves irradiation inducing internal rotations in the material, which contributes to change the geometry of the irradiated area leading to distortions, which depend upon the flow received. The latter may be reversible if we have photochromic material available (which would both induce provisional conversion and be reversible).

1.3.1.1. *Photochemical changes*

All of these processes, apart from photodegradation, are shown in Figure 1.7 stemming from Khim *et al.* [KHI 14]. The curvature of polymeric fibers can be connected to a non-homogeneous division of photoproducts (see [GE 16a, AND 17]).

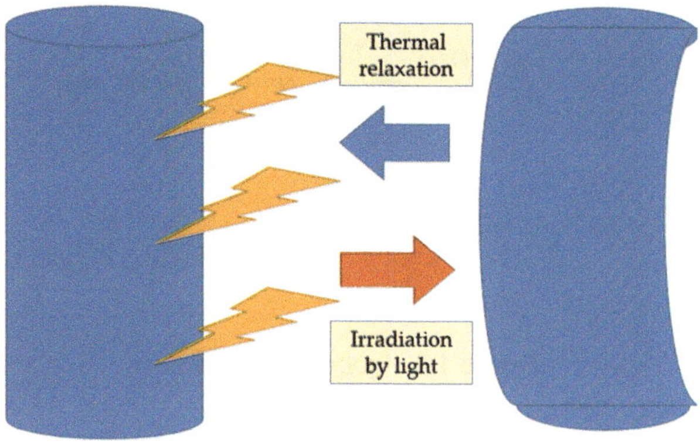

Figure 1.6. *Distortion induced by the light of a photo retractable polymer cylinder. For a color version of the figure, see www.iste.co.uk/andre/printing3.zip*

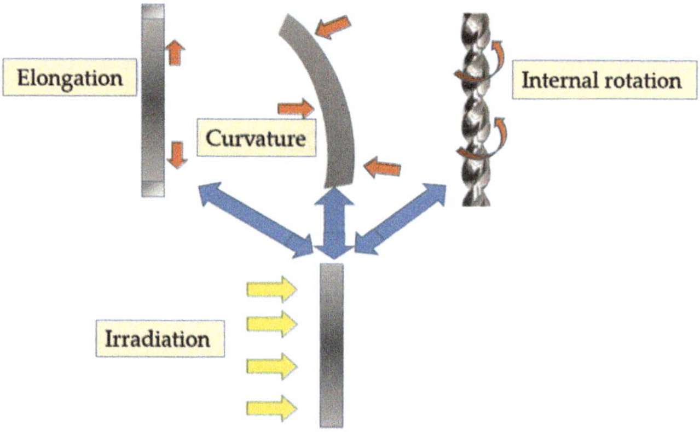

Figure 1.7. *Photomechanics of polymers. For a color version of the figure, see www.iste.co.uk/andre/printing3.zip*

Another example, stemming from Mahimwalla's thesis [MAH 13], is the significant distortion of polymers containing azobenzene groups (see Figure 1.8 and [MAH 12]). These groups through irradiation go from configurations Trans to Cis depending upon the given wavelength used (see Figure 1.9). The effects are noteworthy as can be seen in Figure 1.12 (see also [DES 15, ROS 15]).

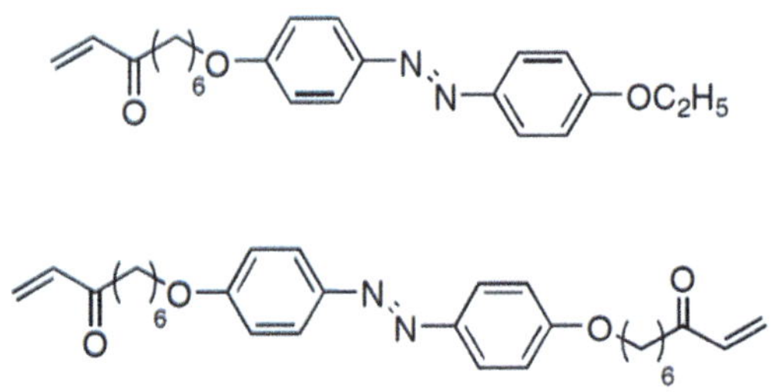

Figure 1.8. *Azobenzene polymers*

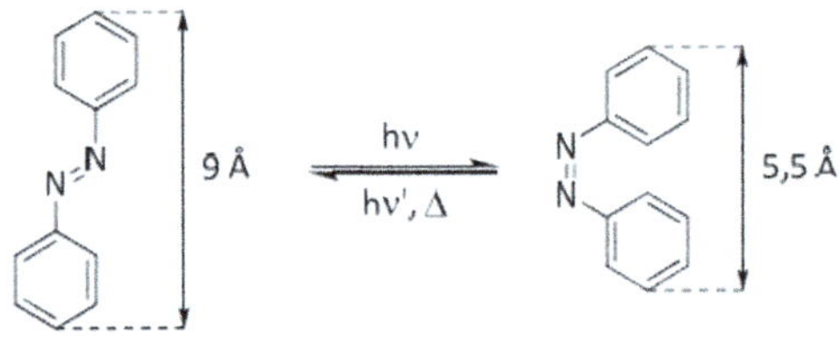

Figure 1.9. *Sizes of two isomers [DEL 07]. The conversion from Trans to Cis operates by ultraviolet means (300–400 nm), returning to waves which are visible (being between 400 and 500 nm)*

The azobenzene molecule is generally found in the Trans conformation, which may be considered as uniaxial. When this molecule is exposed to visible light, it can undergo photoisomerization to a Cis conformation (see Figures 1.8 and 1.9). The given molecule may then return to the Trans conformation consecutively with a thermically or optically induced conversion. Once this cycle is completed, the molecule may be located moving in a new direction. This process may be used to act upon the macroscopic direction of the molecules of a given film. When polarized light is linearly used to activate the cycle of photoreorientation Trans -> *C*is, only molecules for which the electronic absorption transition moment is located within the given direction (or within the vicinity) of the polarization absorb light and undergo a change of direction. In this way, it is possible to reduce the number of molecules oriented parallel to the direction of polarization and to increase the number of molecules oriented at right angles to this direction [ROC 15].

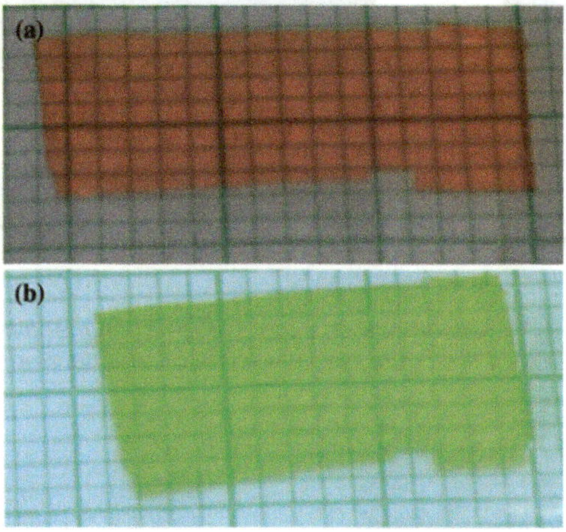

Figure 1.10. *Effect of irradiation on a polymer film containing azobenzene groups: a) before irradiation; b) immediately after irradiation according to Mahimwalla et al. [MAH 13]. For a color version of the figure, see www.iste.co.uk/andre/printing3.zip*

These authors present the fairly spectacular results of Yamada *et al.* [YAM 09], corresponding to the distortion of an induced polymer film, both by UV irradiation (conversion Trans -> Cis) at 366 nm (240 mW.cm^{-2}) and in the visible domain (λ > 540 nm and 120 mW.cm) for the inverse conversion. Figure 1.11 kindly supplied by Kevin Ge [GE 16b] shows a reversible conversion of a polymer which is distorted with temperature. Figure 1.12 by Cheng *et al.* [CHE 10] completes the previous figure: having a double layer of polymers, one of which contains azotolane which can under a given isomerization Cis - > Trans thermic and Trans -> Photochemical Cis under visible light irradiation, it is possible to induce a temporary withdrawal of the irradiated film, followed after the irradiation ceases by a return to the initial position (see also Figure 1.13 by Zhang *et al.* [ZHA 15]). Other works are able to respond to this conversion principle [ERC 10, ILI 11, PET 15, GAO 16]. We may note that in view of the molecular structure of azobenzene and azotolane, it may be possible to take account of the direction of the transition moments of these molecules placed within a polymeric system to make conversions within polarized light determined with the given space, to create programmable diffraction networks (see above). Robotic applications are envisaged by Umedachi, Vikas and Trimmer [UME 13].

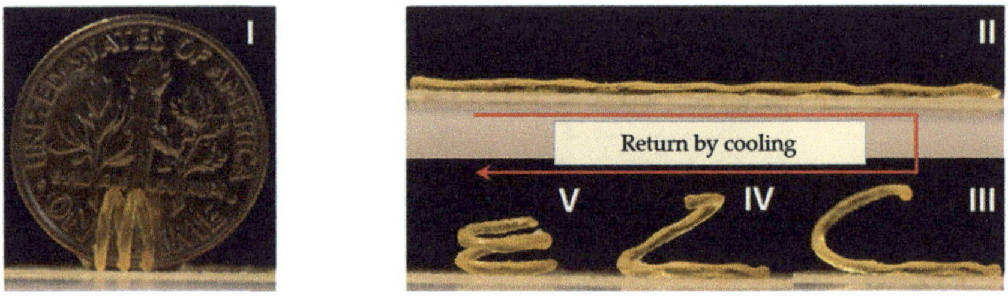

Figure 1.11. *Distortions of a light-induced polymer film. For a color version of the figure, see www.iste.co.uk/andre/printing3.zip*

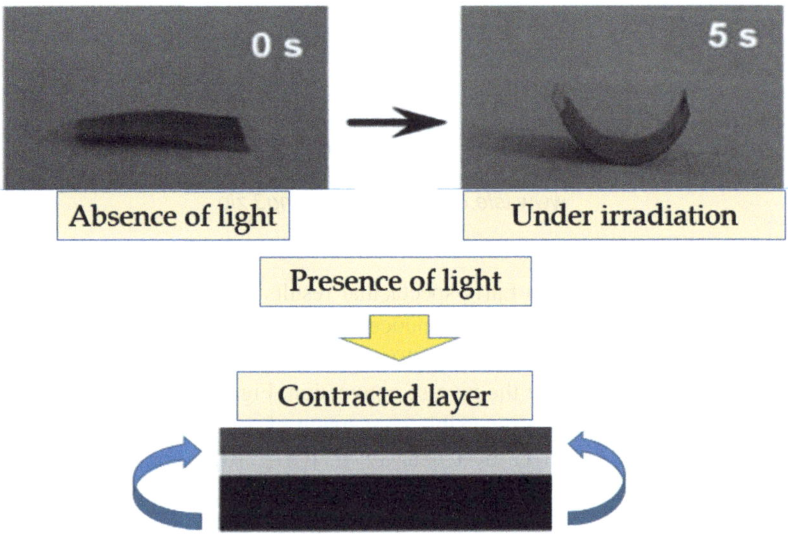

Figure 1.12. *Bimetallic element subject to light irradiation. For a color version of the figure, see www.iste.co.uk/andre/printing3.zip*

Such systems can be produced by 3D printing with the assistance of various passive materials, but highly elastic, for example, with aliphatic acrylic epoxy-based monomer mixes (monofunctional) and urethane-based (bifunctional) substances. Details of these were recently published by Patel *et al.* in 2017 [PAT 17].

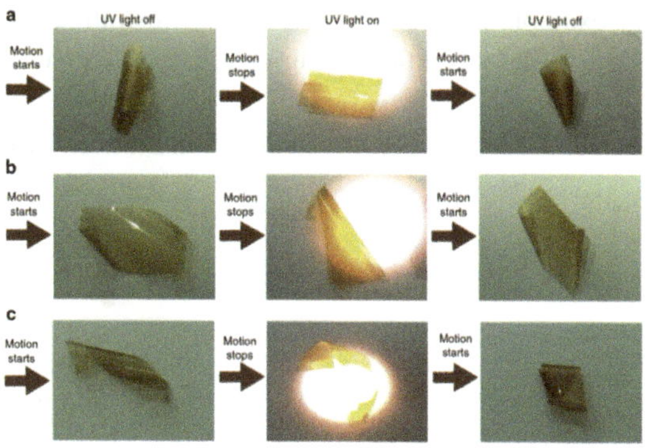

Figure 1.13. *Distortions of a light-induced polymer film according to Zhang et al. [ZHA 15]. For a color version of the figure, see www.iste.co.uk/andre/printing3.zip*

1.3.1.2. *Thermic effects on polymers*

Zhao, Qi and Xie [ZHA 15] said that it is possible to have at one's disposal polymeric materials, the shape of which may either depend on the temperature or on photochemical conversions. These authors in their experiments were seeking to find reversible reactions affecting the space between molecules or macromolecules (see Figure 1.14).

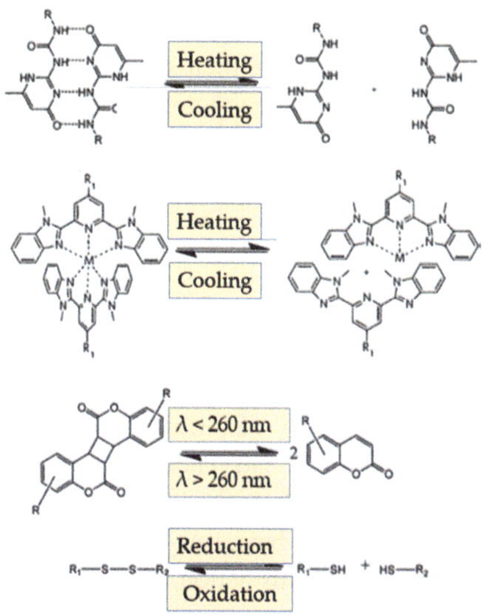

Figure 1.14. *Reversible systems having volume changes*

An explanatory example is given in Figure 1.15 illustrating the change of shape. Other examples are given in this article using various active materials.

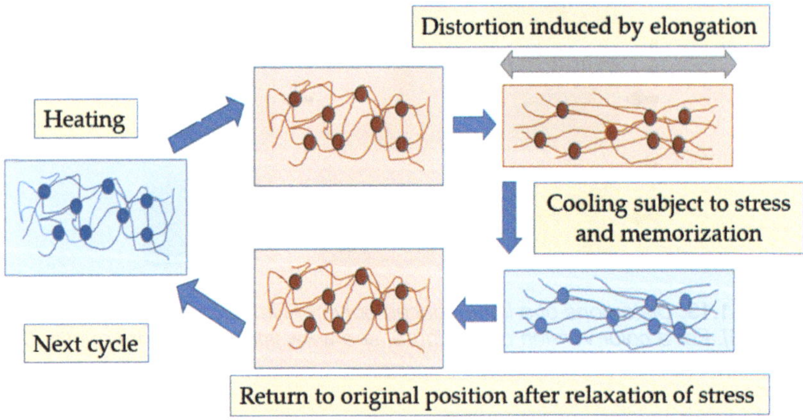

Figure 1.15. *Principle of thermically induced reversible distortions. For a color version of the figure, see www.iste.co.uk/andre/printing3.zip*

Inks produced from hydrogels are capable of inducing distortion by humidity and by temperature [NAF 17]. Le Duigou *et al.* [LED 16] used wood particles as a charge within a given polymer. Oriented within shearing extrusion of this fusible polymer, they retain the memory for their given direction below the material's freezing point and can be seen to change their shape in the presence of humidity (or heating). Finally, from optimization view point, Zhang *et al.* [ZHA 11] added to a material sensitive to heat and light (poly-(N-isopropyl-acrylamide or pNIPAM), charges in carbon nanotubes. This facilitated the transfer of heat within the mass of material and the absorption of light by the given surface.

1.3.1.3. *Other systems*

In the same vein, a "paper" developed by researchers from the Donghua University in China may be qualified as "robot origami" with a composite material capable of reacting to temperature by distorting. A double-layer flat material may be, according to Rochefort [ROC 15], half made up of an oxide of graphite, the other layer of polymerized dopamine or polydopamine, a material that reacts upon contact with water. When exposed to light or heat, this layer shrinks. The other layer, having good mechanical strength, reacts to this action by bending or folding up. By using a high-precision infrared laser, researchers were able to develop tiny pieces of paper, capable of moving on demand. This research may serve to produce artificial muscles. Other examples are shown in Figure 1.16 of polymers sensitive to localized stimuli within space and which can be sensitively distorted. These were researched by Kevin Ge *et al.* [GE 16b] – see also Zark *et al.* [ZAR 15].

Figure 1.16. *Another example of polymers sensitive to given stimuli and validation by additive manufacturing. For a color version of the figure, see www.iste.co.uk/andre/printing3.zip*

For their part, Grossweiler *et al.* [GRO 15] used polymers containing spiropyranes which are sensitive to mechanical stress. These may become distorted and change color with a relaxation time of several minutes (see Figure 1.17). This type of material enables them to produce micro pliers.

There is another example that taken from the works of Chen *et al.* [CHE 17]. Active polymers, ready to respond to a blue-light stimulation (from an LED), containing trithiocarbonates (TTC) groupings with an accordion-like structure may be activated by organic catalysts. In the presence of light, monomers are formed (the inverse process of polymerization), which indeed has the effect of drawing out the macromolecule. The monomers are distributed uniformly throughout the structure. They confer new properties on the material, which then enables us to modify the shape of the object. This complex technique necessitates the specific conditions for use.

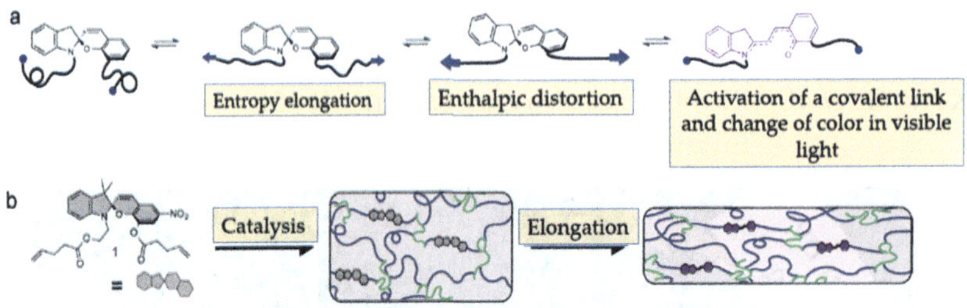

Figure 1.17. *Effect of a mechanical constraint on the chemical conversion of a spiropyrane-based polymer. For a color version of the figure, see www.iste.co.uk/andre/printing3.zip*

Table 1.1, partly extracted from Gao *et al.* [GAO 16], gives indications as to convertible matter and upon the methods of conversion applied within this domain.

Type of materials	Mechanism and commentary	Bibliographical references
Polyethylene glycol; lipid bilayers	Contact with water	Jamal *et al.* [JAM 13], Villar, Graham and Bayley [VIL 13], Villar, Heron and Bayley [VIL 11]
Polymers at low-transition temperature (PNIPAAm; PCL); alumina ceramics	Transition temperature	Peppas *et al.* [PEP 00], Wang *et al.* [WAN 15], Bakarich *et al.* [BAK 15]
Magnetic materials	Precise positioning by the given location of the magnetic field	Kokkinis, Schaffner and Studart [KOK 15]
Polymers containing light-induced isomerizable components	Photochromic compounds	Zhang *et al.* [ZHA 15]

Table 1.1. *Various types of smart matter and associated mechanisms*

1.3.1.3.1. Biological materials

Natural plants demonstrate spontaneous curvature alterations due to the direction of the cellulose, which is subject to an anisotropic swelling when in the presence of water. Gladman *et al.* [GLA 16] imitated this process using a hydrogel containing such rigid cellulose fibrils. They spontaneously become aligned with shear points at the time of their flow through the nozzle of the extruder of the printer, which induces both swelling anisotropies and elastic module anisotropies. With a dual layer positioning, it is possible to achieve swelling differentials, hence a calculable curvature for the given object. By using anisotropy within the material, Qe *et al.* [GE 16b] show in Figure 1.18 how it is possible to develop the form of an object according to the stimulation it receives.

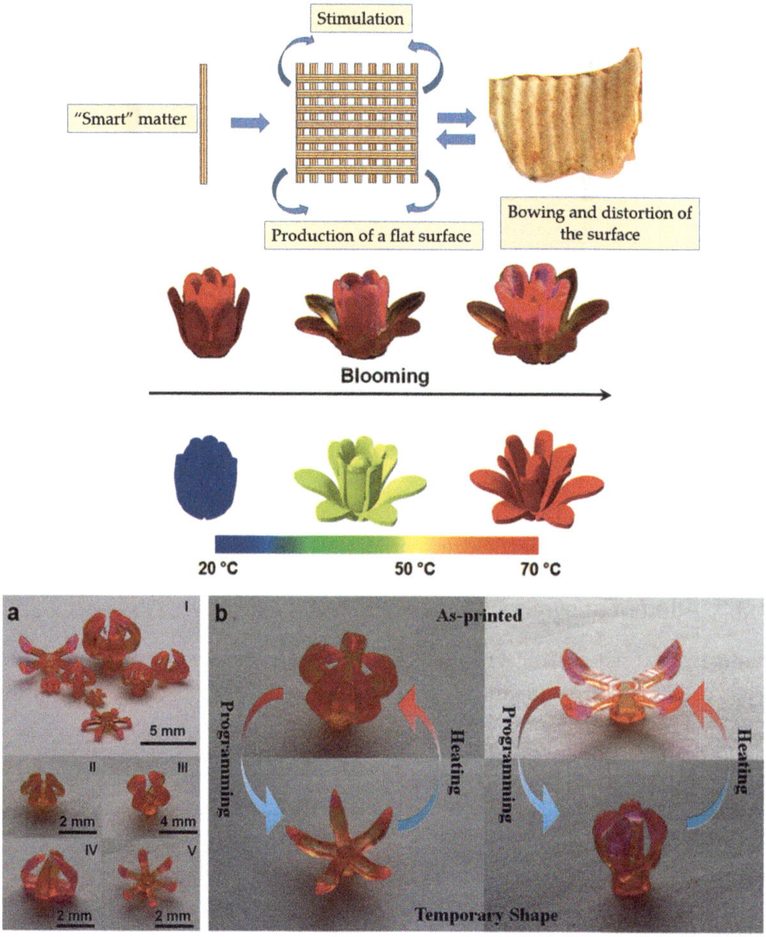

Figure 1.18. *The anisotropy of the material creates a curvature during the temperature change (we could have obtained the same phenomenon with the swelling of bilayer threads charged with cellulose). Complex morphology (here a flower) is generated with developments in the forms of objects. For a color version of the figure, see www.iste.co.uk/andre/printing3.zip*

Lind *et al.* [LIN 16] used, as material able to be activated, the heart's muscle cells which contract under the influence of electrical stimulation. Figure 1.19 shows the phenomena of compression/expansion enabling a plastic material to bend (see also [JAC 16]).

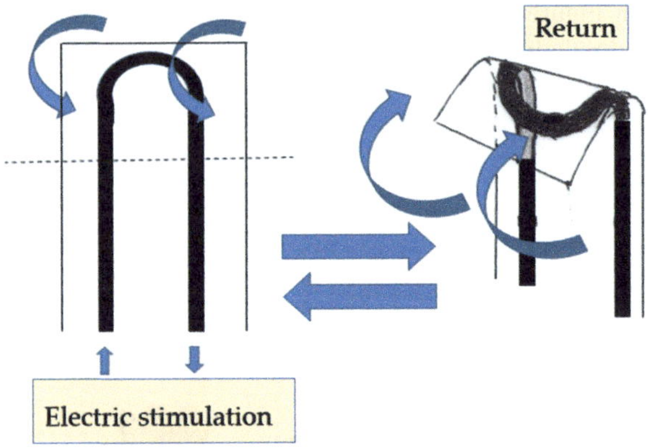

Figure 1.19. *Distortions induced by muscle cells. For a color version of the figure, see www.iste.co.uk/andre/printing3.zip*

1.3.1.3.2. Active mineral materials

Dufaud, Marchal and Corbel [DUF 02] studied the effect of an electric current containing PZTs with the piezoelectric properties aiming to produce actuators.

1.3.1.3.3. Attempt at synthesis

Table 1.2, extracted from Formlab [FOR 15], enables a comparison between the materials used in additive manufacturing and those currently used in 4D printing.

Additive manufacturing	4D printing
Thermoplastics	Self-assembly materials
Metals	Multi-materials
Ceramics	Shape-memory systems, polymer couplings and systems for continuous change in shape (humidity, temperature, light irradiation and other factors)

Table 1.2. *Differences between the materials used in 3D and 4D printing*

1.3.2. *Physical alterations*

A surfactant is a molecule with a hydrophilic head and a hydrophobic tail, the hydrophilic head is directed preferentially to the aqueous phase, contrary to its hydrophobic tail. Thus, generally, the surfactants are positioned preferentially at interfaces water/air or water/oil. "Their presence stabilizes the interfaces, which implies a local reduction in surface tension. As the presence of surfactants on a given surface locally decreases the surface tension, it is possible to create a surface tension slope. From this results a movement of fluids towards the areas of higher surface tension" [MAR 14], as is shown in Figure 1.20. The addition of surfactant (shown in orange), leads to reduction of the surface tension from γ_0 to γ, which induces a distortion of the surface leading to a flow of fluids (blue arrows) towards the strongest areas of surface tension γ_0.

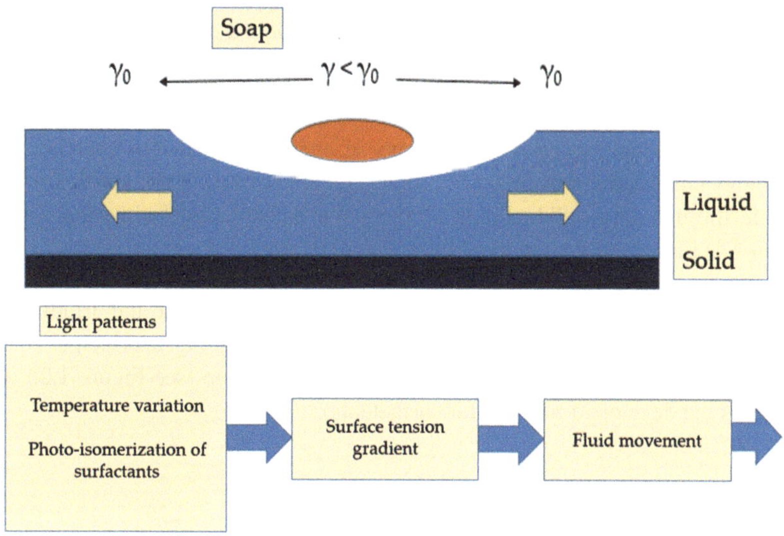

Figure 1.20. *Distortion of fluid induced by surfactants. For a color version of the figure, see www.iste.co.uk/andre/printing3.zip*

If you use an active surfactant, for example, containing azobenzene, the isomerization Trans-Cis is likely to alter the surface tension, which may enable matter flows. In UV irradiation, the concentration in the Trans isomer diminishes, which increases surface tension. According to the light distribution, a surface tension slope may be obtained with the possibility of having a photo-induced Marangoni effect.

Thus, if a drop is placed on a surface, it will be able to move along this by taking advantage, as Figure 1.21 shows, of the light-induced alterations to surface tension [SHI 99].

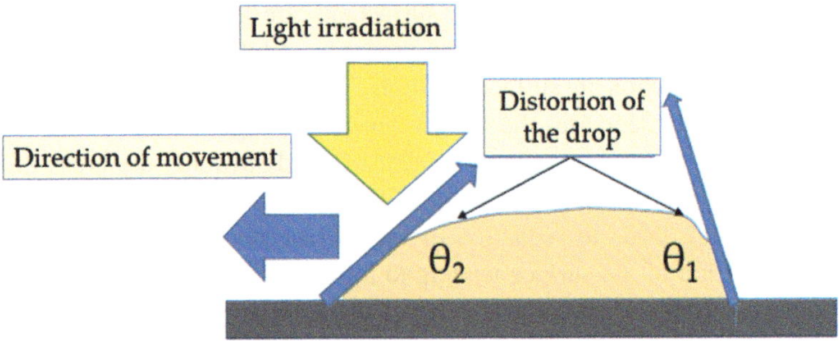

Figure 1.21. *Possibility of guiding the movement of a drop on the surface. For a color version of the figure, see www.iste.co.uk/andre/printing3.zip*

COMMENT 1.– taking account of the forms of energies called into question, if this technology enables the movement of droplets on the horizontal flat surface, it is only possible through immiscible fluids of the same density (see swimming robots in section 1.4) for an application to additive manufacturing.

COMMENT 2.– with the Trans-BTHA [SHI 99] showed that the irradiation of hanging drops may lead to their fall. From there, envisaging a principle of rapid matter deposition using this process, only involves one step (see Figure 1.22 which represents the developed formula for surfactants).

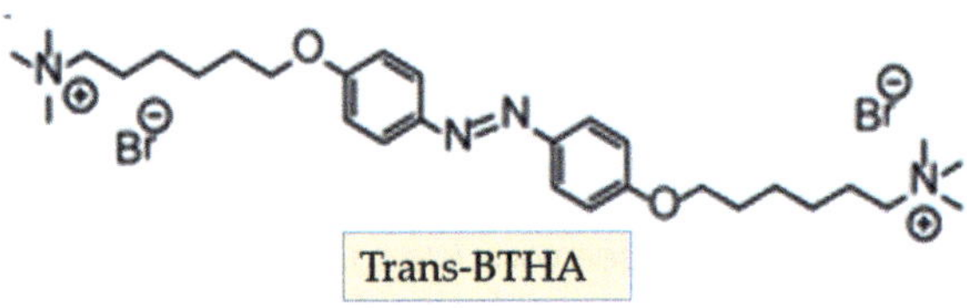

Figure 1.22. *Formula developed for surfactants*

1.3.3. *Distortion of metal parts*

1.3.3.1. *Drops*

By way of an initial example, we quote the article of 3ders.org [3D 14], a sign of the possibility of rotating and moving the droplets of an alloy Ga/In/Se melting at 10°C, floating (due to surface tension) on the water via a magnetic field. The Chinese authors Sheng, Zhang and Liu [SHE 14], of the Chinese Academy of Sciences in Beijing, also succeeded in altering the shapes and sizes of the droplets. It is a matter of seeing how we can access the third dimension. Figure 1.23 shows the effect of an electric field [NEW 15].

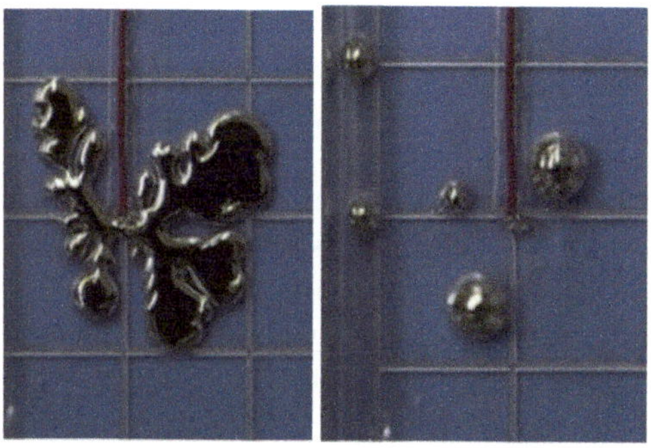

Figure 1.23. *Effect of an electric field (left) on the shape of an alloy droplet*

1.3.3.2. *Magnetic-oriented architectures*

Recently, Martin, Fiore and Erb [MAR 15] linked with photopolymerizable resins for magnetic particles, which can be specifically oriented, by altering the range and the orientation of the given field. This achievement, where two forms of energy are coupled together, enables the production of singular magnetic property structures, at the same time as a possible strengthening of the material by the increase in the sequence of elongated particles in the polymeric matrix. Catheters having neonatal applications are thought to have been produced using this principle.

1.3.3.3. *Shape-memory alloys*

Lastly, shape-memory alloys may be used [BAL 96]. These alloys have several novel properties among metallic materials: the capacity to remember an initial shape and to go back to it even after distortion. There is the possibility of alternating between two previously memorized forms, when its temperature varies around a given critical temperature. Moreover, superelastic behavior enables elongation without permanent distortion greater than that achieved with other metals. Among the main shape-memory alloys, we find a variety of alloys made of nickel and titanium as principal constituents, in almost equal proportions [MEI 12, DAD 14]. Although the name "nitinol" is in fact only the name of one of these "quasi-equiatomic nickel-titanium alloys", this designation is currently used within works in the field to denote all of these alloys, with very similar properties. To a lesser extent, some copper–aluminum alloys also have shape-memory properties [WIK 15]. Although these materials have not actually been used in additive manufacturing, their spheres of application today cover medicine, the aerospace industry, the car industry, women's clothing, the spectacle trade, security and research [NIM 17]. This development was for a long time limited by low durability and their capacity to regain their initial form [CHL 15]. However, everyone remembers an example which is now classic, that of a demonstration by Uri Geller, before an astounded audience of the distortion of spoons produced by such alloys using a thermal effect (L'arrêt public gouverne MENT, [LAR 10]). Several applications in additive manufacturing can be envisaged (by remembering low-volume variations, and also shape changes which may be reasonable). In particular, as Figure 1.24 indicates, metal shape-memory alloys have a specific advantage (relative to polymers). This is their weight/power correlation (hence their use as actuators). This has the potential for exploitation in 4D manufacturing.

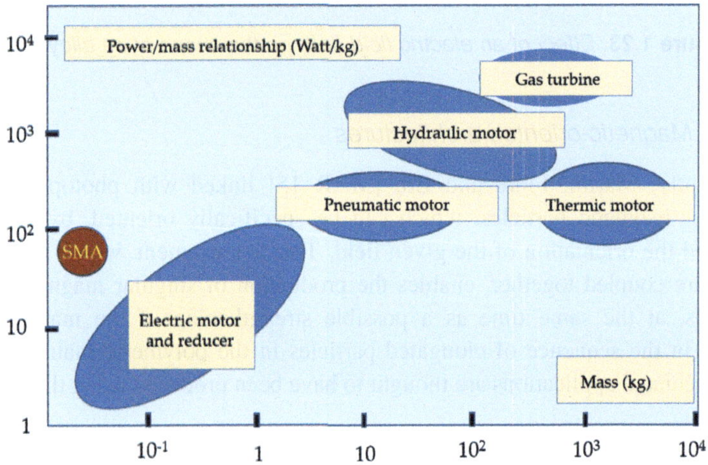

Figure 1.24. *Power–mass relationship of shape-memory alloys (SMA) versus motors according to Patoor [PAT 06]. For a color version of the figure, see www.iste.co.uk/andre/printing3.zip*

1.3.4. *Conclusion*

The original component of this recently emerged sphere for additive manufacturing resides in the principle of “smart” matter to extract spatial functionality from it (rather than falling within spontaneous self-assembly). At this stage, although “photochemical muscles” have already been the subject of works, to the author’s knowledge, the idea which consists of starting from a simple shape, which may easily be produced by traditional techniques (molding and traditional machining) for post-treatment by light (or all other forms of localized energy in space and time) and the idea of obtaining a complex object has not been put forward. It poses the question of having available materials with a high scalability (we recall that for shape-memory alloys, the gap between both forms is generally of the order of a few percent), of being capable of processing the inverse issue. Starting from the basis of achieving a given form, what may in fact be the initial form?

However, although this route appears interesting, we should tackle studies of very simple objects to infer given laws. Evidently, it may be a question of working on surfaces rather than volumes to simplify the situation. For all that there is a risk of the number of influencing parameters to take into consideration being high, leading not to a unique solution for a given spatial resolution, but to a set of solutions. At the same time, we run the risk of the absence of solutions for forms, which are too complex. Thus, starting from a given continuous material, the formation, for example, of a percolating environment could be problematic. These means are explored below.

1.4. A transition to 4D printing: swimming robots

In section 2.2, the possibility of moving a droplet of matter by the provision of external energy was shown. On this principle basis, minuscule swimming robots containing particles of iron oxide linked by chemical bonds are able to move, thanks to a magnetic force [CHE 16]. In this paper, these particles are not intended to self-assemble to create a given object, but from a size of the order of a nanometer, this object should be able to circulate within the vascular system from an external magnetic field ([RAO 15], Figure 1.25). Indeed, the swimming robots’ swimming is well controlled, as they move at a low Reynolds numbers: “Indeed the Reynolds number, in respect of flow is very small, of the order of 0.00001, whilst within usual flows (for example that we ought to consider for a human swimming in water) this number is typically of the order of 100 000” [ALO 12]. A swimming robot made up of three spheres, placed side by side, was studied by Alouges, De Simone and Lefebvre [ALO 08]; Naja and Golestanian [NAJ 04] and Garcia [GAR 13] where it is shown that it can effectively swim (slowly) within a three-dimensional flow with a low Reynolds number.

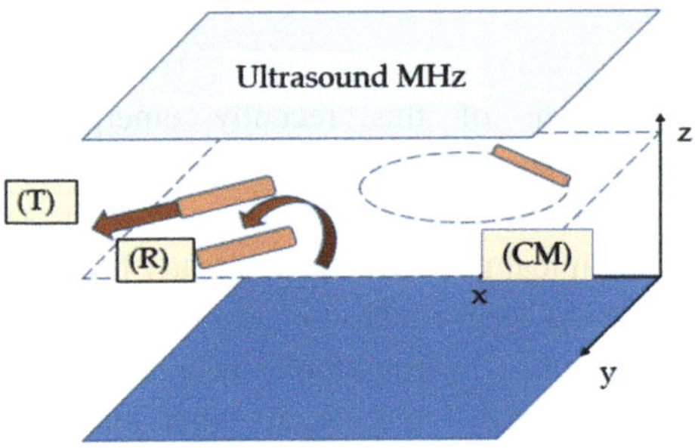

Figure 1.25. *Principle of robot-swimmer movement (various types: translation (T); rotation (R) and circular movement (CM)). For a color version of the figure, see www.iste.co.uk/andre/printing3.zip*

Although the researcher perspective is to carry medicines (cargo effect) or to use swimming robots for emitting obstructed vessels (icebreaker effect), it is envisaged to use for a given voxel or convertible matter which, in a state of weightlessness, may be guided upon a given object undergoing construction. The principle of this construction necessitates accurate guidance of each robot, which is considered as isolated from other swimming robots. Otherwise, apart from the feasibility of a precise command, these entities behave according to their density and flow as Lushia and Peskin [LUS 13] and Liebchen, Cates and Marenduzzo, [LIE 16] showed. Yet, in Volumes 1 and 2, the question of voxel size was widely mentioned as it was linked to manufacturing time. Within this system, by using variable-sized particles, it is possible to envisage the construction of nanometric objects or a more significant size using a "multi-voxel" principle, indeed, multi-materials with cargo voxels of varying sizes (see Figure 1.26).

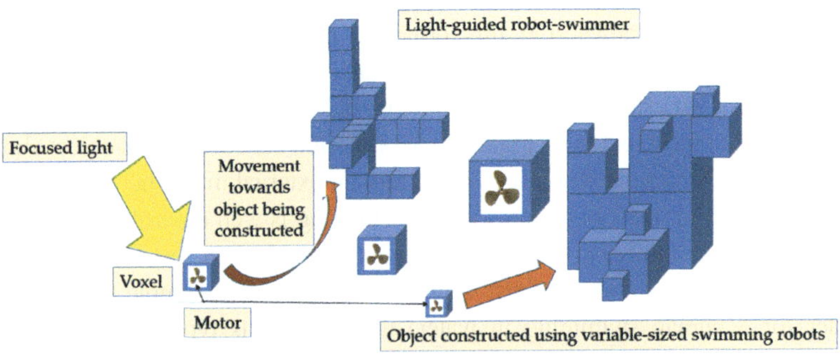

Figure 1.26. *Application of swimming robots to additive manufacturing, for example, light-induced polymerization. For a color version of the figure, see www.iste.co.uk/andre/printing3.zip*

Various guidance techniques may be envisaged. These are presented in Table 1.3, largely stemming from the works of Rao *et al.* [RAO 15].

Locomotion principle	Materials	Commentary	Bibliographical references
Electrophoresis	Microspheres, bimetallic rods, fibers	Known mechanisms, self-generated electric field, sensitivity to the given medium conductivity, high speed > 100 μm.s^{-1}	Paxton *et al.* [PAX 05]; Wheat *et al.* [WHE 10]
Diffusiophoresis	Heterogeneous voxel	Self-generated concentration slope with enrichment of fluid support in substances dissolved in the process (difficult to apply in additive manufacturing)	Hong *et al.* [HON 10]; Ibele *et al.* [IBE 09]; Howse *et al.* [HOW 09] and [HOW 07], Clément *et al.* [CLÉ 16]; Clément [CLÉ 16]
Bubble propulsion	Elongated forms	Catalysis for the decomposition of a reagent producing bubbles, with an uncertain trajectory	Wilson, Nolte and van Hest, [WIL 12]; Gibbs and Zhao, [GIB 09]; Solovev *et al.* [SOL 09]; Zhang *et al.* [ZHA 15a]
Electric field	Microparticles	Localized electrolysis in water producing bubbles, accuracy difficult to achieve and slow movements	Calvo-Marzal [CAL 10]; Loget and Kuhn, [LOG 11]
Magnetic rotation	Flexible stretched out shapes	Swimmer rotation within a magnetic field inducing locomotion, accuracy possible but slow speed	Gao *et al.* [GAO 10] and [GAO 11]; Ghosh and Fischer [GHO 09]; Tottori *et al.* [TOT 12]; Peyer *et al.* [PEY 12]; Li *et al.* [LI 16a]
Ultrasound and acoustic waves	Metallic particles	Conversion of moving energy, uncontrolled behavior, high speed	Wang *et al.* [WAN 12]; Ahmed *et al.* [AHM 16]
Photochromism	Elongated form	Spiropyranes or azobenzene compounds are likely to have light-induced Trans—Cis transitions leading to the movement of flagellum	Li *et al.* [LI 16]; Huang *et al.* [HUA 15] Stanton, Trichet-Paredes and Sanchez [STA 15]

Table 1.3. *Various forms of propulsion of swimming robots*

For magnetic propulsions, Li *et al.* give some indications as to the speed of movement according to means of action. This data is shown in Table 1.4.

Nature of magnetic stimulation	Speed in $\mu m.s^{-1}$	Commentary
Internal particle oscillation	30.9	Movement as if a fish
Oscillation	6-14.4	Boat (scull) movement
Rotary feed	18-40-180-250	Screw movement

Table 1.4. *Speeds of movements obtained by Li et al. [LI 16a]*

In the case of azobenzene Trans–Cis linked to a light-induced transformation, the motor is linked to the voltage induced by the change in form of azobenzene within a sandwich structure, such as that shown in Figure 1.27. There is, with this type of mechanism, a means to control partial rotation of a fixed flagellum on this motor as shown in Figure 1.28 ([HUA 15], see also [EGU 16]). Although the constraints are not homogeneous within the different layers, a moment of forces appears and the thin layers may then curve spontaneously, subject to the action of external forces. Thus, by exploiting these constraints, a new "self-winding" material produced with cross-linked polydimethylsiloxane (PDMS), a transparent and biocompatible elastomer and silicone oil was able to be produced.

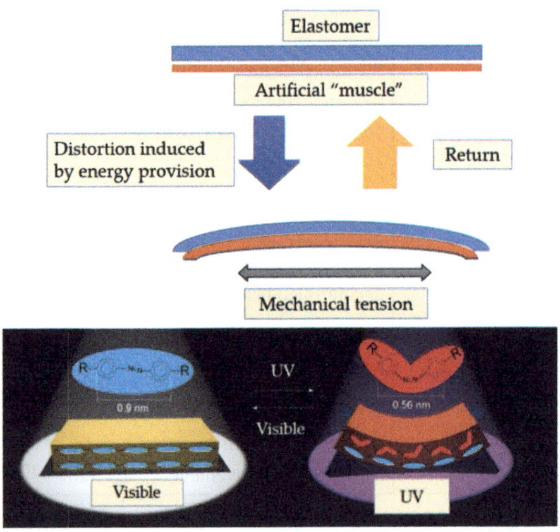

Figure 1.27. *Principle of operation of photochemical motor. For a color version of the figure, see www.iste.co.uk/andre/printing3.zip*

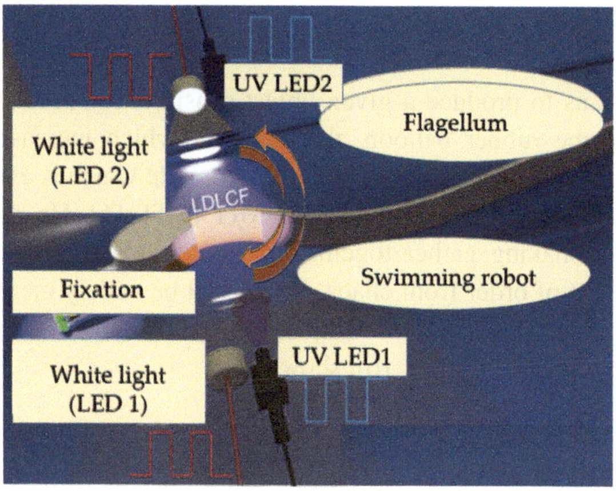

Figure 1.28. *Relationship between photochemical motor and its flagellum. For a color version of the figure, see www.iste.co.uk/andre/printing3.zip*

In their publication, Khoo *et al.* [KHO 15] have produced a synthesis of applicable processes in terms of "smart" matter. A summary of their conclusions is comprised in Table 1.5.

Materials and structures	Approximate resolution	Commentary
Piezoelectric nanocomposites	5 μm	Production of electric charges when a given constraint is both applied (and reciprocated). The support material is a polymer such as the PEGDA (Poly(ethylene glycol) diacrylate)
Memory-shape alloys (NiTi)	/	Change of shape induced by temperature (problem of hysteresis to take into consideration)
Memory-shape polymers	20 μm	Distortion of a bi-component system with various properties (bimetallic strip effect) induced by the temperature or by the light (heating)
Bimaterial system including a dielectric elastomer	30 +/- 10 μm	Electric activation and bimetallic strip effect
Polymeric composite and origami	45 +/- 15	Movements induced by temperature

Table 1.5. *Several examples of applied smart matter in additive manufacturing*

1.5. 4D Printing

A simple means to produce a given object sensitive to stimulation of energetic origin is that of the rubber balloon, the shape of which increases in size as the pressure of the gas filling it increases. Another example which is even better adapted to the demonstration comes from Tibbits at MIT [MIT 16]. He places objects in a flask which after shaking gather together to create another object (the traditional problem of creation of order from chaos, which will be widely mentioned in the part devoted to bio-printing). Figure 1.29 inspired by Tibbits and Thomas [TIB 13], and published by Leist and Zhou [LEI 16], shows this.

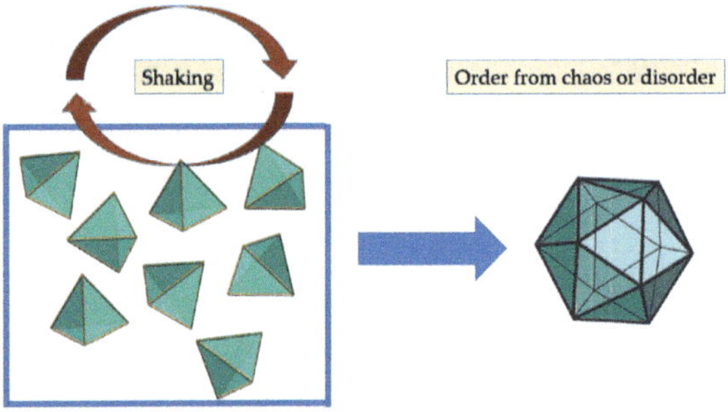

Figure 1.29. *Creation of order through disorder (general concept according to [LEI 16], citing Tibbits and Olson). For a color version of the figure, see www.iste.co.uk/andre/printing3.zip*

From there producing more sophisticated components such as the actuators shown in the previous chapter or folding systems including internal expansion enables interesting developments of shapes, such as those in Figure 1.30 by Yuk *et al.* [YUK 17] which is an object produced by additive manufacturing, the shape of which may be altered by the effect of hydraulic pressure. There are only extensions of this concept. It is the same, for example, with the works of Ballandras *et al.* [BAL 96] where a structure produced by 3D printing was connected to a structure-memory alloy to manufacture an actuator. According to Pimenta [PIM 15], a humanoid on a genuine human scale may be directly printed at home, thanks to a 3D printer. All the parts have been designed to print without any particular difficulty and self-assemble either by using screws or clamps. Other examples of the same type exist (see, for example, [DEM 14, ATL 15, MAR 12, STR 13, MEH 14, YIM 07,

KOT 98, KEA 13, YIN 16, HAR 16, PRA 16, ENG 17]). However, is this 4D printing?

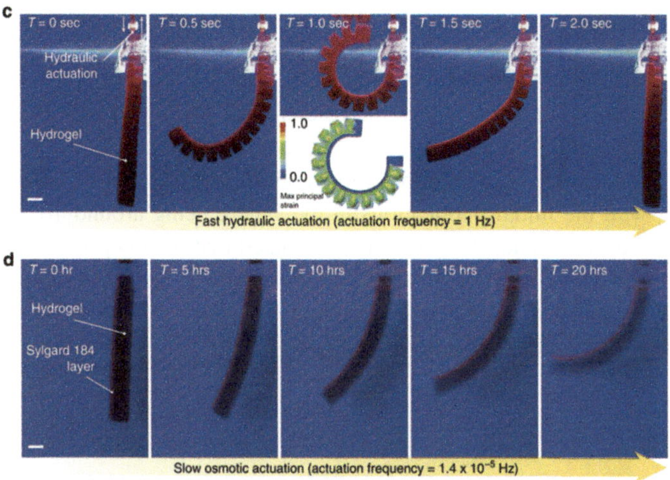

Figure 1.30. *Hydraulic actuator. For a color version of the figure, see www.iste.co.uk/andre/printing3.zip*

The concept of 4D printing hypothesizes that the given object is adaptive through integration and less through the juxtaposition of the various components, even when produced through 3D printing. The fourth dimension is time and/or the given functionality [PIC 17]. The 4D technology concept enables the printing of objects which themselves change shape and self-assemble with time. The dividing line is therefore not so simple to find especially around the first aspect of subject framework: changing shape by itself, as every robot does, even every machine does. Moreover, the German firm BMW [BMW 16] in its research on vehicles of the future has developed a prototype relying upon the concept consisting of spatial conversion of the vehicle structure according, for example, to its speed. The idea of 4D printing tackles the assembly and the installation of products/objects which may be able to self-assemble and their implementation in environments facilitated by "Programmable Matter". As has already been suggested, programmable materials which build themselves thus make superfluous the need for assembly plants and heavy industrial installation procedures. The robotization, the core of productivity gains of the 20th Century, would thus be integrated to the products themselves. Nowadays, even Figaro Magazine (a French Newspaper) is speaking of 4D printing [4DP 15]!

Independently of this problem of definition, it appeared useful to investigate, through the library site of the University of Lorraine, what is produced on the emerging theme. Figure 1.31 shows the results from the key words "4D Printing". Approximately 200 publications make up the available "stock", but what this figure shows is that it is an over-exponential development, indeed illustrating an emergence, with the initial publications in the sphere being around 10 years old or more. Moreover, Leist and Zhou [LEI 16] estimated that the 4D printing market will be worth US$63 million in 2019 and nine times more in 2025 (values that it is appropriate to consider relative to 3D printing estimates of US$5 billion to US$30 billion/year on this latter date).

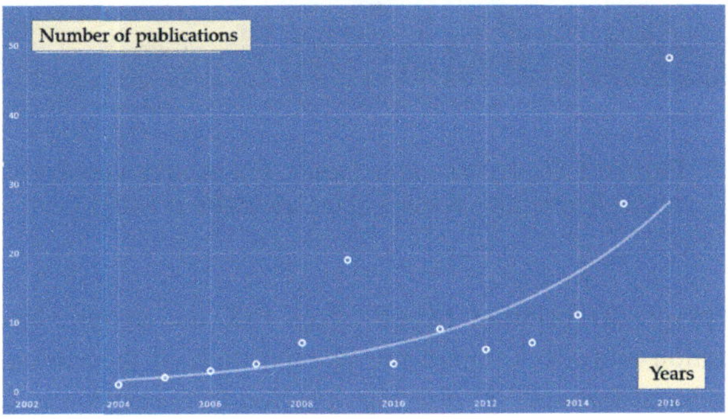

Figure 1.31. *4D printing publications (curve: exponential through given optimal points)*

1.5.1. *Automation and robots*

D. Rus of MIT succeeded in demonstrating the capacity for robotized voxels to self-assemble in a given form, by using a type of distributed algorithm, without a central server, in the manner of cellular automation. His team developed "m-cubes" liable to roll on each other and also to climb on each other [SUS 14]. In the first stage, it is possible to envisage, as MIT [MIT 13a] showed, articulating on the basis of origamis, sets of voxels placed among each other to enable the introduction of temporal development in the overall geometry of a given object. The photographs in Figure 1.32 give examples of such developments.

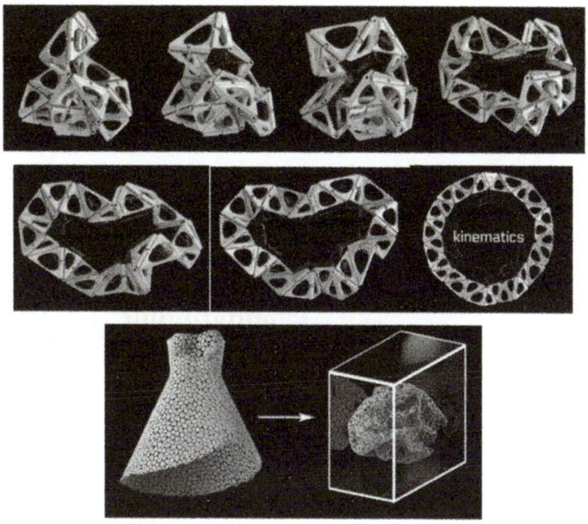

Figure 1.32. *Distortions from movable voxels (with the consent of the organization Nervous System; http://nervo.us)*

The same institute [MIT 13] shows that using simple examples, it is possible to "control" not least by visual recognition. This idea, validated by Figure 1.33, relies upon the movement of cylinders by using, on a two-dimensional basis, a means to take an impression of a cornice or like certain toys which permit three-dimensional construction. If we add a digital command to the movement of each cylinder, there is a means of making matter directly or indirectly "programmable". However, in these two examples, which serve to introduce the subject, matter is inert. The idea consists of going slightly further with the approach by "introducing the 'smart' element" directly into the matter, at finer scales than this figure.

Figure 1.33. *Reversible sculpture toy [CDI 17]*

Figure 1.34 from MIT [MIT 13a] shows the intrinsic complexity of such a system which will not be able to be used at lower scales due to the high number of actuators which it would be necessary to put together within a finished volume. There is also the issue of the size of voxels to take into consideration.

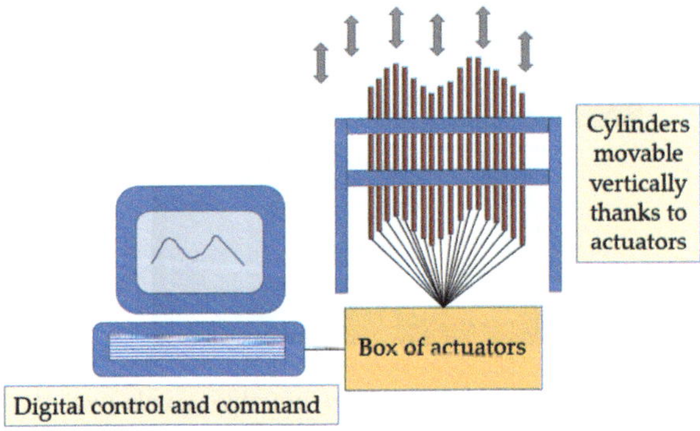

Figure 1.34. *Experimental set-up of the MIT [MIT 13]. For a color version of the figure, see www.iste.co.uk/andre/printing3.zip*

A 4D technology still in the process of assembly of programmed development of robots [GE 13, GE 14, THI 14, BOU 14, LI 15, BAK 15, CRA 15, SAV 14, THO 14, FIS 14, BÜC 15, JIN 11, AZN 12, MAC 10, MAC 12] and other references) may consist of creating “catoms”, capable of self-assembling magnetically to produce any object [GIL 12, KNA 10]. From 2007 onwards, researchers at the MIT had created prototypes from cylinders approximately five centimeters in diameter, which have subsequently been somewhat miniaturized. They have manufactured cylindrical “catoms” of around 1 mm in diameter, capable of communicating and receiving energy. There is one small problem, they do not yet move [SUS 14, HAR 12, HAR 15], but the cubes from the team Rus at MIT meanwhile move to create genuine three-dimensional structures, however, having a number of voxels which are very limited, as Figure 1.35 shows [GIL 12].

In this project, a new “smart” system manufactured from a component material and a software approach enables the creation of a set of programmable objects (taking a “smart” collective shape). The material component is a micro-robot manufactured in series which uses the same IT mechanism, the same energy, communication, and adhesion and locomotion distributions. The software approach enables reconfiguration of a set of micro-robots for a given end.

Figure 1.35. *Example of 4D printing using cubes [SUS 14]*

A demonstration was carried out with the assistance of a limited number of programmed mechanisms. To go further, it extends to the resolution of the following scientific and technological challenges:

1) The change of scale, while maintaining the same levels of movement and adhesion performance, to achieve compatible sizes with a given applicable need, although the number of robots becomes important. We can cite the project of the research unit FEMTO-ST in Besançon (France), which is engaged in this research direction. "For the material, we will create a genuine 3D micro-robot through LSI/MEMS integration between the chip and a distortable substrate. Our main manufacturing process will concern the actuator, which groups together all of the necessary micro-robot functionalities" [BOU 16].

2) The control of the cooperation of movements of a set of robots moves in an autonomous way, without affecting other robots. However, to produce a complex object with an "acceptable" accuracy, it is necessary to have a large number of "nano-robots" engaged within collective manufacturing, with the difficulties of multiple interactions between a multitude of micro-robots before avoiding one another, as the following Figure 1.36 shows [YEO 15].

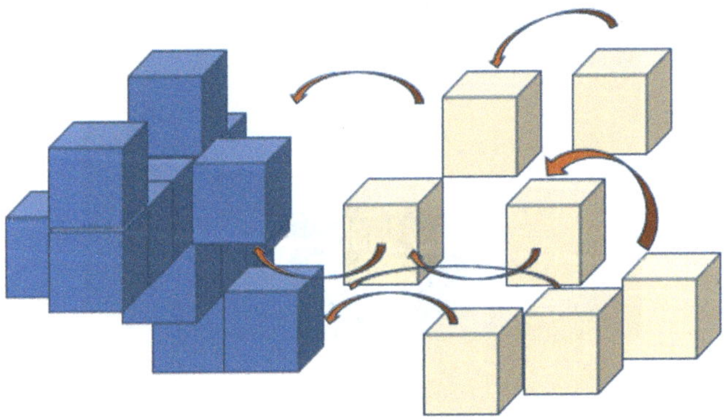

Figure 1.36. *Collective approach to interdependent movements in object production and movements which may be concerted*

What is shown on these figures is a form of tangled causality which may be observed in a number of mechanical systems [SÈV 05], with a nonlinearity and largely unpredictable effects which may appear, especially if the number of independent voxels at the start has increased. "For software the challenge is to intensify the number of components which may be efficiently controlled with a single easy-to-understand program" by the user. "We are going to create Foxel, an invariant descriptive recursive language implemented within a logical programming language. In addition, we will devise 'Meld', to create programs which are naturally distributed, tolerate breakdowns, and which can also progress towards clear evidence. This project follows the 'Claytronics' project initiated by Intel and the University Carnegie Mellon, then co-directed with FEMTO-ST" [BOU 16].

3) The energy supply in increasingly small objects, by reducing the size, we reduce the space devoted to energy storage, with a charge that varies with the cube of the average dimension of the object.

1.5.1.1. *Concluding comments*

"Although the necessary progress as regards hardware is yet to be accomplished, it does benefit from a fairly simple road map: the strategy of miniaturizing! The creation of distributed software is, for its part, even more uncertain. We have been working on this subject for more than two decades. There have been an increasing number of experiments, as much in the real as the virtual world. However this type of process is still an economic study without practical applications" [SUS 14]. Nevertheless, going from a few centimeters to a few hundred μm corresponds to change in scale of approximately 100, or for a three-dimensional object having a factor of 10^6. It is nowadays difficult to imagine a process involving energized micro-robots and having a collective assembly program and moving without causing collisions. Perhaps it is necessary to investigate the movements of a given number of social insects (bees, ants and termites) who know how to move within a complex environment asking the following: "Why is the group moving coherently whilst each individual seems to be autonomous? How are the activities of all individuals coordinated without supervision? Behaviorists studying the behavior of social insects observe cooperation within colonies as being subject to self-organization: often, it results from interactions between individuals" [BON 00]. That which nature knows how to do with ants, man begins to know how to do. However, the application phase, within 3D or 4D additive manufacturing, is only taking tentative steps. Nevertheless, although we can content ourselves with less precise forms, it is not impossible to envisage a very flexible process to produce "smart" objects. It would certainly be useful to monitor this issue.

1.5.2. *Origami*

Sheets of origami when the author was young generally had a square shape (but this was not always the case). The various shapes are obtained by successively folding the given sheet. In theory, it should not be cut. The given origami can represent an animal, a plant or an object, but can also represent geometric forms which are simple or complex: these are the so-called "modular" origamis. Although this Far Eastern expression is introduced into this chapter (apart from DNA origami), it is in order that we can create a three-dimensional shape with an initially flat structure.

For centimeter-sized entities, it is possible to position the traditional actuator at given folds. Their rotation enables the movement of given planes and, consequently, the movement of origami which can, through an adapted control, become a robot-origami. Nisser *et al.* [NIS 16] used origami techniques to create and deploy four robots constructed from a single sheet. These were made from several layers of different materials. This process would enable facilitation of the deployment of robots in swarms for various applications. Shape-memory polymers and the origami principle are used to create robots from a single sheet, enabling precise folding control thanks to mounted sensors. For the moment, the capacities of small origami robots are fairly limited. Their top speed is approximately 10 mm per second and their autonomy approximately a few minutes [ACK 16]. This coupling example between robotics and "smart" matter is indeed not unique.

Very recently, researchers at MIT presented a "robot miniature" prototype capable of adopting various shapes, of moving and of transporting loads by pushing them. This small ultra-light robot, which is remotely controlled, can close in on itself to adopt different postures. These include walking, swimming and moving heavier loads than the robot itself (before finally self-destroying within the acetone substance). What is interesting is that to produce this movable 3D object, it is sufficient to print programmable matter using a traditional 2D printer.

"This machine is only made up of two extremely thin layers of polystyrene (or paper) between which are a magnet and a sheet of PVC measuring with a side measuring 1.7 cm in length. These relatively simple materials enable the inventor to produce remarkable feats when working within a suitable environment. Thus, when the device is placed upon a heated area at the right temperature, the sheet of PVC of which it is made folds in less than a minute according to pre-determined structural lines. Then, thanks to the combined action of four solenoid mechanisms arranged below the surface, the "robot" is able to move in any direction at an average speed of 3 to 4 cm per second. Moreover, it is capable of carrying a weight of 0.6 grams, climbing a slope and cutting a chosen path through various obstacles" [RAN 15].

In the same spirit, other works have been published: Rus at the MIT shown by Hardesty [HAR 15] on origamis; see also Khoo *et al.* [KHO 15], Ge, Qi and Dunn, [GE 13, GE 16a], Ge *et al.* [GE 14]; Hawkes *et al.* [HAW 10]. The following Figure 1.37 shows the potentialities of these microelements distortable due to one or more localized magnet(s) localized around the given object, having cell components with the Peltier effect and controlled movement by magnetization determined at the surface. Optic or thermic control can be used.

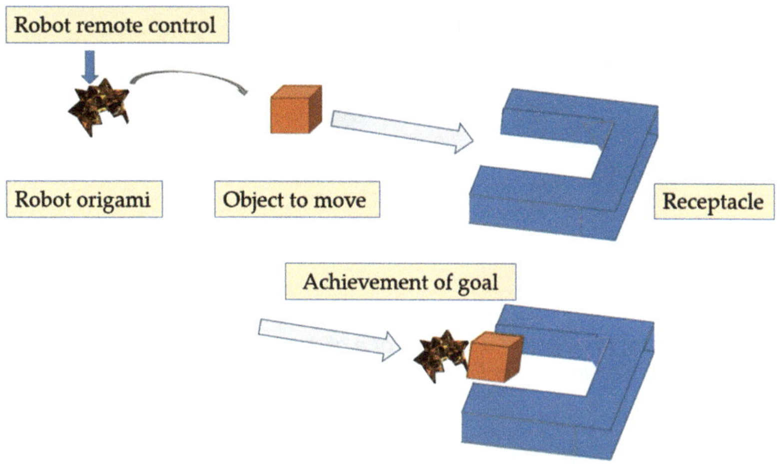

Figure 1.37. *Robot-origami moving a given object (from the start to reaching the objective). For a color version of the figure, see www.iste.co.uk/andre/printing3.zip*

Using this system (which has been subject to other publications by Daniela Rus' team such as: Felton *et al.* [FEL 13, FEL 13a], An and Rus, [AN 14], Hawkes *et al.* [HAW 10], Onal, Wood and Rus, [ONA 13], we find ourselves in the same difficulty of self-assembly collective production, having a three-dimensional object responding to instructions. However, it is possible to use light in Figure 1.7, the possibility of controlling micro-robots to move voxels. This enables the manufacture of an object or its completion using the matter of an existing object. This system might be used advantageously in areas which are difficult to access or even dangerous.

Figure 1.38 shows the principle which stems from Ge *et al.* [GE 14] resumes the same idea with thermomechanical systems. In their publication, the relaxation time is of the order of 0.4 seconds. For an adapted design, there is the possibility of achieving more complex forms provided that the object can be described on a plan before being folded (see Figure 1.40 stemming from Kimionis *et al.* [KIM 15], see also Tenzeris *et al.* [TEN 16].

Examples stemming from Py *et al.* [PY 07] use the influence of water and surface tension illustrating the concept within another framework, that of altering surface tension (see Figure 1.39). Others use porous high-power displaced polymer materials distorted by an electric field [HAN 04].

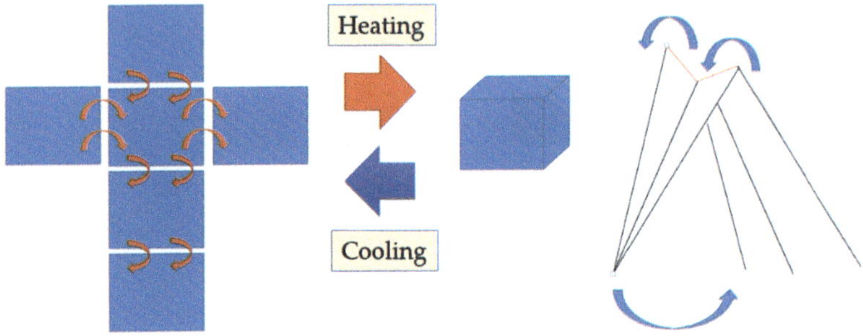

Figure 1.38a. *Principle of operation of a given thermomechanical origami. For a color version of the figure, see www.iste.co.uk/andre/printing3.zip*

Figure 1.38b. *RF antenna produced by additive manufacturing and multi-materials producing origami to achieve a 3D structure (with the permission of M.M. Tenzeris). For a color version of the figure, see www.iste.co.uk/andre/printing3.zip*

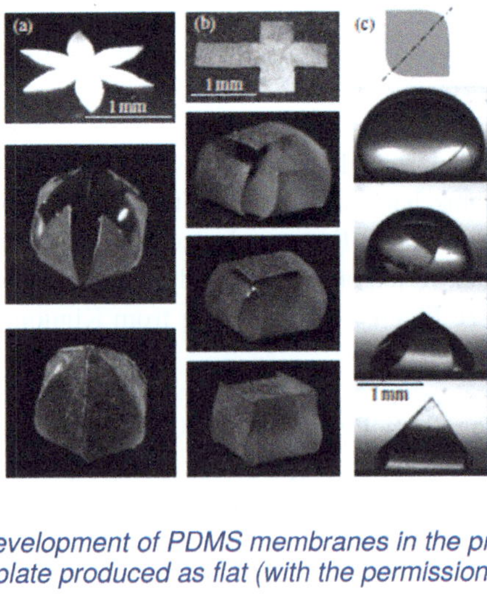

Figure 1.39. *Development of PDMS membranes in the presence of water per the initial template produced as flat (with the permission of Benoît Roman)*

1.5.3. *Octobot*

Mazzolai and Mattoli [MAZ 16]; Wehner *et al.* [WEH 16] printed the body of a silicone octopus in 3D. They used tanks of oxygenated water and catalytic reactors which store localized hydrogel-based inks in their tentacles. These contain particles of platinum, which catalyze the controlled decomposition of oxygenated water which frees the oxygen serving as a thruster. Pistons and gates regulate the movements of gas to propel the robot. "Being less than 2 centimeters in thickness and almost the size of a human palm, the Octobot can operate for 4 to 8 minutes before getting to the end of its supply of oxygenated water. Larger versions of this robot (shown in Figure 1.40), may be capable of remaining active for longer time periods upon each replenishment" [DEV 16].

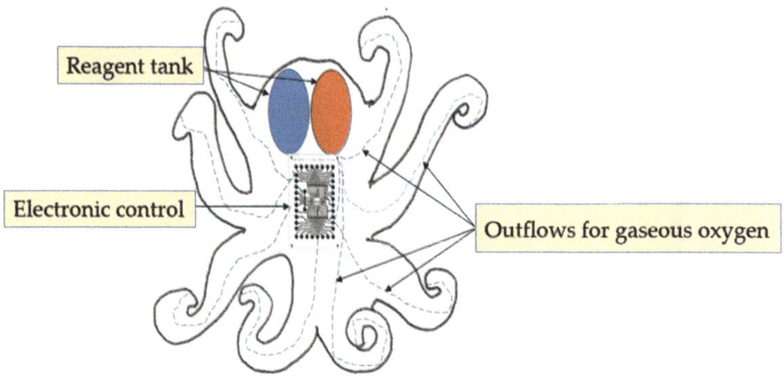

Figure 1.40. *Approximate drawing of an "Octobot". For a color version of the figure, see www.iste.co.uk/andre/printing3.zip*

1.5.4. *Massive objects*

Figure 1.41 stemming from an idea of Yang *et al.* [YAN 16] and Ge *et al.* [GE 16a] highlights the aptitude of memory-shape polymers involved in additive manufacturing to produce actuators (see also [SEN 16, LU 17, ZAR 15, YAN 16]). This is completed by the actual demonstration of the idea on the figure as shown at the bottom.

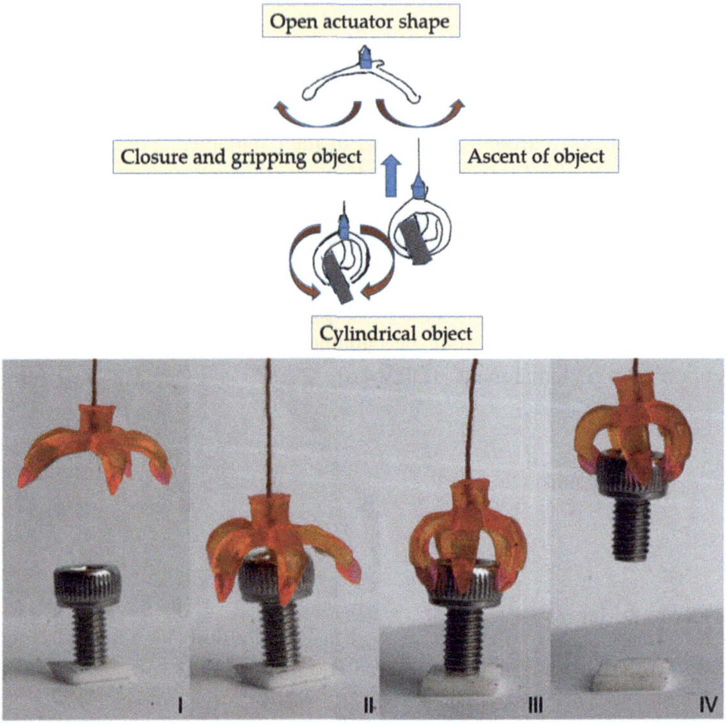

Figure 1.41. *Actuator thermically agitated by a shape-memory polymer (image supplied by Kevin Ge [KEV 16]). For a color version of the figure, see www.iste.co.uk/andre/printing3.zip*

Although it is possible to calculate the shape of a 3D object fairly easily and to envisage it developing, the experiments shown to illustrate the significance of this emerging domain essentially concern elongated objects or those in the form of layers. Concerning massive objects, such as that of the chosen provocative example in Figure 1.42 to illustrate the idea, the problem to process becomes more delicate, as it involves:

– a radical change of shape;

– a radical change in functionality.

Since it does not appear possible to the author to resolve the complex problem of the conversion in this figure, posing questions linked to the nature of the material (so that it becomes distorted and that it has good machining properties), energy provision (within the particular material) and the process, it appears at this stage necessary to commit to a "minimalist" demonstration. This happens by choosing an apparently simple example of that of the passage from a sphere to a cube, as shown in Figure 1.42. To simplify the problem, we could content ourselves with a form of surface modeling of these two entities (which are then assumed to be defined by their surface). In making up the base object with line segments (actuators, the length of which may be activated by the provision of remote energy), it is already possible to consider a given number of interrelationships to take into account between these various segments. This is because the reduction of a segment (pixel) affects others, being those which are the closest and less and less the others which are the furthest away.

In a recent article [ALA 16], the conversion to apply to each pixel of a cube to achieve a sphere, or the inverse conversion, can be calculated fairly easily using morphing methods which the authors have developed. "The work carried out shown in this publication indeed enables us to go from one shape to another using the same meshing: we mesh the so-called 'reference' shape. They calculate the conversions to apply to it to obtain the final desired form. The advantage from a digital viewpoint is that both forms have the same meshing, that is to say the same number of nodes and the same connectivity matrix, which enables us to deal with a digital problem without having to re-input the same data each time" [BEL 16]. From a principle viewpoint where there is the preservation of volume, the work of mathematical modeling around the relevant distortion to apply to each surface component was produced by Rohmer [ROH 14] and Rohmer, Hahmanna and Cania [ROH 15]. Sophistication linked to the non-interpenetration of surfaces (the example of an arm folding) has been developed by Vaillant *et al.* [VAI 12]. This scientific activity takes its origin, at the time of the animation of virtual characters of the calculation of a credible skin distortion at the joints. The production phase of a distortion is usually called "skinning".

There is thus, for this "very simple" example, the possibility of modeling the local energy path (for each pixel) to achieve a given conversion (with at the end of the chain, the possibility of analyzing the feasibility of an object from a given form).

This work is to be undertaken before addressing massive objects, which may however, as a first approach, be represented by a meshing close to that proposed on the surface of the object, by using a three-dimensional network.

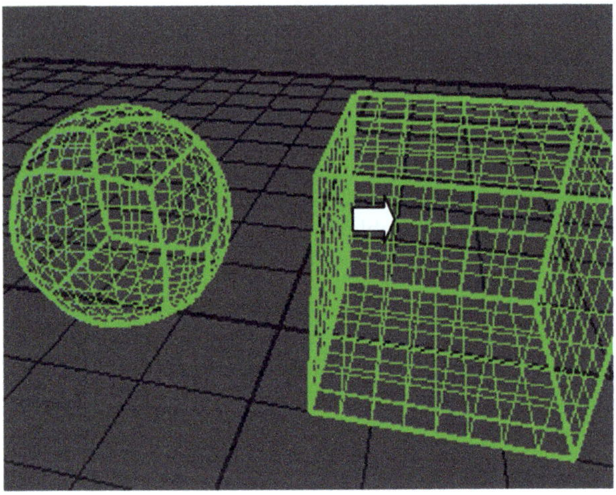

Figure 1.42. *Desired transition from a sphere to a cube. For a color version of the figure, see www.iste.co.uk/andre/printing3.zip*

1.6. Conclusion

The original component of this emerging sphere lies in the principle of voluntarily making the matter "smart", so as to extract a spatial functionality (rather than falling, at the present time, however exciting these concepts are within robotic self-assembly. This poses the question of the size of voxels involved and their cooperativeness in the given movements). At this stage, if "photochemical muscles" have already been the subject of works (which may be advantageously considered in depth), to the knowledge of the author, the concept which consisted of starting from a simple shape, easily produced by conventional techniques (for example, molding and conventional machining) so as, after processing by localized energy in time and space (for example, with light), to achieve a complex object. This is yet to be explored. It poses the question of having available materials with a high expandability (we recall that for shape-memory alloys, the gap between both forms is generally of the order of a few percent). It is then a matter of being able to process the converse issue. Starting from the basis of achieving a given shape, what might the initial shape actually be?

However, although this route appears interesting, we might acknowledge the need to tackle studies of very simple objects to infer given laws. Seemingly, it may initially be a question, as has been suggested, of working on surfaces rather than volumes to simplify the situation. For all that, the number of influencing parameters to take into consideration risks being higher leading not to a unique solution to a given spatial resolution, but indeed a set of solutions (and, at the same time, the absence of solutions). Thus, starting from a continuous material, the formation, for example, of a percolating environment may be problematic. However, it should be studied.

Table 1.6 collates all of the elements to master to develop this process (classified as + for "easy", = "feasible", - "existence of an impediment", -- "sticking point") with commentary.

Issue	Attainability	Commentary
Processes	??	Spatial transformation of matter induced by energy (electromagnetic, photochemical, heat and other forms of energy). Ideas are yet to be researched, the sphere is practically uncharted (outside of aspects of robot self-assembly) with the emergence of work on the so-called "swimming robots" (micrometric space and biomedical applications)
Flexibility	-	Possibility of using reversible reactions functioning by analogical means (photo-mechanic effects around polymers) or digital (shape-memory alloys)
Materials	=	Function of process (or processes)
Modeling	--	Outside of author competence (see the inverse problem)
Converse problem	--	Complex research avenue to explore, probably starting by studies of surface distortion rather than volume. Ideas are sought in this area
Interdisciplinarity	-	Collaborative encounter between photonics, photochemistry, mechanics, automation engineers, and other domains to move towards applications (or other processes)

Table 1.6. *Elements to bring under control for development of the 4D process*

To reach the objective of a good control of this futuristic field, there is a need for individuals originally from unconnected disciplines to explore these new boundary objects which appear capable of a genuine future application. The expected convergence (for which associated lines of questioning will be subject to discussion in Part 1 of Volume 3) necessitates that we depart from "conventional" means of innovation research. In exploring this axis, "accurately", nowadays, capable of illustrating its capacity for proofs of creative concepts, but still limited in number, these methods appear to have reached their limits. Around such an original and key theme, still uncharted, it may be more interesting to consider what enables "disruptive" innovation. This is the only way to think otherwise about scientific developments around new applications. Nowadays, we are therefore witnessing types of research "Uberisation" and 4D printing deserves better.

France Stratégie has stated that "A wave of innovations is materializing which is creating new markets, and is overturning the economy in numerous sectors. Faced with these disruptive innovations, the state must define its position: a wait-and-see stakeholder possibly having a considerable economic and social cost which could lead to a loss of sovereignty" [CHA 17]. However, within too many management teams, risk-taking is too often perceived as an interesting topic of conversation, not often appropriate, even esoteric, "particularly for researchers and academics, but with no real significance on daily activities in terms of strategies, and without any impact on their decision-taking processes" [LOU 17]. This idea is confirmed by R. Quirion, the President of the Comité de l'Évaluation (Assessment Committee) of CNRS who wrote in [CNR 16]: "At the CNRS, as elsewhere, interdisciplinarity is a genuine paradox. Although it is one of the most pluridisciplinary institutions to exist, [CNRS] must nevertheless make substantial efforts to stimulate interdisciplinary research. This indicates that there exists even within the institution, barriers which must be lowered if we wish to take advantage of the wealth and diversity of science generated".

To conclude for the time being, let us leave the responsibility to Alain Le Méhauté [LEM 14] to provide his contribution for this chapter (which actually he is not aware of as such). He wrote: "We estimate the economic potential of this future technology to be several hundred billion dollars. We know that hundreds of billions of dollars equates to between 1 and 4 million jobs... besides knowing that the control of this technique induces multiple derivative creations, including conceptual ones. We cannot regret the intellectual poverty of clichés, and the absence of ambition which we make clear in our public and private research through result-based programs. It is often very sad to read the content of guidelines into which the research will fall within and compromise the future, if it wishes to be financed".

It is a question of creating a little piece of nature to do so, and then to do so with the so-called "programed" objects, rather than remaining does so, by only operating in respect of opportunities (even if that only gets a foot in the door). Upon an open passionate subject, French academic research cannot set aside this accessible domain which bears a number of applications that are as yet unknown. Let us therefore try to prove Alain wrong.

1.7. Bibliography

[3D 14] 3DERS.ORG, "Chinese test self-forming liquid metal seeing a Terminator like future", available at: http://www.3ders.org/articles/20140304-chinese-test-self-forming-liquid-metal-seeing-a-terminator-like-future.html, 2014.

[ACK 16] ACKERMAN E., "Swarm of origami robots can self assemble out of a single sheet", available at: http://spectrum.ieee.org/automaton/robotics/robotics-hardware/harvard-self-folding-origami-robots, 2016.

[AHM 16] AHMED D., BAASCH T., JANG B. *et al.*, "Artificial swimmers propelled by acoustically activated flagella", *Nano Letters*, vol. 16, pp. 4968–4974, 2016.

[ALA 16] AL AKHRAS H., ELGUEDJ T., GRAVOUIL A. *et al.*, "Towards an automatic isogeometric analysis suitable trivariate models generation – Application to geometric parametric analysis", *Computer Methods in Applied Mechanics and Engineering*, vol. 316, pp. 623–645, 2016.

[ALO 08] ALOUGES F., DE SIMONE A., LEFEBVRE A., "Swimming at low Reynolds number at optimal strokes: an example", *Journal of Nonlinear Science*, vol. 18, pp. 277–302, 2008.

[ALO 12] ALOUGES F., "La natation dans le miel ou l'Hélium liquide: un problème de contrôle linéaire et sans drift", available at: http://smai.emath.fr/canum2012/resumesPDF/alouges/Abstract.pdf, 2012.

[ALV 04] ALVARO M., "Auto-organisation, autonomie et identité", *Revue Internationale de Philosophie*, vol. 228, pp. 135–150, 2004.

[AN 14] AN B., RUS D., "Designing and programming self-folding sheets", *Robotics & Autonomous Systems*, vol. 62, pp. 976–1001, 2014.

[AND 17] ANDRÉ J.C., "Impression 3D: niches applicatives porteuses", *Techniques de l'Ingénieur*, BM7970 V1, p. 17, 2017.

[ATL 15] ATLANTIC 3D, "Découvrez nos applications", available at: http://www.atlantic-3d.fr/, 2015.

[AZN 12] AZNAR F., PUJOL M., RIZO R., "Macroscopic definition of distributed swarm morphogenesis", *Connection Science*, vol. 24, pp. 162–192, 2012.

[BAK 15] BAKARICH S.E., GORKIN III R., IN HET PANHUIS M. *et al.*, "4D printing with mechanically robust, thermally actuating hydrogels", *Macromolecular Rapid Communications*, vol. 36, pp. 1211–1217, 2015.

[BAL 96] BALLANDRAS S., CALIN M., ZISSI S. *et al.*, "Actionneurs miniatures à alliages à mémoire de forme fabriqués par micro-stéréo-photolithographie", *Journal de Physique III France*, vol. 6, pp. 1759–1774, 1996.

[BAQ 04] BAQUIAST J.P., "La morphogénèse", available at: http://www.automatesintelligents.com/echanges/2004/jan/morphogenese.html, 2004.

[BEA 15] BEAUX-ARTS, *Les secrets des chefs-d'œuvre de la BD d'humour*, TIM, Paris, 2015.

[BEL 16] BEL-BRUNON A., Private communication, 2016.

[BIZ 11] BIZZARRI M., GIULANI A., CUCINA A. *et al.*, "Fractal analysis in a systems biology approach to cancer", *Seminars in Cancer Biology*, vol. 21, pp. 175–182, 2011.

[BMW 16] BMW, "BMW vision Next 100", available at: http://www.next100.bmw/en/topics/bmw-vision-next-100.html, 2016.

[BON 00] BONABEAU E., THÉRAULAZ G., "L'intelligence en essaim", available at: https://www.researchgate.net/profile/Eric_Bonabeau/publication/230660873_L'intelligence_en_essaim/links/53ff53f00cf2194bc29a4abd.pdf, 2000.

[BOU 14] BOUFFARON P., Impression 3D: Les prémisses d'une nouvelle (r)évolution industrielle, French and United States Embassy, available at: http://sf.france-science.org/wp-content/uploads/2014/11/SMM14_025.pdf, p. 48, 2014.

[BOU 16] BOURGEOIS J., "Matériel et logiciel pour créer de la matière programmable", available at: http://www.agence-nationale-recherche.fr/projet-anr/?tx_lwmsuivibilan_pi2%5BCODE%5D=ANR-16-CE33-0022, 2016.

[BRU 76] BRUTER C.P., "La théorie du système général; théorie de la modélisation", PUF, Paris, available at: http://www.mcxapc.org/inserts/ouvrages/0609tsgtm.pdf, 1976.

[BÜC 15] BÜCKMANN T., KADIC M., SCHITTNY R. *et al.*, "Mechanical metamaterials with anisotropy and negative effective mass density tensor made from one constituent material", *Physica Status Solidi*, vol. 252B, pp. 167–174, 2015.

[CAL 10] CALVO-MARZAL P., SATTAYASAMITSATHIT S., BALASUBRAMANIAN S. *et al.* (eds), "Chemical sensing based on catalytic nano-motors: motion-based detection of trace silver", *Chemical Communications*, vol. 46, pp. 12082–12083, 2010.

[CAM 74] CAMPBELL D.T., "Downward causation in hierarchically organised biological systems", in AYALA F., DOBZHANSKY T. (eds), *Studies in the Philosophy of Biology*, University of California Press, Berkeley, 1974.

[CDI 17] CDISCOUNT, "Jouet magique avec impression forme 3D", available at: http://www.cdiscount.com/juniors/jeux-de-societe-cartes/jouet-magique-avec-impression-forme-3d/f-12079-gen6922263311393.html, 2017.

[CÉR 11] CÉRÉZUELLE D., *La technique et la chair: essai de philosophie de la technique*, Parangon/VS E., Lyon, 2011.

[CÉR 17] CÉRÉZUELLE D., "Pour en finir avec le dogme de l'Immaculée Conception de la science", available at: http://sciences-critiques.fr/pour-en-finir-avec-le-dogme-de-limmaculee-conception-de-la-science/, 2017.

[CHA 06] CHAUVET G., *Comprendre l'organisation du vivant et son évolution vers la conscience*, Vuibert, Paris, 2006.

[CHA 17] CHARRIÉ J., JANIN L., "Répondre à l'innovation disruptive", http://francestrategie1727.fr/actions/repondre-a-linnovation-disruptive/, 2017.

[CHE 10] CHENG F., ZHANG Y., YIN R. *et al.*, "Visible light induced bending and unbending behavior of crosslinked liquid-crystalline polymer films containing azotolane moieties", *J. Mater. Chem.*, vol. 20, pp. 4888–4896, 2010.

[CHE 16] CHEANG U.K., KIM M.J., "Fabrication and control of simple low Reynolds number micro-swimmers", *Applied Physics Letters*, vol. 109, p. 034101, 2016. Doi: http://dx.doi.org/10.1063/1.4954946.

[CHE 17] CHEN M., GU Y., SINGH A. *et al.*, "Living additive manufacturing: transformation of parent gels into diversely functionalized daughter gels made possible by visible light photoredox catalysis", *ACS Central Science, Article ASAP*, 2017. Doi: 10.1021/acscentsci.6b00335, available at: http://pubs.acs.org/doi/full/10.1021/acscentsci.6b00335.

[CHI 11] CHIU C.Y., LI Y., RUAN L. *et al.*, "Platinum nanocrystals selectively shaped using facet-specific peptide sequences", *Nature Chemistry*, vol. 3, pp. 393–399, 2011.

[CHL 15] CHLUBA C., GE W., LIMA DE MIRANDA R. *et al.*, "Ultralow-fatigue shape memory alloy films", Science, vol. 348, pp. 1004–1007, 2015

[CHO 96] CHOSSAT P., *Les symétries brisées, Sciences d'avenir*, Pour la Science/Belin, Paris, 1996.

[CHO 15] CHOSSAT P., "Les mathématiques de la morphogénèse: l'autre grande contribution d'Alan Turing (I)", available at: http://images.math.cnrs.fr/Les-mathematiques-de-la.html, 2015.

[CLE 16a] CLÉMENT E., *The Microfab/microlab (MiFaMiLab) Platform at ESPCI*, Singapore, 1–2 December Nanoscale 3D Printing Workshop, 2016.

[CLE 16b] CLÉMENT E., LINDNER A., DOUARCHE C. *et al.*, "Bacterial suspensions under flow", *European Physical Journal Special Topics*, vol. 225, p. 2389, 2016.

[CNR 16] CNRS, "Jean-Pierre Sauvage, un Nobel pour les machines moléculaires", available at: https://lejournal.cnrs.fr/articles/jean-pierre-sauvage-un-nobel-pour-les-machines-moleculaires, 2016.

[COL 01] COLLIN J.P., DIETRICH-BUCHECKER R., SAUVAGE J.P., "Vers des machines et des moteurs moléculaires", *Actualité Chimique*, vol. 6, pp. 27–32, 2001.

[CRA 15] CRAWFORD M., "4D printing, the next level of additive manufacturing", available at: https://www.asme.org/engineering-topics/articles/manufacturing-processing/4d-printing-next-level-additive-manufacturing, 2015.

[DAD 14] DADBAKHSH S., SPEIRS M., KRUTH J.P. *et al.*, "Effect of SLM parameters on transformation temperatures of shape memory nickel titanium parts", *Advanced Engineering Materials*, vol. 16, pp. 1140–1146, 2014.

[DEM 14] DE MATHAREL L., "Poppy, le robot open source fabriqué avec une imprimante 3D", available at: http://www.usine-digitale.fr/article/poppy-le-robot-open-source-fabrique-avec-une-imprimante-3d.N249523, 2014.

[DES 15] DE SIMONE A., GIDONI P., NOSELLI G., "Liquid crystal elastomer strips as soft crawlers", *Journal of the Mechanics and Physics of Solids*, vol. 84, pp. 254–272, 2015.

[DEL 07] DELONCLE R., Dendrimères phosphorés à motifs azo-benzène: vers des matériaux photo-adaptatifs, Thesis, Université Paul Sabatier, Toulouse, France, 2007.

[DEV 16] DEVOUGE A., "Voici l'Octobot, le premier robot entièrement mou au monde", available at: http://dailygeekshow.com/octobot-robot-mou/?utm_source=newsletter&utm_medium=e-mail&utm_campaign=Newsletter_Journaliere_2016-09-03, 2016.

[DIO 92\93] DIONNET V., HÉLIODORE F., LE MÉHAUTÉ A. *et al.*, Procédé de fabrication d'un objet fractal par stéréo-lithographie et objet fractal obtenu par un tel procédé, Patent nos 9206652 (French), 93401389.7 (European), 1992\1993.

[DUF 02] DUFAUD O., MARCHAL P., CORBEL S., "Rheological properties of PZT suspensions for stereolithography", *Journal of the European Ceramic Society*, vol. 22, pp. 2081–2092, 2002.

[EAS 09] EASTES R.E., DARRIGA C., BATAILLE X. *et al.*, "Des vers dans les jardins chimiques pour (re)mettre la science en culture", *Alliage*, vol. 64, pp. 73–89, 2009.

[EGU 16] EGUNOV A.I., KORVINKB J.G., LUCHNIKOV V.A., "Polydimethylsiloxane bilayer films with an embedded spontaneous curvature", *Soft Matter*, vol. 12, pp. 45–52, 2016. Doi: 10.1039/C5SM01139F, extract from: http://www.cnrs.fr/inc/communication/direct_labos/luchnikov.htm.

[ENG 17] ENG K., "A peek into the brave new world of programmable materials", available at: http://ideas.ted.com/a-peek-into-the-brave-new-world-of-programmable-materials/, 2017.

[ERC 10] ERCOLE F., DAVIS T.P., EVANS R.A., "Photo-responsive systems and biomaterials: photochromic polymers, light-triggered self-assembly, surface modification, fluorescence modulation and beyond", *Polymer Chemistry*, vol. 1, pp. 37–54, 2010.

[FEL 13a] FELTON S.M., TOLLEY M.T., SHIN B. *et al.*, "Self-folding with shape memory composites", *Soft Matter*, vol. 7, pp. 7688–7694, 2013.

[FEL 13b] FELTON S.M., TOLLEY M.T., ONAL C.D. *et al.*, "Robot self-assembly by folding: a printed inchworm robot", *2013 IEEE International Conference on Robotics & Automation (ICRA)*, Karlsruhe, Germany, IEEE Transactions, pp. 277–282, 2013.

[FIG 15] FIGARO MAGAZINE, "L'impression 4D, la nouvelle dimension", *Le Figaro*, p. 30, 8 May 2015.

[FIS 14] FISCHER J., THIEL M., WEGENER M., "Matter made to order", *IEEE Spectrum*, vol. 51, pp. 34–58, 2014.

[FOR 15] FORMLABS, "Design, adoption, and readiness: the future of 3D printing and additive manufacturing", available at: http://www.liebertpub.com/lpages/3dp-formlabs-learning-center/178/, 2015.

[FRE 94] FREUD S., *Essai de psychanalyse*, Payot, Paris, 1994.

[GAO 10] GAO W., SATTAYASAMITSATHIT S., MANESH K.M. *et al.*, "Magnetically powered flexible metal nanowire motors", *Journal of the American Chemical Society*, vol. 132, pp. 14403–14406, 2010.

[GAO 11] GAO W., SATTAYASAMITSATHIT S., OROZCO J. *et al.*, "Highly efficient catalytic microengines: template electrosynthesis of polyaniline/platinum microtubes", *Journal of the American Chemical Society*, vol. 133, pp. 11862–11864, 2011.

[GAO 16] GAO B., YANG Q., ZHAO X. *et al.*, "4D bio-printing for biomedical applications", *Trends in Biotechnology*, vol. 34, pp. 746–756, 2016. Available at: http://dx.doi.org/10.1016/j.tibtech.2016.03.004.

[GAR 13] GARCIA M., Hydrodynamique de micro-nageurs, Thesis, University of Grenoble, France, available at: https://tel.archives-ouvertes.fr/tel-00952771/document, 2013.

[GE 13] GE Q., QI J., DUNN M.L., "Active materials by four dimensions printing", *Applied Physics Letters*, available at: http://dx.doi.org/10.1063/1.4819837, vol. 103, p. 131901, 2013.

[GE 14] GE Q., DUNN C.K., QI H.J. *et al.*, "Active origami by 4D printing", *Smart Materials and Structures*, vol. 23, p. 094007, 2014. doi: 10.1088/0964-1726/23/9/094007.

[GE 16a] GE Q., QI H.J., DUNN M.L., "Active materials by four-dimension printing", *Applied Physics Letters*, vol. 103, p. 131901, 2016. doi: available at: http://dx.doi.org/10.1063/1.4819837.

[GE 16b] GE Q., SAKHAEI A.H., LEE H. *et al.*, "Multimaterial 4D printing with tailorable shape memory polymers", *Scientific Reports 6*, p. 31110, 2016. doi: 10.1038/srep31110.

[GHO 09] GHOSH A., FISCHER P., "Controlled propulsion of artificial magnetic nano-structured propellers", *Nano Letters*, vol. 9, pp. 2243–2248, 2009.

[GIB 09] GIBBS J.G., ZHAO Y.P., "Autonomously motile catalytic nano-motors by bubble propulsion", *Applied Physics Letters*, vol. 94, p. 163104, 2009.

[GIL 12] GILPIN K., KNAIAN A., RUS D., "Robot pebbles: one centimeter modules for programmable matter through self-dissassembly", available at: http://cba.mit.edu/docs/papers/10.05.knaian.ICRA.pdf, 2012.

[GLA 16] GLADMAN A.S., MATSUMOTO E.A., NUZZO R.G. *et al.*, "Biomimetic 4D printing", *Nature Materials*, vol. 15, pp. 413–418, 2016.

[GRO 15] GROSSWEILER G.R., BROWN C.L., HEWAGE C.B. *et al.*, "Mechano-chemically active soft robots", *ACS Applied Materials & Interfaces*, vol. 7, pp. 22431–22435, 2015.

[HAM 14] HAMPEL C., *Eléments d'épistémologie*, Armand Colin, Paris, 2014.

[HAN 04] HANSON D.F., WHITE V., "Converging the capabilities of EAP artificial muscles and the requirements of bio-inspired robotics", *Proceedings of SPIE 5385 Smart Structures and Materials: Electroactive Polymer Actuators and Devices (EAPAD)*, 27 July, vol. 5385, pp. 29–41, 2004.

[HAR 12] HARDESTY L., "Self-sculpting sand", available at: http://newsoffice.mit.edu/2012/smart-robotic-sand-0402, 2012.

[HAR 15] HARDESTY L., "Centimeter-long origami robot", available at: http://news.mit.edu/2015/centimeter-long-origami-robot-0612, 2015.

[HAR 16] HARVARD UNIV., "A 3-D material that folds, bends and shrinks on its own", available at: https://www.youtube.com/watch?v=maKILHxcGAE, 2016.

[HAW 89] HAWKING S., *Une brève histoire du temps*, Flammarion, Paris, 1989.

[HAW 10] HAWKES E., AN B., BENBERNOU N.M. *et al.*, "Programmable matter by folding", *Proceedings of the National Academy of Sciences*, vol. 107, pp. 12441–12445, 2010.

[HON 10] HONG Y., DIAZ M., CÓRDOVA-FIGUEROA U.M. *et al.*, "Light-driven titanium-dioxide-based reversible microfireworks and micromotor/micropump systems", *Advanced Functional Materials*, vol. 20, pp. 1–9, 2010.

[HOW 07] HOWSE J., JONES R., RYAN A. *et al.*, "Self-motile colloidal particles: from directed propulsion to random walk", *Physical Review Letters*, vol. 99, p. 048102, 2007.

[HUA 15] HUANG C., LU J.A., TIAN X. *et al.*, "Miniaturized swimming soft robot with complex movement actuated and controlled by remote light signals", *Scientific Reports,* vol. 5, Article number: 17414, 2015.

[IBE 09] IBELE M., MALLOUK T.E., SEN A., "Schooling behavior of light-powered autonomous micromotors in water", *Angewandte Chemie International Edition*, vol. 48, pp. 3308–3312, 2009.

[ILI 11] ILIC-STOJANOVIC S., NIKOLIC L., NIKOLIC V. *et al.*, "Stimuli-sensitive hydrogels for pharmaceutical and medical applications", *Facta Universitatis*, vol. 9, pp. 37–56, 2011.

[JAC 16] JACKSON B., "3D printing advances end to animal testing with Harvard's heart on a chip", available at: https://3dprintingindustry.com/news/3d-printing-advances-end-animal-testing-harvards-heart-chip-97338/, 2016.

[JAM 13] JAMAL M., KADAM S.S., XIAO R. *et al.*, "Bio-origami hydrogel scaffolds composed of photo-crosslinked PEG bilayers", *Advanced Healthcare Materials*, vol. 2, pp. 1142–1150, 2013.

[JIN 11] JIN Y., MENG Y., "Morphogenetic robotics: an emerging new field in developmental robotics", *IEEE Transactions On Systems, Man and Cybernetics*, vol. C41, pp. 145–160, 2011.

[JOS 12] JOSHI S.D., DAVIDSON L.A., "Epithelial machines of morphogenesis and their potential application in organ assembly and tissue engineering", *Biomechanics and Modeling in Mechanobiology,* vol. 11, pp. 1109–1121, 2012.

[KEA 13] KEATING S., OXMAN N., "Compounds fabrication: a multi-functional robotic platform for digital design and fabrication", *Robotics & Computer Integrated Manufacturing*, vol. 29, pp. 439–448, 2013.

[KEK 12] KEKENES-HUSKEY P.M., GILLETTE A., HAKE J. *et al.*, "Finite-element estimation of protein–ligand association rates with post-encounter effects: applications to calcium binding in troponin C and SERCA", *Computational Science & Discovery,* vol. 5, available at: http://dx.doi.org/10.1088/1749-4699/5/1/014015, 2012.

[KEK 15] KEKENES-HUSKEY P.M., EUN C., MCCAMMON J.A., "Enzyme localization, crowding, and buffers collectively modulate diffusion-influenced signal transduction: insights from continuum diffusion modeling", *Journal of Chemical Physics*, available at: http://dx.doi.org/10.1063/1.4929528, vol. 143, p. 094103, 2015.

[KHI 14] KHIM T., ZHU L., AL-KAYSI R.O. *et al.*, "Organic photomechanical material", *ChemPhysChem*, vol. 15, pp. 400–414, 2014.

[KHO 15] KHOO Z.X., MEI TEOH J.E., LIU Y. *et al.* , "3D printing of smart materials: a review on recent progresses in 4D printing", *Virtual and Physical Prototyping*, available at: http://dx.doi.org/10.1080/17452759.2015.1097054, vol. 10, pp. 103–122, 2015.

[KIM 15] KIMIONIS J., ISAKOV M., KOH B.S. *et al.*, "3D-printed origami packaging with inkjet-printed antennas for RF harvesting sensors", *IEEE Transactions On Microwave Theory and Technique*, vol. 63, pp. 4521–4532, 2015.

[KNA 10] KNAIAN A.N., Electropermanent magnetic connectors and actuators: devices and their application in programmable matter, PhD Thesis, MIT, Boston, 2010.

[KOK 15] KOKKINIS D., SCHAFFNER M., STUDART A.R., "Multimaterial magnetically assisted 3D printing of composite materials", *Nature Communications*, vol. 6, p. 8643, 2015.

[KOT 98] KOTAY K., RUS D., VANA M. *et al.*, "The self-reconfiguring robotic molecules: desogn and control algorithm", Workshop on algorithmics, available at: http://citeseerx.ist.psu.edu/viewdoc/download?doi=10.1.1.42.6594&rep=rep1&type=pdf, 1998.

[KUK 13] KUKSENOK O., BALAZS A.C., "Modeling the photoinduced reconfiguration and directed motion of polymer gels", *Advanced Functional Materials*, vol. 23, pp. 4601–4610, 2013.

[LAG 12] LAGADEC P., *Du risque majeur au mégachocs*, Préventique, Bordeaux, 2012.

[LAR 10] L'ARRÊT PUBLIC GOUVERNEMENT, "Portrait d'un charlatan", available at: http://legosanstrique.blogspot.fr/2010_08_01_archive.html, 2010.

[LAR 15] LARGER L., *Groupe de veille INSIS – contribution*, INSIS- CNRS, Paris, 2015.

[LAX 12] LAXMINARAYANA G.K., Approches colloïdale et bio-inspirée en nano-plasmonique, Thesis, Paul Sabatier University, Toulouse, available at: http://thesesups.ups-tlse.fr/1723/1/2012TOU30103.pdf, 2012.

[LED 10] LEDUC S., *Théorie physico-chimique de la vie et générations spontanées*, Poinat, Paris, 1910.

[LED 12] LEDUC S., *La biologie synthétique: étude de biophysique*, Poinat, Paris, 1912.

[LED 16] LE DUIGOU A., CASTRO M., BEVAN R. *et al.*, "3D printing of wood fibre biocomposites: from mechanical to actuation functionality", *Materials & Design*, vol. 96, pp. 106–114, 2016.

[LEH 02] LEHN J.M., "Toward self-organization and complex matter", *Science*, vol. 295, pp. 2400–2403, 2002.

[LEM 14] LE MÉHAUTÉ A., "Interview d'Alain Le Méhauté – L'un des pères de l'impression 3D", available at: http://www.priximprimante3d.com/inventeur/, 2014.

[LES 13] LESNE A., "Une conjonction de réactions chimiques et de diffusion: Alan Turing, les motifs et les structures du vivant", available at: https://interstices.info/jcms/int_71868/alan-turing-les-motifs-et-les-structures-du-vivant, 2013.

[LEI 16] LEIST S.K., ZHOU J., "Current status of 4D printing technology and the potential of light-reactive smart materials as 4D printable materials", *Virtual and Physical Prototyping*, vol. 11, pp. 249–262, 2016.

[LEW 16] LEWIS C., *Too Fast to Think: How to Reclaim Your Creativity in a Hyper-connected Work Culture*, Kogan Page, Paris, 2016.

[LI 16a] LI W., WU X., QIN H. *et al.*, "Light-driven and light-guided micro-swimmers", *Advanced Functional Materials*, vol. 26, pp. 3164–3171, 2016.

[LI 16b] LI T., LI J., ZHANG H. *et al.*, "Magnetically propelled fish-like nano-swimmers", *Small*, vol. 12, pp. 6098–6105, 2016.

[LIE 16] LIEBCHEN B., CATES M.E., MARENDUZZO D., "Pattern formation in chemically interacting active rotors with self-propulsion", *Soft Matter*, vol. 12, pp. 7259–7264, 2016. Doi: 10.1039/C6SM01162D.

[LIN 16] LIND J.U., BUSBEE T.A., VALENTINE A.D. *et al.*, "Instrumented cardiac microphysiological devices via multimaterial three-dimensional printing", *Nature Materials*, 2016. Doi:10.1038/nmat4782.

[LOG 11] LOGET G., KUHN A., "Electric field-induced chemical locomotion of conducting objects", *Nature Communications*, vol. 2, p. 535, 2011. Doi: 10.1038/ncomms1550.

[LOU 17] LOUISOT J.M., "Optimiser la prise de risques", *Risk-Assur*, avalable at: https://www.riskassur-hebdo.com, vol. 474, pp. 28–37, 2017.

[LU 17] LU L., GUO P., PAN Y., "Magnetic-field-assisted projection stereolithography for 3D printing of smart structures", *ASME*, forthcoming, vol. 139, p. 071008-1, 2017. Doi: 10.1115/1.4035964.

[LUS 13] LUSHIA E., PESKIN C.S., "Modeling and simulation of active suspensions containing large numbers of interacting micro-swimmers", *Computers & Structures*, vol. 122, pp. 239–248, 2013.

[MAC 10] MACLENNAN B.J., "Morphogenesis as a model for nano-communication", *Nano Communication Networks*, vol. 1, pp. 1199–1208, 2010.

[MAC 12] MacLennan B.J., "Molecular coordination of hierarchical self-assembly", *Nano Communication Networks* , vol. 3, pp. 116–128, 2012.

[MAH 12] Mahimwalla Z., Yager K., Mamiya J.I. *et al.*, "Azobenzene photomechanics prospects and potential applications", *Polymer Bulletin*, vol. 69, pp. 967–1006, 2012.

[MAH 13] Mahimwalla Z., Characterization of photo-induced photo-mechanical responses in azobenzene polymers, PhD Thesis, McGill University, Montreal, 2013.

[MAN 05] Mano N., Heller A., "Bio-electrochemical propulsion", *Journal of the American Chemical Society*, vol. 127, pp. 11574–11575, 2005.

[MAR 12] Martinez, R.V., Fish C.R., Chen X. *et al.*, "Elastomeric origami: programmable paper-elastomer composites as pneumatic actuators", *Advanced Functional Materials*, vol. 22, pp. 1376–1384, 2012.

[MAR 14] Marques-Serra A.V., "Matière molle programmable par voie chimique ou optique: compaction de l'ADN, activité protéique et opérations microfluidiques", available at: https://tel.archives-ouvertes.fr/tel-01142069/file/pdf2star-1429012631-these_archivage_3159641.pdf, 2014.

[MAR 15] Martin J.J., Fiore B.E., Erb R.M., "Designing bioinspired composite reinforcement architecture via 3D magnetic printing", *Nature Communications*, vol. 6, 2015. Doi: 10.1038/ncomms9641.

[MAY 09] Maye M.M., Nykypanchuk D., Cuisinier M. *et al.*, "Stepwise surface encoding for high-throughput assembly of nanoclusters", *Nature Materials*, vol. 8, pp. 388–391, 2009.

[MAZ 16] Mazzolai B., Mattoli V., "Robotics: generation soft", *Nature*, vol. 536, pp. 400–401, 2016.

[MEH 14] Mehta A.M., Rus D., "An end-to-end system for designing mechanical structures for print-to-folds robots", *ICRA/IEEE International Conference on Robotics and Automation*, 2014. Doi: 10.1109/ICRA.204.6907044.

[MEI 12] Meier H., Haberland C., Frenzel J., "Structural and functional properties of NiTi shape memory alloys produced by Selective Laser Melting", *Innovative Developments in Virtual and Physical Prototyping*, pp. 291–296, 2012.

[MIT 13a] MIT, "Amazing technology invented by MIT – tangible media", available at: https://www.youtube.com/watch?v=lvtfD_rJ2hE and https://www.youtube.com/watch?v=lCARHatJQJA&feature=iv&src_vid=lvtfD_rJ2hE&annotation_id=annotation_4171896729, 2013.

[MIT 13b] MIT, "This is why 4D printing is cool", available at: https://www.youtube.com/watch?v=yKTVsDPGQgg, 2013.

[MUG 08] Mugnier L., "Introduction aux problèmes inverses: application à l'imagerie optique à haute résolution en astronomie, C9", in Léna P. (ed.), *Des données à l'objet: le problème inverse*, EDP Sciences/CNRS, Paris, available at: http://laurent.mugnier.free.fr/ftp/SupOpt/methodesPhy-LM.pdf, 2008.

[NAF 17] NAFICY S., GATELY R., GORKIN III R. *et al.*, "4D printing of reversible shape morphing hydrogel structures", *Macromolecular Materials and Engineering*, vol. 302, p. 1600212, 2017. Doi: 10.1002/mame.201600212.

[NAJ 04] NAJA A., GOLESTANIAN R., "Simple swimmer at low Reynolds number: Three linked spheres", *Physical Review*, vol. E69, p. 062901, 2004.

[NAK 10] NAKANO H., "Direction control of photomechanical bending of a photochromic molecular fiber", *Journal of Materials Chemistry*, vol. 20, pp. 2071–2074, 2010.

[NEW 15] NEW SCIENTIST, "Liquid metal brings shape-shifting robot a step closer", available at: https://www.newscientist.com/article/dn27127-liquid-metal-brings-shape-shifting-robot-a-step-closer/, 2015.

[NIC 77] NICOLIS G., PRIGOGINE I., *Self-Organization in Non-equilibrium Systems: From Dissipative Structures to Order through Fluctuations*, Inter-Science, New York, 1977.

[NIK 10] NIKITIN M.P., ZDOBNOVA T.A., LUKASH S.V. *et al.*, "Protein-assisted self-assembly of multifunctional nanoparticles", *Proceedings of the National Academy of Sciences of the United States of America*, vol. 107, pp. 5827–5832, 2010.

[NIM 17] NIMESIS, "Exemples d'applications", available at: http://www.nimesis.com/faq/exemples-dapplications/, 2017.

[NIS 16] NISSER M.E.W., FELTON S.M., TOLLEY M.T. *et al.*, "Feedback-controlled self-folding of autonomous robot collectives", *IROS 2016 – IEEE/RSJ International Conference on Intelligent Robots and Systems*, Daejeon, South Korea, pp. 1254–1261, 9–14 October 2016.

[NYK 08] NYKTER M., PRICE N.D., ALDANA M. *et al.*, "Gene expression dynamics in the macrophage exhibit criticality", *Proceedings of the National Academy of Sciences USA*, vol. 105, pp. 1897–1900, 2008.

[ONA 13] ONAL C.D., WOOD R.J., RUS D., "An origami-inspired approach in worm robots", *IEEE/ASME Mechatronics Transactions*, vol. 18, pp. 430–438, 2013.

[PAT 17] PATEL D.K., SAKHAEI A.K., LAYANI M. *et al.*, "Highly stretchable and UV curable elastomers for digital light processing based 3D printing", *Advanced Materials*, 2017. Doi: 10.1002/adma.201606000.

[PAT 06] PATOOR E., "Comportement et applications des alliages à mémoire de forme", Séminaire Matériaux IN2P3-CNRS, available at: http://www.in2p3.fr/actions/formation/Materiaux06/AMF_IN2P3.pdf, 2006.

[PAX 04] PAXTON W.F., KISTLER K.C., OLMEDA C.C. *et al.*, "Catalytic nano-motors: autonomous movement of striped nano-rods", *Journal of the American Chemical Society*, vol. 126, pp. 13424–13431, 2004.

[PEP 00] PEPPAS N.A., BURES P., LEOBANDUNG W. *et al.*, "Hydrogels in pharmaceutical formulations", *European Journal of Pharmaceutics and Biopharmaceutics*, vol. 50, pp. 27–46, 2000.

[PET 15] PETR M., HELGESON M.E., SOULAGES J. *et al.*, "Rapid viscoelastic switching of an ambient temperature range photo-responsive azo-benzene side chain liquid crystal polymer", *Soft Matter*, 2015, doi: 10.1039/b000000x.

[PEY 12] PEYER K.E., QIU F., ZHANG L. *et al.*, "Movement of artificial bacterial flagella in heterogeneous viscous environments at the microscale", available at: https://infoscience.epfl.ch/record/180951/files/Kathrin-Nelson_Movement__Microscale_2012.pdf, 2012.

[PIC 17] PICKETT I., "For shape-changing smart materials, applications abound", *Appliance Design*, vol. 65, pp. 17–19, 2017.

[PIM 15] PIMENTA J., "InMoov, le premier robot humanoïde que vous pourrez imprimer en 3D directement depuis chez vous", available at: http://soocurious.com/fr/inmoov-le-premier-robot-humanoide-que-vous-pourrez-imprimer-en-3d-directement-depuis-chez-vous/, 2015.

[PRA 16] PRAVANEE M., "4D Printing in action", available at: https://www.youtube.com/watch?v=qMJXzPiJkH8, 2016.

[PRI 94] PRIGOGINE I., *Les lois du chaos*, Flammarion, Paris, 1994.

[PY 07] PY C., REVERDY P., DOPPLER L. *et al.*, "Capillary origami: spontaneous wrapping of a droplet with an elastic sheet", *Physical Review Letters*, vol. 98, p. 156103, 2007.

[QUI 16] QUIRION R., "President, Evaluation scientifique du CNRS", CNRS, Paris, available at: http://www.cnrs.fr/fr/organisme/docs/espacedoc/comite-consultatif-oct-2016.pdf, 2016.

[RAN 15] RANOSA T., "MIT tiny origami robot can fold itself and self-destruct: perfect profile for future microsurgeon?", available at: http://www.techtimes.com/articles/56678/20150530/mit-tiny-origami-robot-can-fold-itself-and-self-destruct-perfect-profile-for-future-micro-surgeon.htm, 2015.

[RAO 15] RAO K.J., LI F., MENG L. *et al.*, " A force to be reckoned with: a review of synthetic micro-swimmers powered by ultrasound", *Small*, vol. 11, pp. 2836–2846, 2015.

[ROC 15a] ROCHEFORT M., "Bougeant au contact de la chaleur ou de la lumière, ce papier en graphène pourrait révolutionner la robotique", available at: http://soocurious.com/fr/graphene-papier-bouge/?utm_source=newsletter&utm_medium=e-mail&utm_campaign=Newsletter_Journaliere_2015-11-13, 2015.

[ROC 15b] ROCHON P., "Description des intérêts actuels en matière de recherche", available at: http://www.rmc.ca/fr/physique/dr-paul-rochon-bsc-phd-peng-professeur, 2015.

[ROH 14] ROHMER D., "Les surfaces gagnent du volume", *Quadrature*, available at: https://hal.inria.fr/hal-01017885/document, 2014.

[ROH 15] ROHMER D., HAHMANN S., CANI M.P., "Real-time continuous self-replicating details for shape deformation", *Computers and Graphics, Elsevier, International Conference Shape Modeling International*, available at: https://hal.inria.fr/hal-01152928/document, vol. 51, pp. 67–73, 2015.

[ROS 15] ROSALES A.M., MABRY K.M., NEHLS E.M. *et al.*, "Photo-responsive elastic properties of azo-benzene-containing poly-(ethylene-glycol)-based hydrogels", *Biomacromolecules*, vol. 16, pp. 798–806, 2015.

[SAN 13] SANCHEZ C., Enseignement: La chimie mésoscopique: du nanomatériau au mésocristal, Annuaire du Collège de France; Résumé des cours et travaux 113e année, pp. 225–250, 2012–2013.

[SAR 03] SARIKAYA M., TAMERLER C., JEN A.K.Y. *et al.*, "Molecular bio-mimetics: nanotechnology through biology", *Nature Materials*, vol. 2, pp. 577–585 2003.

[SAV 14] SAVAGE N., "Time for a change", *ACM Communications*, vol. 57, pp. 16–18, 2014.

[SÈV 05] SÈVE L., "De quelle culture logico-philosophique la pensée du non-linéaire a-t-elle besoin?", in J. GUESPIN-MICHEL (ed.), "*Emergence, complexité et dialectique*", Odile Jacob, Paris, 2005.

[SEN 16] SENATOV F.S., NIAZA K.V., ZADOROZHNYY M.Y. *et al.*, "Mechanical properties and shape memory effect of 3D-printed PLA-based porous scaffolds", *Journal of the Mechanical Behavior of Biomedical Materials*, vol. 57, pp. 139–148, 2016.

[SHE 14] SHENG L., ZHANG J., LIU J., "Diverse transformations of liquid metals between different morphologies", *Advanced Materials*, 2014. doi: 10.1002/adma.201400843, available at: https://www.researchgate.net/profile/Lei_Sheng2/publication/262773664_Liquid_Metals_Diverse_Transformations_of_Liquid_Metals_Between_Different_Morphologies_Adv_Mater_342014/links/546e9d950cf2bc99c2155a1b.pdf.

[SHA 15] SHALYGO Y., "The kinetic basis of morphogenesis", *13th European Conference on Artificial Life*, New York, pp. 122–129, 2015.

[SHI 99] SHIN J.Y., ABBOTT N.L., "Using light to control dynamic surface tensions of aqueous solutions of water soluble surfactants", *Langmuir*, vol. 15, pp. 4404–4410, 1999.

[SIM 11] SIMS P., *How Breakthrough Ideas Emerge from Small Discoveries*, Business Book, London, 2011.

[SLO 10] SLOCIK J.M., NAIK R.R., "Probing peptide–nanomaterial interactions", *Chemical Society Reviews*, vol. 39, pp. 3454–3463, 2010.

[SOL 09] SOLOVEV A., MEI Y., BERMÚDEZ UREÑA U. *et al.*, "Catalytic micro-tubular jet engines self-propelled by accumulated gas bubbles", *Small*, vol. 5, pp. 1688–1692, 2009.

[STA 11] STAETZLER K., SONNENSCHEIN C., SOTO A.M., "Systems biology beyond networks: generating order from disorder through self-organization", *Seminars in Cancer Biology,* vol. 21, pp. 165–174, 2011.

[STA 15] STANTON M.M., TRICHET-PAREDES C., SANCHEZ S., "Applications of 3D printing for micro-swimmers and bio-hybrid robotics", *Lab on a Chip*, vol. 15, pp. 1634–1637, 2015.

[STE 13] STENGERS I., *Une autre science est possible! Manifeste pour un ralentissement des sciences*, Les empêcheurs de penser en rond, Paris, 2013.

[STR 13] STROUD A.B., MORRIS M., CAREY K. *et al.*, "MU-L8: the design architecture and 3D printing on a teen-sized humanoid soccer robot", *Proceedings of 8th Workshop on Humanoids Soccer Robots, 13th IEEE-RAS International Conference on Humanoids Robots*, available at: http://www.eng.mu.edu/abwilliams/heirlab/papers/HSR13_Williams.pdf, p. 5, 2013.

[SUS 14] SUSSAN R., "Retour sur la matière programmable", available at: http://www.internetactu.net/2014/07/11/retour-sur-la-matiere-programmable/, 2014.

[TAM 07] TAMERLER C., SARIKAYA M., "Molecular bio-mimetics: utilizing nature's molecular ways in practical engineering", *Acta Biomaterialia*, vol. 3, pp. 289–299, 2007.

[TEN 16] TENZERIS M.M., BAHR R.A., PAPADOPOULOS A.D. *et al.*, "Multi-domain modeling of 3D printed, nanotechnology and morphing/origami-based RF modules", *Microwave Symposium (IMS), 2016 IEEE MTT-S International*, 2016. doi: 10.1109/MWSYM.2016.7540101.

[TIB 13] TIBBITS S., TOMAS A.F., "Biomolecular, chiral and irregular self-assemblies", *Acadia 2013 Adaptative Architecture*, available at: http://papers.cumincad.org/data/works/att/acadia13_261.content.pdf, pp. 261–268, 2013.

[THI 14] THILMANY J., "4D printing", *Mechanical Engineering*, pp. 22–24, February 2014.

[TIB 16] TIBBITS S., *Self-Assembly Lab: Experiments in Programming Matter*, Routledge, New York, 2016.

[THO 83] THOM R., *Paraboles et catastrophes*, Flammarion, Paris, 1983.

[THO 14] THOMAS C., TIBBITS S., BANNING G., "The programmable world", *Scientific American*, vol. 311, pp. 60–65, 2014.

[THU 80] THUILLIER P., *Le petit savant illustré*, le Seuil, Paris, 1980.

[TOT 12] TOTTORI S., ZHANG L., QIU F. *et al.,* "Magnetic helical micromachines: fabrication, controlled swimming, and cargo transport", *Advanced Materials*, vol. 24, pp. 811–816, 2012.

[TUR 52] TURING A., "The chemical basis of morphogenesis", *Philosophical Transactions of the Royal Society of London*, vol. B237, pp. 37–72, 1952.

[TUR 15] TURK & DEGROOT, *Léonard; cadeau de génie*, Editions du Lombard, Brussels, Belgium, 2015.

[UME 13] UMEDACHI T., VIKAS V., TRIMMER B.A., "Highly deformable 3D printed soft robot generating inching and crawling locomotion with variable friction legs", *IEEE/RSJ International Conference on Intelligent Robots and Systems (IROS)*, pp. 4590–4593, 2013.

[VAI 12] VAILLANT R., BARTHE L., GUENNEBAUD G. *et al.*, "Déformation de la peau d'un personnage avec prise en compte des contacts", *Revue Electronique Francophone d'Informatique Graphique, Association Française d'Informatique Graphique*, vol. 6, pp. 37–49, 2012.

[VIL 11] VILLAR G., HERON A.J., BAYLEY H., "Formation of droplet networks that function in aqueous environments", *Nature Nanotechnology*, vol. 6, pp. 803–808, 2011.

[VIL 13] VILLAR G., GRAHAM A.D., BAYLEY H., "A tissue-like printed material", *Science*, vol. 340, pp. 48–52, 2013.

[WAN 12] WANG W., CASTRO L.A., HOYOS M. *et al.*, "Autonomous motion of metallic micro-rods propelled by ultrasound", *ACS Nano*, vol. 6, p. 6122, 2012.

[WAN 15] WANG X., SUN Y., PENG C. *et al.*, "Transitional suspensions containing thermosensitive dispersant for three-dimensional printing", *ACS Applied Materials & Interfaces*, vol. 7, pp. 26131–26136, 2015.

[WEH 16] WEHNER M., TRUBY R.L., FITZGERALD D.J. *et al.*, "An integrated design and fabrication strategy for entirely soft, autonomous robots", *Nature*, vol. 536, pp. 451–455, 2016.

[WHE 10] WHEAT P.M., MARINE N.A., MORAN J.L. *et al.*, "Rapid fabrication of bimetallic spherical motors", *Langmuir*, vol. 26, pp. 13052–13055, 2010.

[WIK 15] WIKIPEDIA, "Alliage à mémoire de forme", available at: https://fr.wikipedia.org/wiki/Alliage_%C3%A0_m%C3%A9moire_de_forme, 2015.

[WIL 12] WILSON D.A., NOLTE R.J., VAN HEST J.C., "Autonomous movement of platinum-loaded stomatocytes", *Nat. Chem.*, vol. 26, pp. 268–274. doi: 10.1038/nchem.1281, 2012.

[YAM 09] YAMADA K., SONE J., CHEN J., "Three-dimensional photochemical microfabrication of conductive polymers in transparent polymer sheet", *Optical Review*, vol. 16, pp. 208–212, 2009.

[YAN 16] YANG Y., CHEN Y., WEI Y. *et al.*, "3D printing of shape memory polymer for functional part fabrication", *International Journal of Advanced Manufacturing Technology*, vol. 84, pp. 2079–2095, 2016.

[YEO 15] YEOM K., "Morphological approach for autonomous and adaptative system: the construction of three dimensional artificial model based on self-reconfigurable agents", *Neurocomputing*, vol. 148, pp. 100–111, 2015.

[YIM 07] YIM M., SHN W.M., SALEMI B. *et al.*, "Modular self-reconfigurable robot systems: challenges and opportunities for the future", *IEEE Robotics Automation Magazine*, pp. 43–52, 2007.

[YIN 16] YINLING E., 4D printing: towards biomimetic additive manufacturing, PhD Thesis, MIT, Boston, 2016.

[YOS 10] YOSHINO T., KONDO M., MAMIYA J. *et al.*, "Three dimensional photo-mobility of crosslinked azo-benzene liquid crystalline polymer fiber", *Advanced Materials*, vol. 22, pp. 1361–1363, 2010.

[YU 03] YU Y., NAKANO M., IKEDA T., "Photo-mechanics: directed bending of a polymer film by light", *Nature*, vol. 425, p. 145, 2003.

[YU 15] YU K., RITCHIE A., MAO Y. *et al.*, "Controlled sequential shape changing components by 3D printing of shape memory polymer multimaterials", *Procedia IUTAM*, vol. 12, pp. 193–203, 2015.

[YUK 17] YUK H., LIN S., MA C. *et al.*, "Hydraulic hydrogel actuators and robots optically and sonically camouflaged in water", *Nature Communications*, vol. 8, 2017. doi: 10.1038/ncomms14230.

[ZAR 15] ZAREK M., LAYANI M., COOPERSTEIN I. *et al.*, "3D printing of shape memory polymers for flexible electronic devices", *Advanced Materials*, 2015. doi: 10.1002/adma.201503132.

[ZAR 17] ZAREK M., MANSOUR N., SHAPIRA S. *et al.*, "4D printing of shape memory-based personalized endoluminal medical devices", *Macromolecular Rapid Communications*, 2017. doi: 10.1002/marc.201600628.

[ZAM 89] ZAMBRANO M., *Les clairières du bois*, Editions de l'éclat, Paris, 1989.

[ZEN 13] ZENIL H., "Turing patterns with Turing machines: emergence and low-level structure formation", *Natural Computing*, vol. 12, pp. 291–303, 2013.

[ZHA 11] ZHANG X., PINT C.L., LEE M.H. *et al.*, "Optically- and thermally-responsive programmable materials based on carbon nanotube-hydrogel polymer composites", *Nano Letters*, vol. 11, pp. 3239–3244, 2011.

[ZHA 15a] ZHANG L., LIANG H., JACOB J. *et al.*, "Photo-gated humidity-driven motility", *Nature Communications*, 2015. doi: 10.1038/ncomms8429, available at: www.nature.com/naturecommunications.

[ZHA 15b] ZHANG J., YAO Y., SHENG L. *et al.*, "Self-fueled biomimetic liquid metal mollusk", *Advanced Materials*, vol. 27, pp. 2648–2655, 2015.

[ZHA 15c] ZHAO Q., QI H.J., XIE T., "Recent progress in shape memory polymer: new behavior, enabling materials, and mechanistic understanding", *Progress in Polymer Science*, vols 49–50, pp. 79–120, 2015.

PART 2

Live "Smart" Matter and (Bio-printing)

Introduction to Part 2

"To create an artificial life that simulates natural life, it may be necessary in theory to know what constitutes natural life". [COD 68]

"The epistemic status of artificial life currently appears to oscillate between the intermediate and the encroached upon forms of life. Artificial life is intermediate in the sense that it appears to prove that there are a certain number of dynamic and encryption processes that do not belong to natural life in their own right. Such processes constitute a common basis of operations for the living and machines, whatever their substrate. In this sense, artificial life plays the role of all modeling processes". [PAR 94]

"Everything that the most subtle and ingenious of men can do to imitate or aid Nature, which is accustomed to be understood as Magic, to the extent that we have discovered various resources and means that they practice to overcome these extraordinary processes. What we can particularly note amongst us regarding the invention of Canons and Printing, and upon the discovery of this 'new world', are the people who principally believe that our ships were made by Magic, our archways by enchantment, and that the Spaniards were devils who came to destroy them with the thunder and lightning of their harquebuses and pistols". [NAU 45]

"Extropy is the exact inverse of entropy. It is faith in the possibility of endlessly increasing the power, the faith in the perpetual progress of science and humanity". [ROM 10]

"God does not mass-produce, does not shape two things using the same mold, things are never so similar that one rose is similar to another rose. God has not spared his own power which is absolute. He can therefore renew the world in the emergence of infinite shapes and, in truth, he does this at every moment". [ALF 89]

"What changes radically, is that we no longer combine, initially, the knowledge of each discipline, but we start by collecting together what we do not know. 'The state of the non-art', in which the discipline cannot possibly provide all of the answers to the issues that the subject poses, is what we seek to both know and put together. Each discipline looks at what the object cannot deal with through its given concepts and logic. This questions scientific practice more deeply because these non-knowledge aspects force us to see interdisciplinarity as something other than as an additive procedure. It obeys a subtraction logic (without lacking anything) which leads each discipline to be reinterpreted by the others. Mastery is no longer at the center of the process. Non-knowledge is no longer at the margins but, in fact, at the center of the process". [SCH 10]

"The question becomes one of yes or no. Are living organisms 'simply' or exclusively physicochemical systems? However what exactly is a positive response to this question?". [HEM 14]

"Einstein is one of the most original thinkers that I have ever met. As his research goes in every direction, it is only to be expected that the majority of avenues that he takes turn out to be dead ends". (Henri Poincaré in a letter on Albert Einstein applying for a permanent position in France, quoted by De Gaulejac [DEG 12])

"However whilst it keeps looking for the obvious things, soon science runs dry. There may be given fashions, and it is those who may follow them who are likely to take the credit. Those who might not follow them may not have them and indeed simply learn to follow these fashions in their turn". [SÉG 09]

"May heaven spare us from snobbery which not only acknowledges the possibility of a lean and superficial form of work, but protests, in a spirit of anxious arrogance, against the competition for the vigor and ideas where we find them". [WIE 50]

I.1. Introduction

Bio-printing is a technique, for additive manufacturing, enabling 3D printing of successive layers of cells in biomatrices to regenerate the structure of an entire organ through replication [AND 16, AND 17, GUÉ 17]. Bio-printing has already been used to produce parts of skin and bones, and for vascular grafting and tracheal splints, as well as cardiac tissue and cartilaginous structures. Although still very recent, this method is highly promising, not only in the sphere of regenerative medicine, but also for the discovery of medical drugs and research in toxicology (see [CHU 15, TUR 15, ATA 14, SHA 14, R&D 15, NAK 10, GUI 10, MEL 12, FER 13, ATA 15, PAU 15, KPC 13, TED 14, VEN 14, DOU 14, RIM 15, MA 15, LEE 15, STE 16] and, indeed, others).

"The classic image of the world was well described by Feynman [FEY 65]... when he compared nature to an immense game of chess. Each movement, taken in isolation, would be simple and the complexity just like irreversibility would simply result from the large number of pieces in play. However, today it is difficult to accept this image as a valid comparison" [PRI 94]. In the chapters which follow, besides technological aspects, predictability will naturally be mentioned. How are we to stand out from the crowd as regards technological aspects? On the one hand, instability (chaos) – probability – irreversibility and the search for determinism (at least for the given intended application). On the other hand, or put another way, the incidence of consubstantial instability of the living on actual determinism?

Figure I.1 resumes that of the previous chapter from Ramirez-Ferrero [RAM 15]. The latter sets out the positioning of additive manufacturing technologies on a Hype diagram (the Hype cycle is a representation of emergent technologies in development of a given moment in history). For this author, this is indeed an emerging phenomenon.

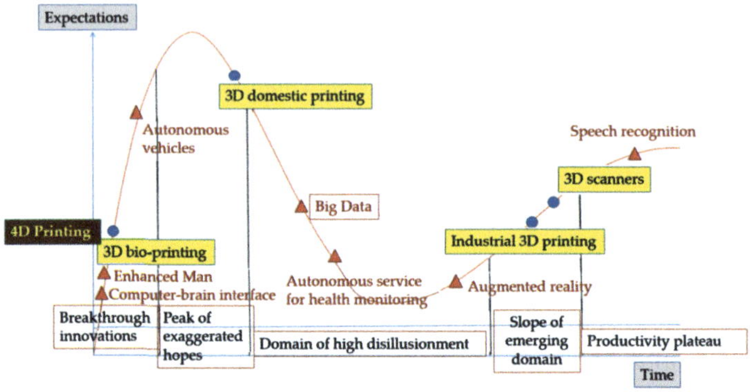

Figure I.1. *Hype representation and 3D technologies. For a color version of the figure, see www.iste.co.uk/andre/printing3.zip*

This curve is completed in Figure I.2 [EXP 13] which still leaves a little time for the development of bio-printing relative to the techniques using "inert" matter.

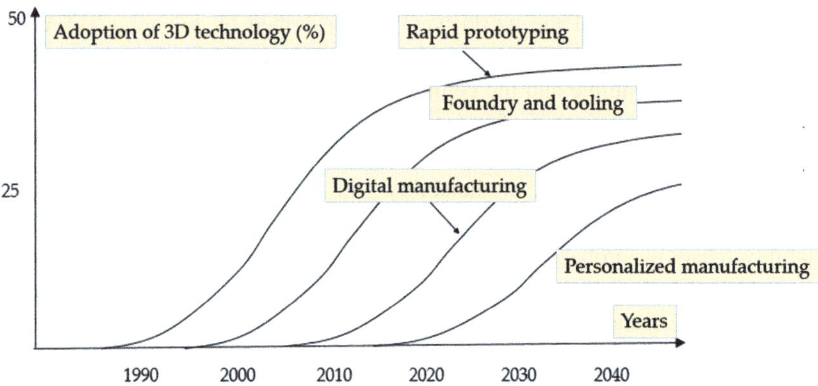

Figure I.2. *Time lags in the various targets for additive manufacturing*

"Bernard Stiegler, in line with Leroi-Gourhan, thus lays down the assumption of the coupling of Man and matter, the coupling between anthropogenesis and technogenesis: the technique is 'the pursuit of life by means other than life itself' [STI 94]. This premise induces an essential theoretical consequence: the need to think of the majority of social, cultural, political and cognitive phenomena… from this co-construction, this so-called 'Man-Technology' combination, and notably using memory technologies" [SER 08].

According to [DID 06], stem cells are the precursors which multiply and pass from the primitive lack of differentiation to actual differentiation taking place. At the first stage, "totipotent" cells are capable of reconstituting all tissues and even returning to the embryo phase. They then become "pluripotent", still capable of differentiating themselves, without a possible return to the embryo stage. Finally, they are then "multipotent", with a capacity for a more modest differentiation. It is indeed this capacity to recreate tissues from stem cells, which constitutes one of the bases for the fascination with bio-printing. It thus makes man both the master and owner of nature. Thus, Descartes, and more specifically his theories, are therefore not altogether dead.

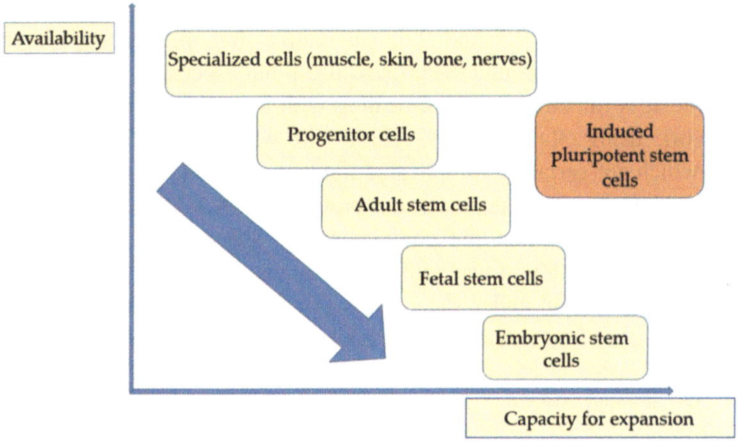

Figure 2.25. *Cells used in bio-printing operate through their potential for both differentiation and growth. For a color version of the figure, see www.iste.co.uk/andre/printing3.zip*

However, Rackham *et al.* [RAC 16] may have demonstrated the possibility of reprogramming all types of human cells into another type of cell, without resorting to stem cells. The bioinformatics system, known as "Mogrify", enables the prediction of the best reprogramming process. This is according to the cell type we start with and the cell type we wish to obtain. The biological process, called transdifferentiation, is in large part the source of hopes and advances in regenerative medicine. This is an IT algorithm to predict cell factors involved in cell transformation.

That said, the choices of materials and cells being made (see the "review" paper of Shao, Sang and Fu, [SHA 15]), arise, as has already been shown within the bio-printing processes. Specifically relevant is the choice of method for introducing cells into the object under construction, being one of the following:

– individual cells in solution;

– individual cells placed in hydrogel;

– cells placed upon a surface;

– "encapsulated" cells.

2.3.4. *Process–material couplings*

Upon reading Tables 2.5 and 2.6, we realize that the criteria for choosing materials will play a role around the process itself, but especially around desired cell development. It is thus a question of dealing with a significant barrier, without the need to work using 3D machines. These fall within the domain of technical problems, which are certainly complicated but less complex. Figure 2.26 shows this complex coupling between materials and processes [SKA 15]. This is because we must take into consideration the complementary aspects of rheology, viscosity, absorption and evaporation of water and temperature. Additional relevant factors are the cross-linking reaction under shear stress (the issue of relaxing constraints), the potential presence of free radicals able to inhibit cell growth and other factors.

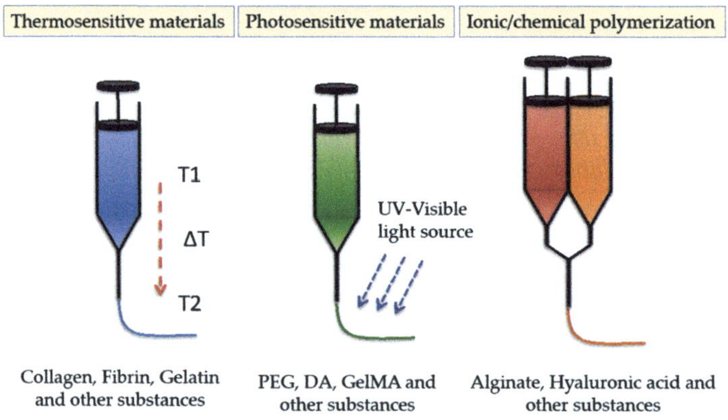

Figure 2.26. *Materials–processes coupling within bio-printing [GUÉ 17]. For a color version of the figure, see www.iste.co.uk/andre/printing3.zip*

Chua and Yeong [CHU 15] proposed families of materials likely to be of relevance to the sphere of bio-printing. Table 2.7 sets out these various families.

Family	Examples
Polymers	Poly-L-lactic acid (PLLA), poly-glycolic acid (PGA), polycaprolactone (PCL), poly-(lactide-co-glycolide) PLGA and others
Ceramics and glass	Hydroxylapatite (HAP), alumina, bioactive lenses (amorphous lenses containing silica)
Hydrogels	Collagen, gelatin, fibrin, alginate, chitin, chitosan, hyaluronic acid, synthetic hydrogels (poly-(2-hydroxyethyl methacrylate) (PHEMA), polyvinyl alcohol (PVA), polyethylene glycol (PEG)

Table 2.7. *Families of materials used in bio-printing*

Table 2.8, extracted from Chua and Yeong's publication [CHU 15], indicates material–process relationships, which are generally used in bio-printing. The relationship from the works of Skardal and Atala is set out in Table 2.9.

Materials	Printing processes
Natural hydrogels	
Alginate	Injection (IP – Inject Printing), LIFT, pneumatic extrusion, biological laser printing (BLP)
Matrigel	LIFT, BLP, pneumatic extrusion
Collagen	IP, LIFT, BLP
Gelatin	BLP, LIFT, IP
Gel-MA	Projection by optical means stereolithography (or stereolithography apparatus – SLA)
Hyaluronic acid (HA)	Extrusion
Synthetic hydrogels	
PEG	Acoustic ejection, SLA, projection by optical means
Hybrid hydrogels	
PCL, alginate	Fusion printing/solidification, pneumatic extrusion
PCL, fibrinogen, collagen	IP, electrospinning
PCL, PLGA, HA, gelatin, collagen	Pneumatic extrusion, fusion-based printing/solidification

Table 2.8. *3D methods and bio-printing materials*

Materials	Activation method	Commentary
Polyethylene glycol (PEG)	Photopolymerization	Timeliness (minutes), mechanical properties which are easy to control (size of monomers and polyfunctionalities), non-adhesion of cells except if a particular procedure is carried out, encapsulation of cells, cross-linking of other polymers
Collagen	Hydrophobic bonds	Good cell adhesion, slow cross-linking (1 hour); implants, skin substitutes, encapsulations
Hyaluronic acid	Michaelis addition which is a pH function Photopolymerization Thyramine + H_2O_2	Good cell adhesion, slow cross-linking except in the presence of light (the function of intensity), modest mechanical properties, geometry which is difficult to control (with H_2O_2); filling components, encapsulation, bio-printing
Gelatin	Hydrophobic bonds according to temperature Glutaraldehyde	Good cell adhesion, unstable material except for cross-linking with glutaraldehyde; encapsulation, bio-printing, scaffolds (sequential population if involves glutaraldehyde).
Alginate	Exchange of Na^+ ions by Ca^{++} ions	Rapid process, geometry difficult to control; microspheres and encapsulation
Fibrin	Thrombin fibrinogen	Rapid, cross-linking, good cell adhesion, imprecise geometry; encapsulation and cell freedom
Polycaprolactone (PCL)	Fusion processing	Good mechanical properties, high temperature; no encapsulation, production of scaffolds

Table 2.9. *Materials used with their origin and their significance*

2.3.5. *Subsequent cell growth*

When the tissue is bio-printed, it is generally envisaged that it should be left to grow in terms of cell population, and to make biodegradable scaffold components disappear before a potential transplant takes place. Modeling such a development is a difficult process, which it will be necessary to tackle if we wish to be able to transplant a bioconstruct of any given organ or set of organs. This is in response to the particular practitioner's instructions [MCC 14]. In the case of transplants, independently of the issue of bio-printing, vascularized systems of biological additions must be part of the given patient's system.

Professor Lutolf's team at the EPFL (École Polytechnique Fédérale de Lausanne), in Switzerland, is using support materials likely to be locally degraded enabling cell growth. Figure 2.27 shows this process [GJO 15]. Irradiated areas create free spaces in which cells and nutriments can be found, in theory leading to the objective of cell growth being achieved. This type of technology is vital in order to design specific spaces for cell development, but for all that, the issue of determinism, the size of spaces and their spatial distribution as such remains to be dealt with. Other scientific activities of this team must be cited for their quality and their top-down approach to the theme (see, for example, [RAN 14, MOS 14, ROC 13, GOB 11, GIL 10, LUT 09, LUT 09]).

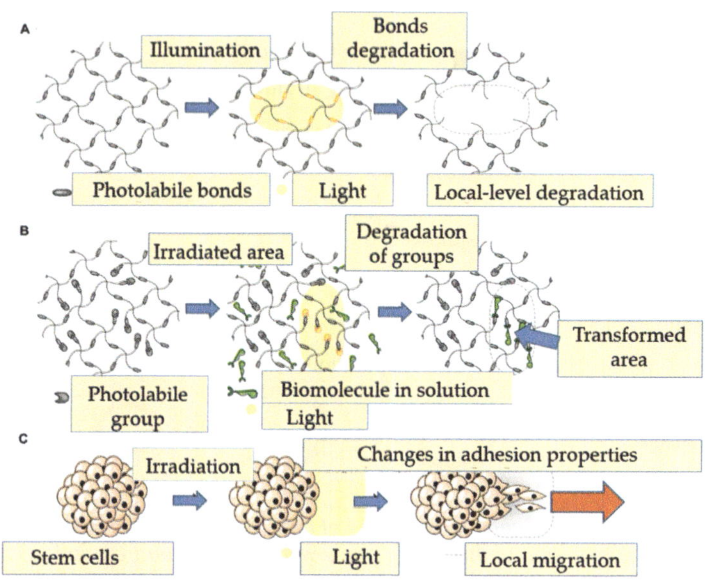

Figure 2.27. *Creation of free spaces for cell growth. For a color version of the figure, see www.iste.co.uk/andre/printing3.zip*

To go a little further in terms of complexity, according to [MAL 15, GUÉ 17], vascularized bone tissue remains a significant technological issue, owing to efficiencies limited in time. According to [WAN 14], it was possible to produce a structure by extrusion of a composed precursor of calcium phosphate, hydroxyapatite and tricalcium phosphate. On this basis, they injected cells (phages coupled to peptides) which behave as endothelial cells from the point of view of migration and adhesion. See also Rasskazova *et al.* [RAS 15] for compounds of hydroxyapatites and oligomers of lactic acid, enabling the manufacture of bone graft substitutes. This process will enable a means to produce vascularized bone structures, for which we may devise regenerative medicine applications.

2.4. Comment: 4D bio-printing

In a recent article, An, Chua and Mironov [AN 16] defined the concept of 4D bio-printing. It is defined in Figure 2.28. In fact, it is "conventional" 4D bio-printing given that this method is already a means of 4D manufacturing. This is because the time has come to take into consideration, along with the aspects of cell growth, both the differentiation and progressive elimination of supports.

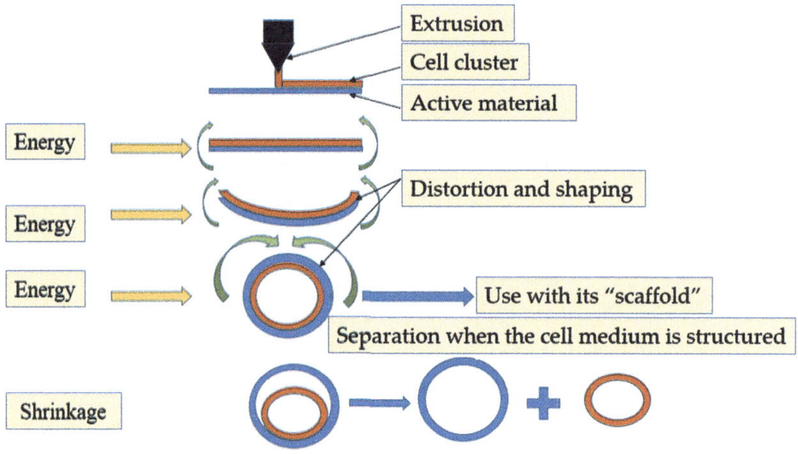

Figure 2.28. *Example of means of realization of 4D bio-printing. For a color version of the figure, see www.iste.co.uk/andre/printing3.zip*

2.5. Other applications

2.5.1. *Biological applications*

Aside from the bio-printing aspect for the manufacture of tissues, a large part of short-term activities (during cell survival) is oriented towards studies of toxicology, cosmetics and applications for medical research (see, for example, [KIN 17, RAY 15, POI 16, KNO 15]). In addition, taking account of the possibility of using the given patient's own biological cells, it will be increasingly possible to find customized treatments. Works on this subject include Knowlton *et al.* [KNO 15]; Santoro *et al.* [SAN 15], Peng *et al.* [PEN 16], Charbe, McCarron and Tambuwala [CHA 17], Ozbolat, Peng and Ozbolat; [OZB 16]. Table 2.10, from MADEELI, an agency for development, export and innovation [MAD 17], sets out the 3D technologies in the process of development to invest in this sphere with their time estimate for economic access for market. Figure 2.29, from the work of Asghar *et al.* [ASG 15], sets out 3D methods applied to the study of cancers with the aim of customized medicine.

Process	Time for pharmaceutical market access	Time for cosmetics market access
Ink jets (acoustic and thermic-based mechanisms)	2017	2018
Extrusion (pneumatics and mechanics)	2017	2018
LIFT (laser assistance)	2019	2018

Table 2.10. *Bio-printing technologies used in the pharmaceutical and cosmetics industries, with their year of entry to market estimated by MADEELI*

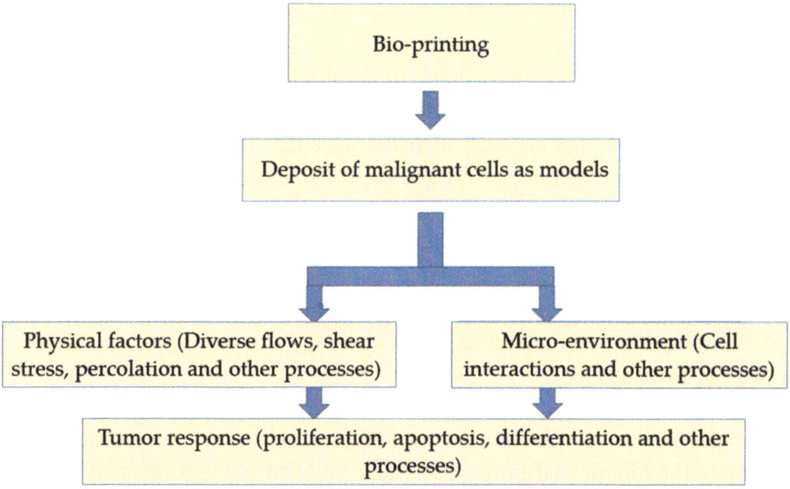

Figure 2.29. *Research into cancer and bio-printing*

2.5.2. *Is it possible to feed ourselves thanks to bio-printing?*

The company Modern Meadow, created in 2011 [DIL 15, CLA 16], has the objective of producing leather from stem cells which are "100% ethical" through bio-printing. For his part, Mark Post in Maastricht [POS 14] is already producing meat culture based upon stem cells. This remains highly costly, but these activities focus on means of manufacture close to bio-printing.

2.5.3. *Bioluminescence and electronics*

The Costa team [NIK 17] produced a flat screen made up of bio-LEDs for the system of backlighting, as well as color filters with the assistance of bioluminescent proteins deposited in a polymer matrix. Color filters are manufactured thanks to the use of 3D printing techniques, while maintaining the luminescent properties of proteins and optimum stability. They may offer an alternative to the filters in the current liquid crystal display screens, which have the increased costs and the contrast and brightness limits. With these new systems, recycling could become less problematic.

2.5.4. *Bio-printed Bio-bots or "soft robots" produced by additive manufacturing*

"Soft-robots" [CAR 13] are made up of deformable materials (fluids, elastomers, gels and others) which imitate the elastic and rheological properties of biological tissues and organs. In theory, such a robot can adapt its shape and its locomotion strategy according to its environment and the tasks it has to accomplish. This sphere emerging from "bioinspired" machines constitutes the result of a highly interdisciplinary form of research, within the line of sight of numerous biological applications. The production of the elements of these machines can take place using additive manufacturing techniques [HIL 12]. However, the number of publications around the coupling "soft robot" and "3D printing" is more modest. This is even more marked for bio-bots.

However, as the introduction of the chapter indicated, processing the use of living things for robotics (from a conceptual point of view, a form of reverse bio-printing), the creation of machines may be envisaged from the elements shown in Figure 2.30 (see [KAM 14, AND 17]). This includes the increase in original natural living components.

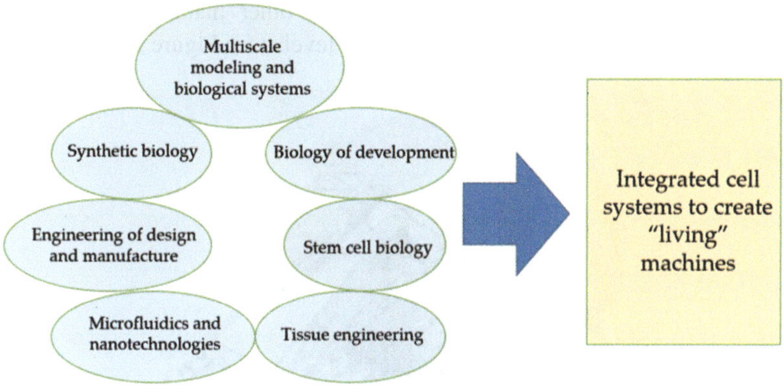

Figure 2.30. *Inherent elements for the production of a bio-bot*

In a review paper, Ahadjian *et al.* [AHA 13] described the need for production of living elements. Examples are muscles for various applications including bio-bots, and repair aspects which have been broadly mentioned. Muscle cells included in a hydrogel can contract by electrical excitation, which enables the proposal of machine movements using muscle cells, as Figure 2.31 shows [AND 17a]. Several studies have already been completed of muscles applicable in the sphere (see, for example, [KIM 14, RAM 17, LIN 16]).

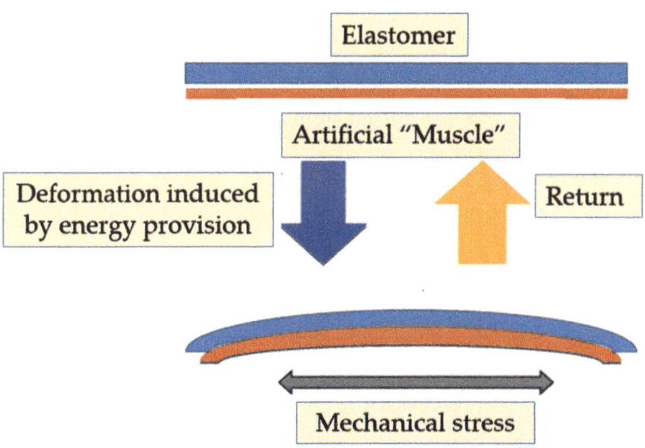

Figure 2.31. *Movement induced by muscle excitations. For a color version of the figure, see www.iste.co.uk/andre/printing3.zip*

The production of microactuators is possible as the works of Sreetharan *et al.* [SRE 12] and Raman, Cvetkovic and Bashir [RAM 17] attest. The size reduction aspect may constitute a way forward, because other manufacturing techniques cannot provide similar or superior performance levels (see Figure 2.32).

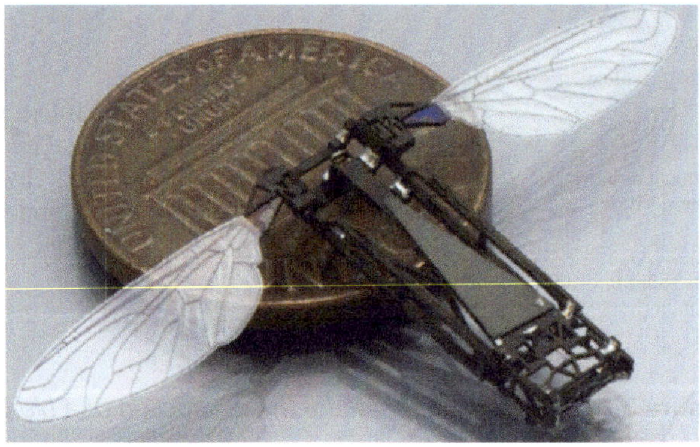

Figure 2.32. *Winged Microrobot – the body is produced by additive manufacturing (reproduced with the permission of P.S. Sreetharan)*

Table 2.11 attempts to compare bio-printing and bio-bots, which in theory have common points of reference.

Properties	Bio-printing	Bio-bot
Combination of living/artificial aspects	Yes	Yes
Cells	Specific functionalities	Muscle cells
Nature of cells	Customized stem cells	"Standard" cells
"Right" to mistakes for a given application (Risk analysis)	No	Possible
Lifespan	Long	May be short
Self-perpetuation	Necessary	Seldom critical
Information	Necessary	Seldom critical
Functionality	Necessary	Necessary
Development of support or scaffold	Programmed for tissue development	Stable over time
Manufacturing	Uniqueness	Eventually envisaging collective manufacturing

Table 2.11. *Comparative elements between bio-printing and bio-bots*

Evidently, these two axes have common elements, even if the purposes are different. Yet, as has largely been shown, the complexity of bio-printing processes constitutes a barrier which risks limiting the emergence of this technology. The prior exploration of the sphere of bio-bots which relies on the same cultural substance but with more modest constraints, could constitute a more confident basis for action to prove the possibility of examining the theme.

2.6. Conclusion

Bio-printing is a technique which enables three-dimensional printing of successive layers of cells around biomatrices, so as to identically regenerate the structure of an entire organ [AND 16]. Bio-printing has already been used to produce elements of skin, bones, vascular transplants, tracheal splints, cardiac tissues and cartilaginous structures. Although very recent, this method is highly promising not only in the sphere of regenerative medicine, but also the discovery of drugs and research into toxicology.

Tissue engineers try to produce, in the laboratory, human tissue which is correctly vascularized and fairly robust to replace damaged tissue. They endure numerous failures before succeeding in a result which is described as a proof of concept. Other teams played the bio-printing card, but had to be limited to the smallest dimension by producing tissue which was extremely fine. The attempts at printing thick layers have been somewhat negative, cells sandwiched together having no access to either oxygen or nutriments in a sufficient quantity. They do not have the possibility of properly removing carbon dioxide and other waste. They thus end up dying. This means of vascularization is therefore important to control the promotion of additive biomanufacturing.

To work around this problem, nature (in its genuine form) has a network of minuscule blood vessels with thin walls, which feed the entirety while evacuating waste. The vessels can be reproduced using 3D printing. Printing tissue constructions that obey such a diagram, several functional "bioinks" possess biological properties which are both useful and complementary. These contain key elements necessary for the formation of living tissue and can be used [MOO 14], but do we know how to combine them in the optimal manner?

What has just been mentioned, if we wish to make a comparison with "standard" additive manufacturing, is that we add supplementary barriers, of which some fall within conceptual notions:

– a complex choice of supports for the development of systems using living cells;

– cells of variable origin, dependent upon the selected purpose;

– developments of complex cellular function systems of spatial, chemical and mechanical factors and interactions with the given support, various transfers and other aspects;

– the intermediate growth of cells prior to deposition (see [GUI 10a]);

– interdependences between these three major spheres.

Today, it does not seem possible to control the bio-printing system based solely on knowledge of its components, hence understandable trends for a global approach such as engineers practice, taking into account the phenomena judged as predominant in the modelling. It is therefore necessary to come out of the "materialism of the promise". According to the holistic theses, each cell influences the whole organ being constructed, which itself, recursively, intervenes in turn, on the elementary entity. "There is a sequence of actions and reactions. That is to say, we have a system of equilibrium between various forces, rather than a phenomenon which can be reduced to a single cause and to the effects of this cause" [PAR 65]. However, how then can we be sure that we are concerned only with the principle variables, if complex local interdependences (between cells and the environment and vice versa) exist? Yet, bio-printing objects, such as they are presented, are composite, systemic, unstable and subject to constraints (some of which are still unknown to us). Thus, for the author, whatever scales are considered, bio-printing is probably inaccessible to single monodisciplinary approaches ("contingent structuring produced by a particular historic process involving the fragmentation of science", according to [LEG 02]).

It only remains to verify this type of claim: "3D bio-printing, invented in 2009 by Australian and American engineers, will be capable of ensuring organ transplants by 2030, thanks to cutting-edge Russian technologies, which have already been tested. The Russian company 3D Bio-printing Solutions, which was established at the innovation center Skolkovo, successfully produced 3D printing trials. These enabled the reproduction of a thyroid before implanting it into a mouse, and doing so only within the space of two years. This was stated by the company's partner Youssef Khesouani in an interview granted to Sputnik" [SPU 16]. We will attempt to illustrate this in the following chapter by using representative examples which have been subject to scientific publications.

2.7. Appendix: 3D printing for biological applications

In this chapter, bioink printing containing living matter is described. Yet, the current market is moreover somewhat characterized by numerous niches ranging from dentistry to various repairs, via the pharmaceutical and prosthetics industries. It has therefore been thought useful in this appendix to mention the provision of additive manufacturing in these spheres, relying on "traditional" uses. The aim is to produce parts so as to simulate a surgical procedure [OST 15, OST 16b], stimulating brain actuators [KOS 15], biological and medical repairs [ZOP 13, MAR 15, GUI 17]), or dental repairs (ceramics; see, for example, [MIT 12, KAC 10]). In these different applications for conventional additive manufacturing, we seek to make available biocompatible materials to best effect, having specific properties.

Other examples exist for various repairs (bones, tissues, vessels). These are detailed by Shapira, Kim and Dvir [SHA 14], Fischer [FIS 13], Kucukgul *et al.* [KUC 13], Chandler [CHA 15], 3D Printing Buzz [3DP 15], Gilpin [GIL 15], Chua and Yeong [CHU 15] and Lavergne [LAV 14]. Further uses include biotechnologies [LOD 15], the dispensing of medicines [SHE 15b] and research [TAS 13, KER 11, 3DP 15, KAO 15, WTE 15, MOT 15]. Further applications include scaffolds for two-photon stereolithography (Selimis, Mironov and Farsari [SEL 15]) and "bone" parts containing nanoparticles of mechanically strengthened carbon (nanotubes) [GON 16, IM 12]. Further uses are blood vessels (Artivacs [ART 15]); scaffolds produced from droplets [TAN 14]; vascular systems [PAU 15]; for artificial bones [TAN 16, OST 16] and other applications.

In Volume 1 [AND 17b], the production of a skull was set out to enable the simulation of a surgical procedure in the brain [GIA 09]. Similarly, it is now possible to manufacture phantoms for use in medicine (fingerprints in forensic medicine, artificial hands for calibration of medical instruments [ARO 13] and cartilage [HAR 14]). Other applications are prostheses for fingers and hands [FOR 15], applications in dentistry ([DIG 15, CAR 15, DEN 15, BRA 16], respectively) with a market in a profound growth phase ([SMA 15] and Figure 2.33); the metal parts for the kneecap [GAO 15] and an artificial sternum [ARA 15]. Other applications for the development of foot prosthetics are also cited [NGU 13].

The list of publications is very significant with numerous permitted medical applications.

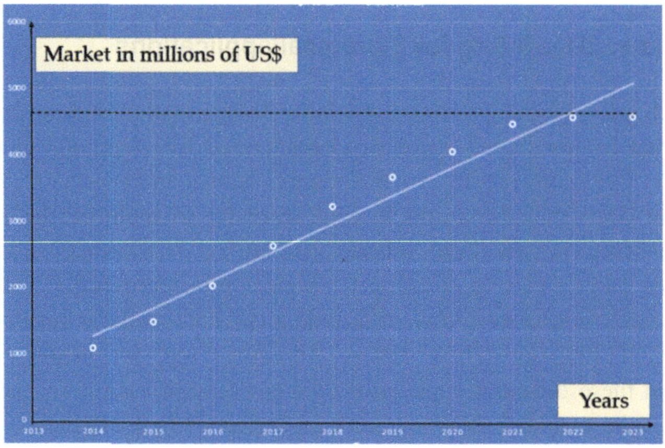

Figure 2.33. *Over ten years, the American additive manufacturing market in dental fixing has tripled [SMA 15]*

Among the recent scientific advances considered by Malaquin [MAL 15], we cite inner ear treatment, skin substitutes, aortic valves and the use of two-photon processes to produce high-precision supports, the development of bionics and other technologies. Also of relevance is the capacity to produce complex cell assembly respecting the morphology and tissue anatomy [LEE 14] for the ear; Michael *et al.* [MIC 13] for the skin; Duan *et al.* [DUA 13] for valves; Weiss *et al.* [WEI 09] for supports; Watts [WAT 16] for cardiac valves; Mannor *et al.* [MAN 13] for bionic aspects and other facets.

COMMENT.– Natural scaffold system

Massachusetts General Hospital and the Harvard Medical School have mastered the regeneration of human hearts from donor cells [MAG 16]. By using cells which cannot be used for transplants, their immersion in *ad hoc* solutions can eliminate "cells able to cause a negative immune response, the organs no longer being made up of their most simple structure. Scientists then introduced into hearts Induced Pluripotent Stem Cells (iPSCs), from cells of adult human skin, reprogrammed to reproduce functional muscular cardiac tissue. Placed in human organs and plunged into a nutriment solution, iPSCs then developed as two types of cardiac cells. Thus, after two weeks, cultivated cellular networks started to resemble those of a correctly structured heart, even if they were not completely formed. Lastly, once fed by an electric shock, the organs started to beat."

This process using cardiac cells from the same organism is less likely, as for an implanted heart, to be rejected by the recipient of the transplant. Although the study enabled the manufacture of 500 million cardiac cells, producing an entire heart would necessitate "tens of billions" of such cells. However, this view where 3D printing is ultimately excluded appears to be an interesting avenue to examine.

It is just the same for bio-printing: we make headway, but a long road must be traveled before reaching our ultimate goal.

2.8. Bibliography

[3DP 15] 3D PRINTING BUZZ, "Archives for the category bio-printing organs/human tissue cells", available at: http://textually.org/3DPrinting/cat_printable_organs.html, 2015.

[ABB 15] ABBOTT R.D., KAPLAN D.L., "Strategies for improving the physiological relevance of human engineered tissues", *Trends in Biotechnology*, vol. 33, pp. 401–407, 2015.

[AHA 13] AHADJIAN S., OSTROVIDOV S., HOSSEINI V.A. *et al.*, "Electrical stimulation as a biomimicry tool for regulating muscle cell behavior", *Organogenesis*, vol. 9, pp. 87–92, 2013.

[ALL 15] ALLEN P.B., ZHAING Z., SCMIDT C.E. *et al.*, "3D printing with nucleic acid adhesives", *ACS Biomaterials Science & Engineering*, vol. 1, pp. 19–27, 2015.

[AN 16] AN J., CHUA C.K., MIRONOV V., "A perspective on 4D Bio-printing", *International Journal of Bio-printing*, vol. 2, pp. 3–5, 2016.

[AND 16] ANDRÉ J.C., LARGER L., Approche épistémologique du bio-printing pour essayer de moins mutiler la recherche sur un objet-frontière, Report to CSI-INSIS – CNRS, Paris, 2016.

[AND 17a] ANDRÉ J.C., Impression 3D: niches applicatives porteuses, Techniques de l'Ingénieur, ref. BM7970 V1, 2017.

[AND 17b] ANDRÉ J.-C., *From Additive Manufacturing to 3D/4D Printing 1*, ISTE Ltd, London, John Wiley & Sons, New York, 2017.

[AND 17c] ANDRÉ J.-C., *From Additive Manufacturing to 3D/4D Printing 2*, ISTE Ltd, London, John Wiley & Sons, New York, 2017.

[ANS 16] ANSM, "Tissus ou organes d'origine humaine ou animale (THA)", available at: http://ansm.sante.fr/Produits-de-sante/Tissus-ou-organes-d-origine-humaine-ou-animale, 2016.

[ARA 15] ARANDA J.L., JIMÉNEZ M.L., RODRÍGUEZ M. *et al.*, "Tridimensional titanium-printed custom-made prosthesis for sternocostal reconstruction", *European Journal of Cardiothoracic Surgery*, vol. 48, pp. 1–3, 2015.

[ARC 11] ARCAUTE K., MANN B.K., WICKER R.B., "Practical use of hydrogels in stereolithography for tissue engineering applications", in BARTOLO P.J. (ed.) *Stereolithography; Materials, Processes and Applications*, pp. 299–331, Springer, New York, available at: http://www.biotech-now.org/health/2014/06/thinking-about-printing-in-the-4th-dimension, 2011.

[ARO 13] ARORA S.S., CAO K., JAIN A.K. *et al.*, 3D fingerprint phantoms, MSU Technical Reports, MSU-CSE-13-12, 2013.

[ART 15] ARTIVACS, Artificial vascularized scaffolds for 3D tissue regeneration, available at: http://www.artivasc.eu/, 2015.

[ASG 15] ASGHAR W., EL ASSAL R., SHAFIEE H. *et al.*, "Engineering cancer microenvironments for in vitro 3-D tumor models", *Materialstoday*, vol. 18, pp. 539–553, 2015.

[ASP 15] ASPECT BIOSYSTEMS, "Human tissues on demand", available at: http://aspectbiosystems.com/, 2015.

[ATE 16] ATERA LAB, "Advanced tissue engineering for research applications", available at: http://ateralabs.com/products/, 2016.

[AUG 11] AUGÉ M., *Où est passé l'avenir?*, Le Seuil, Paris, 2011.

[BAC 77] BACHELARD G., *La formation de l'esprit scientifique*, Vrin, Paris, 1977.

[BAJ 14] BAJAJ P., SCHWELLER R.M., KHADEMHASSEINI A. *et al.*, "3D fabrication strategies for tissue engineering and regenerative medicine", *Annual Review of Biomedical Engineering*, vol. 16, pp. 247–276, 2014.

[BAR 11a] BARTOLO P.J. (ed.), *Stereolithography; Materials, Processes and Applications*, Springer, New York, 2011.

[BAR 11b] BARTOLO P.J., "Stereolithographic process", in BARTOLO P.J. (ed.), *Stereolithography; Materials, Processes and Applications*, Springer, New York, pp. 1–36, 2011.

[BAR 11c] BARTOLO P.J., GIBSON I., "History of stereolithographic process", in BARTOLO P.J. (ed.), *Stereolithography; Materials, Processes and Applications*, Springer, New York, pp. 37–56, 2011.

[BAU 07] BAUMAN Z., *La Décadence des intellectuels. Des législateurs aux interprètes*, Chambon – Actes Sud, Paris, 2007.

[BER 11] BERTSCH A., RENAUD P., "Microstereolithography", in BARTOLO P.J. (ed.), *Stereolithography; Materials, Processes and Applications*, Springer, New York, 2011.

[BHA 13] BHARADWAG A., "Will 3D change everything?", available at: http://3dprintingindustry.com/2013/12/04/video-will-3d-printing-change-everything/, 2013.

[BIK 16] BIKAS H., STAVROPOULOS P., CHRYSSOLOURIS G., "Additive manufacturing methods and modelling approaches: a critical review", *International Journal of Advanced Manufacturing Technologies*, vol. 83, pp. 389–406, 2016.

[BIL 12] BILLIET T., VANDENHAUTE M., SCHELFHOUT J. *et al.*, "A review of trends and limitations in hydrogel-rapid prototyping for tissue engineering", *Biomaterials*, vol. 33, pp. 6020–6041, 2012.

[BOU 15] BOURC'HIS D., "Epigénétique", available at: http://www.inserm.fr/thematiques/genetique-genomique-et-bioinformatique/dossiers-d-information/epigenetique, 2015.

[BRA 16] BRAIAN M., JIMBO R., WENNERBERG A., "Production tolerance of additive manufactured polymeric objects for clinical applications", *Dental Materials*, vol. 32, pp. 853–861, 2016.

[BRU 16] BRUCE K., KRISTOFIC V., "The importance of 3D printing mesenchymal stem cells to tissue engineering", available at: http://slideplayer.com/slide/10947437/, 2016.

[BUR 16] BURKS H.E., PHAMDUY T.B., AZIMI M.S. *et al.*, "Laser direct-write onto live tissues: a novel model for studying cancer cell migration", *Journal of Cellular Physiology*, vol. 231, pp. 2333–2338, 2016.

[CAR 13] CARMEL M., "Soft robotics: a perspective – current trends and prospects for the future", *Soft Robotics*, vol. 1, pp. 5–11, 2013.

[CAR 15] CARBON 3D, "Incredible new 3D printing technique looks like Sci-Fi", available at: http://www.popularmechanics.com/technology/a14586/carbon3d-3d-printer-resin/, 2015.

[CEL 15] CELLINK, "The newly optimized desktop bioprinter for the ultimate bio-printing of human tissues and the 3D cell culturing", available at: http://www.cellink3d.com/, 2015.

[CHA 11] CHANG C.C., BOLAND E.D., WILLIAMS S.K. *et al.*, "Direct-write bio-printing three-dimensional biohybrid systems for future regenerative therapies", *Journal of Biomedical Materials Research. Part B, Applied Biomaterials*, vol. 98, pp. 160–170, available at: https://www.ncbi.nlm.nih.gov/pmc/articles/PMC3772543/#R35, 2011.

[CHA 15] CHANDLER D.L., "Tough bio-gel structures produced by 3D printing", available at: https://newsoffice.mit.edu/2015/3-d-printing-tough-biogel-0601, 2015.

[CHA 16] CHAUHAN N.P.S., AMETA C., PUNJABI P.B., *Scaffolds and Scaffolding Materials in Encyclopedia of Biomedical Polymers and Polymeric Biomaterials*, Taylor and Francis, New York, pp. 6999–7015, 2016.

[CHA 17] CHARBE N., MC CARRON P.A., TAMBUWALA M.M., "Three-dimensional bio-printing: A new frontier in oncology research", *World Journal of Clinical Oncology*, ESPS, vol. 8, pp. 21–36, 2017.

[CHN 12] CHNEIWEISS H., *L'Homme réparé; espoirs, limites et enjeux de la médecine régénérative*, Plon, Paris, 2012.

[CHU 15] CHUA C.K., YEONG N.Y., *Bio-printing: Principles and Applications*, World Scientific, Singapore, available at: http://www.worldscientific.com/worldscibooks/10.1142/9193#t=toc 2015.

[CLA 16] CLARKE C., Australian RIPPA could be the future of 3D printing and food, available at: https://3dprintingindustry.com/news/australian-rippa-future-3d-printing-food-99819/, 2016.

[COL 14] COLLINS S.F., "Bio-printing is changing regenerative medicine forever", *Stem Cells and Development*, vol. 23, Supp. 1, pp. 79–82, 2014.

[CVE 14] CVETKOVIC C., RAMAN R., WILLIAM B.J. *et al.*, "Three-dimensionally printed biological machines powered by skeletal muscle", *Proceedings of the National Academy of Sciences*, vol. 111, pp. 10125–10130, 2014.

[CYF 15] CYFUSE, Bio 3D printer Regenova with Kenzan method, available at: http://www.cyfusebio.com/en/regenova.html, 2015.

[DE 47] DE BEAUVOIR, *Pour une morale de l'ambigüité*, Gallimard, Paris, 1947.

[DEN 15] DENTCA, CAD/CAM denture, available at: http://www.dentca.com/, 2015.

[DES 16] DESRUS H., Etude expérimentale de procédés de bio-impression assistés par laser femto-seconde, PhD thesis, University of Bordeaux, available at: https://tel.archives-ouvertes.fr/tel-01362396/document, 2016.

[DIG 15] DIGICAD, Imprimantes 3D dédiées aux applications dentaires, available at: http://www.digicad.fr/imrpimante-3d-application-dentiste-prothesiste, 2015.

[DIL 15] DILAWAR A., Is company's claim of meat without murder too good to be true?, available at: http://www.theguardian.com/business/2015/sep/05/meat-without-murder-modern-meadow, 2015.

[DOR 15a] DORAY P., "Construction sociale des technologies", in PRUD'HOMME J., DORAY P., BOUCHARD F. (eds), *Sciences, technologis et sociétés de A à Z*, Presses de l'Université de Montréal, Canada, pp. 57–61, 2015.

[DOR 15b] DORAY P., MILLERAND F., "Déterminisme technologique", in PRUD'HOMME J., DORAY P., BOUCHARD F. (eds), *Sciences, technologies et sociétés de A à Z*, Presses de l'Université de Montréal, pp. 66–69, 2015.

[DUA 13] DUAN B., HOCKADAY L.A., KANG K.H. *et al.*, "3D bioprinting of heterogeneous aortic valve conduits with alginate/gelatin hydrogels", *Journal of Biomedical Materials Research Part A*, vol. 101, pp. 1255–1264, 2013.

[DUP 02] DUPUY J.P., *Pour un catastrophisme éclairé – quand l'impossible est certain*, Le Seuil, Paris, 2002.

[FAH 14] FAHARANI R.D., CHIZARI K., THERRIAULT D., "Three-dimensional printing of freeform helical microstructures: a review", *Nanoscale*, vol. 6, pp. 10470–10485, 2014.

[FAN 14] FANG Y.L., CHEN X.G., GODBEY W.T., "Gene delivery in tissue engineering and regenerative medicine", *Journal of Biomedical Materials Research*, vol. 103, no. 8, pp. 1679–1699, 2014.

[FER 10] FERRARO P., COPPOLA S., GRILLI S. *et al.*, "Dispensing nano-pico droplets and liquid patterning by pyro-electrodynamic shooting", *Nature Nanotechnology,* vol. 5, pp. 429–435, 2010.

[FIS 13] FISCHER S., "The body printed: how 3D printing could change the face of modern medicine and why future is still so far away", *IEEE Pulse*, vol. 4, pp. 27–31, 2013.

[FOR 15] FORMLABS.COM, 3D printing custom prosthetics, available at: http://formlabs.com/stories/3d-printed-prosthetics/, 2015.

[GAO 15] GAO, 3D Printing: opportunities, challenges and policy; implications of additive manufacturing, available at: http://www.gao.gov/assets/680/670960.pdf, 2015.

[GIA 09] GIANNOTSIS J., DEDOUSSIS V., "Additive fabrication technologies applied to medicine and health care: a review", *Manufacturing Technology*, vol. 40, pp. 116–127, 2009.

[GIL 10] GILBERT P.M., HAVENSTRITE K.L., MAGNUSSON K.E.G. *et al.*, "Substrate elasticity regulates skeletal muscle stem cell self-renewal in culture", *Science*, vol. 329, pp. 1078–1081, 2010.

[GIL 15] GILPIN L., New 3D bio-printer to reproduce human organs changes the face of healthcare: the inside society, available at: http://www.techrepublic.com/article/new-3d-bioprinter-to-reproduce-human-organs/, 2015.

[GJO 15] GJOREVSKI N., RANGA A., LUTOLF M.P., "Bioengineering approaches to guide stem cell-based organogenesis", *Development*, vol. 141, pp. 1794–1804, 2015.

[GOB 11] GOBAA S., HOEHNEL S., ROCCIO M. *et al.*, "Artificial niche microarrays for probing single stem cell fate in high throughput", *Nature Methods*, vol. 8, pp. 949–955, 2011.

[GON 16] GONÇALVEZ E.M., OLIVERA F.J., SILVA R.F. *et al.*, "3D printed PCL-hydroxyapatite scaffolds filled with CNTs for bone cell growth stimulation", *Journal of Biomedical Materials Research. Part B, Applied biomaterials*, vol. 104, pp. 1210–1219, 2016.

[GRA 12] GRAVITZ L., "Printing muscle", *Technology Review*, vol. 115, pp. 80–82, 2012.

[GUD 16] GUDAPATI H., DEY M., OZBOLAT I., "A comprehensive review on droplet-based bio-printing: past, present and future", *Biomaterials*, vol. 102, pp. 20–42, 2016.

[GUE 17] GUÉDON E., MALAQUIN L., ANDRÉ J.C., Bio-printing, *Techniques de l'Ingénieur*, ref. RE268 V1, p. 23, 2017.

[GUI 10a] GUILLEMOT F., MIRONOV V., NAKAMURA M., "Bio-printing is coming of age: report from the International Conference on Bio-printing and Biofabrication in Bordeaux", *Biofabrication*, vol. 2, p. 010201, 2010.

[GUI 10b] GUILLEMOT F., SOUQUET A., CATROS S. *et al.*, "High throughput laser printing of cells and biomaterials for tissue engineering", *Acta Biomateralia*, vol. 7, pp. 2494–2500, 2010.

[GUI 10c] GUILLEMOT F., SOUQUET A., CATROS S. *et al.*, "Laser assisted cell printing: principle, physical parameters versus cell fate and perspectives in tissue engineering", *Nano-medicine*, vol. 5, pp. 507–515, 2010.

[GUI 14] GUILLEMOT F., "Bio-printing: imprimer du vivant ne relève plus de la science-fiction", available at: http://lesclesdedemain.lemonde.fr/sante/bio-printing-imprimer-du-vivant-ne-releve-plus-de-la-science-fiction_a-11-3707.html, 2014.

[GUI 17] GUIBERT N., DIDIER A., MORENO B. *et al.*, "Treatment of post-transplant complex airway stenosis with a three-dimensional, computer-assisted customized airway stent", *American Journal of Respiratory and Critical Care Medicine*, available at: http://dx.doi.org/10.1164/rccm.201611-2361IM, 2017.

[GUO 14] GUO S., Solvent-cast three dimensional printing of polyactide multifunctional microsystems, Montréal University, PhD thesis, Canada, available at: http://publications.polymtl.ca/1430/, 2014.

[GUV 15] GUVEN S., CHEN P., INCI F., "Multiscale assembly for tissue engineering and regenerative medicine", *Trends in Biotechnology*, vol. 33, pp. 269–279, 2015.

[HAR 14] HARRISON B., Bio-printing: the impossible can be possible, available at: http://www.3ders.org/articles/20140305-benjamin-harrison--tedx-talk-bio-printing-the-impossible-can-be-possible.html, 2014.

[HAR 16] HARVARD, Scaling up tissue engineering, available at: https://www.seas.harvard.edu/news/2016/03/scaling-up-tissue-engineering, 2016.

[HE 09] HE P., Conception et réalisation d'un système micro-fluidique pour la réalisation de gouttes calibrées et leur encapsulation, PhD thesis, Université de Technologie de Compiègne, 2009.

[HIL 12] HILLER J., LIPSON H., "Automatic design and manufacture of soft robots", *IEEE Transactions on Robotics*, vol. 28, pp. 457–466, 2012.

[HOF 02] HOFFMAN A.S., "Hydrogels for biomedical applications", *Advanced Drug Delivery Reviews*, vol. 54, pp. 3–12, 2002.

[HRI 14] HRIBAR K.C., SOMAN P., WARNER J. *et al.*, "Light assisted direct-write of 3D functional biomaterials", *Lab on Chip*, vol. 14, pp. 268–275, 2014.

[HUT 15] HUTMACHER D.W., HOLZAPFEL B.M., DE-JUAN-PARDO E.M. *et al.*, "Convergence of regenerative medicine and synthetic biology to develop standardized and validated models of human diseases with clinical relevance", *Current Opinion in Biotechnology*, vol. 35, pp. 127–132, 2015.

[HYR 16] HYREL 3D, "Reliability is our core value", available at: http://www.hyrel3d.com/, 2016.

[IHE 15] IHEST, Les biotechnologies en Chine: thérapie génique, médecine régénérative, plantes transgéniques, available at: http://www.ihest.fr/IMG/pdf/les_biotechnologies_en_chine_maj_2015_final.pdf, 2015.

[IM 12] IM O., LI J., WANG L.G. *et al.*, "Biomimetic 3D nanocrystalline hydroxyapatite and magnetically synthetized single wall CNT chitosan nanocomposite for bone regeneration", *International Journal of Nanomedicine*, vol. 7, pp. 2087–2099, 2012.

[INS 15] INSIS, Groupe de veille "morphogénèse du CNRS", Communication privée CNRS, Paris, 2015.

[IRV 16] IRVINE S.A., VENKATRAMAN S.S., "Bio-printing and differentiation of stem cells", *Molecules*, vol. 21, p. 1188, 2016. doi: 10.3390/molecules21091188.

[ITO 15] ITOH M., NAKAYAMA K., NOGUCHI R. *et al.*, "Scaffold-Free Tubular Tissues Created by a Bio-3D Printer Undergo Remodel-ing and Endothelialization when Implanted in Rat Aortae", *PLoS ONE*, available at: http://journals.plos.org/plosone/article?id=10.1371/journal.pone.0136681, vol. 10, p. e0136681, 2015. doi:10.1371/journal.pone.0136681.

[JAK 16] JAKUS A.E., RUTZ A.L., JORDAN S.W. *et al.*, "Hyperelastic "bone": a highly versatile, growth factor–free, osteoregenerative, scalable, and surgically friendly biomaterial", *Science Translational Medicine*, vol. 8, 358ra127, 2016. doi: 10.1126/scitranslmed.aaf7704.

[JAN 15] JANA S., LERMAN A., "Bio-printing a cardiac valve", *Biotechnology Advances*, vol. 33, pp. 1503–1521, 2015.

[JER 14] JERONIMIDIS G., "Materials strategies and systems for soft robotics", *Robotsoft Plenary Meeting (31 March–1 April) Pisa*, available at: http://www.robosoftca.eu/public_downloads/plenary1/RoboSoft_GeorgeJeronimidis.pdf, 2014.

[JEU 16] JEUKEN R.M., ROTH A.K., PETERS R.J.R.W. *et al.*, "Polymers in cartilage defect repair of the knee: current status and future prospects", *Polymers*, vol. 8, p. 219, 2016. doi:10.3390/polym8060219.

[JOH 15] JOHNSON B.N., LANCASTER K.Z., ZHEN G. *et al.*, "3D printed anatomical nerve regeneration pathways", *Advanced Functional Materials*, vol. 25, no. 39, pp. 6205–6217, 2015. doi:10.1002/adfm.201501760.

[KAC 10] KACHALIA P.R., GEISSBERGER M.J., "Dentistry a la carte: in-office CAD/CAM technology", *Journal of the California Dental Association*, vol. 38, pp. 323–330, 2010.

[KAH 05] KAHN A., PAPILLON F., *Le secret de la salamandre; la médecine en quête d'immortalité*, NIL, Paris, 2005.

[KAM 14] KAMM R.D., BASHIR R., "Creating living cellular machines", *Annals of Biomedical Engineering*, vol. 42, pp. 445–459, 2014.

[KAN 16] KANG H.W., LEE S.J., KO I.K. *et al.*, "A 3D bio printing system to produce human-scale tissue constructs with structural integrity", *Nature Biotechnology*, vol. 34, pp. 312–319, 2016.

[KAO 15] KAO C.T., LIN C.C., CHEN Y.N. *et al.*, "Poly(dopamine) coating of 3D printed ply(lactic acid) scaffolds for bone tissue engineering", *Materials Science and Engineering*, vol. 56, pp. 165–173, 2015.

[KER 11] KER E.D.F., NAIN A.S., WEISS L.E. *et al.*, "Bio-printing of growth factors onto aligned sub-micron scaffolds for simultaneous control of cell differentiation and alignment", *Biomaterials*, vol. 32, pp. 8097–8107, 2011.

[KHO 13a] KHODA A.K.M., OZBOLAT I., KOC B., "Designing heterogeneous porous tissue scaffold for additive manufacturing process", *Computer-Aided Design*, vol. 45, pp. 1507–1523, 2013.

[KHO 13b] KHODA A.K.M., KOC B., "Functionally heterogeneous porous scaffold design for tissue engineering", *Computer-Aided Design*, vol. 45, pp. 1276–1293, 2013.

[KIM 14] KIM H.Y., Effect of mechanical stretching on the maturation of 3D fascicle-like muscle tissue, PhD thesis, MIT, Boston, 2014.

[KIM 16] KIM J.H., YOO J.J., LEE S.J., "Three-dimensional cell-based bio-printing for soft tissue regeneration", *Tissue Engineering and Regenerative Medicine*, vol. 13, p. 647, 2016. doi: 10.1007/s13770-016-0133-8.

[KIN 17] KING S.M., HIGGINS J.W., NINO C.R. *et al.*, "3D proximal tubule tissues recapitulate key aspects of renal physiology to enable nephrotoxicity testing", *Frontiers in Physiology*, available at: https://www.ncbi.nlm.nih.gov/pmc/articles/PMC5340751/pdf/fphys-08-00123.pdf, vol. 8, p. 123, 2017. doi: 10.3389/fphys.2017.00123.

[KLE 88] KLEBE R.J., "Cytoscribing: a method for micropositioning cells and the construction of two- and three-dimensional synthetic tissues", *Experimental Cell Research*, vol. 179, pp. 362–373, 1988.

[KNO 15] KNOWLTON S., ONAL S., YU C.H. *et al.*, "Bio-printing for cancer research", *Trends in Biotechnology*, vol. 33, pp. 504–513, 2015.

[KOL 14] KOLEVSKY D.B., TRUBY R.L., GLADMAN A. *et al.*, "3D bio-printing of vascularized, heterogeneous cell-laden tissue constructs", *Advanced Materials*, vol. 26, pp. 3124–3130, 2014.

[KOS 15] KOSLOW T., 3D printing brings closure to deep brain stimulation devices, available at: http://3dprintingindustry.com/2015/08/29/3d-printing-brings-closure-to-deep-brain-stimulation-microdevices/?utm_source=3D+Printing+Industry+Update&utm_medium=email&utm_campaign=ab7b6cdee2-RSS_EMAIL_CAMPAIGN&utm_term=0_695d5c73dc-ab7b6cdee2-64557777, 2015.

[KOS 16] KOSLOW T., "Bio-printing breakthrough: researchers 3D print living tissue for functional implants", available at: http://3dprintingindustry.com/2016/02/16/bio-printing-breakthrough-researchers-3d-print-living-tissue-for-functional-implant-use/?utm_source=3D+Printing+Industry+Update&utm_medium=email&utm_campaign=147f79fba4-RSS_EMAIL_CAMPAIGN&utm_term=0_695d5c73dc-147f79fba4-64557777, 2016.

[KUC 13] KUCUKGUL C., OZLER S.B., KARAKAS H.E. *et al.*, "3D hybrid bio-printing of nanovascular structures", *Procedia Engineering*, vol. 59, pp. 183–192, 2013.

[KUC 15] KUCUKGUL C., OZLER S.B., INCI I. *et al.*, "3D bio-printing of biomimetic aortic vascular constructs with self-supporting cells", *Biotechnology and Bioengineering*, vol. 112, pp. 811–821., 2015.

[LAN 10] LANG M., WANG W., CHEN X. *et al.*, "Integrated system for 3D assembly of bone scaffolds and cells", *6th IEEE on Automation Science and Engineering*, Toronto, Canada, pp. 786–791, August 2010.

[LAR 17] LARRÈRE C, LARRÈRE R., *Bulles technologiques*, Wildproject, Paris, 2017.

[LAT 07] LATOUR B., *L'espoir de Pandore*, La Découverte, Paris, 2007.

[LAU 13] LAURENT M., *Systèmes biologiques à dynamique non linéaire*, Ellipses, Paris, 2013.

[LAV 14] LAVERGNE B., LAVERGNE M., *L'imprimante 3D: une révolution en marche*, Favre, Paris, 2014.

[LE 06] LE MOIGNE J.L., La théorie du système général, available at: ftp://ftp-developpez.com/jean-louislemoigne/TGS-TM.pdf, 2006.

[LEE 14] LEE J.S., HONG J.M., JUNG J.W. *et al.*, "3D bio-printing of composite tissue with complex shape applied to ear regeneration", *Biofabrication*, vol. 6, p. 024103, 2014.

[LEG 02] LEGRAND P., "De quelle discipline relève l'étude des objets environnementaux?", in KOURILSKY F. (ed.), *Ingénierie de l'interdisciplinarité,* L'Harmattan, Paris, pp. 65–73, 2002.

[LEI 16] LEI M., WANG X., "Biodegradable polymers and stem cells for bio-printing", *Molecules*, vol. 21, p. 539, 2016.

[LEO 15] LEONARD A., "Inside the weird world of 3D printed body parts", available at: https://medium.com/backchannel/where-s-my-bioprinted-kidney-already-25d8c2a5d676, 2015.

[LEO 16] LEONARD F., GODIN B, "3D in vitro model for breast cancer research using magnetic levitation and bio-printing method", *Methods in Molecular Biology*, vol. 1406, pp. 239–251, 2016.

[LIN 13] LINNENBERGER A., BADINE M.I., FIELDER C. *et al.*, "Three dimensional live cell lithography", *Optics Express*, vol. 21, pp. 10269–10277, 2013.

[LIN 16] LIND J.U., BUSBEE T.A., VALENTINE A.D. *et al.*, "Instrumented cardiac micro-physiological devices via multi-material three-dimensional printing", *Nature Materials*, vol. 16, pp. 303–308, 2016.

[LIP 14] LIPSON H., KURMAN M., *Impression 3D: la prochaine révolution*, First Editions, Paris, 2014.

[LOD 15] LODE A., KRUJATZ F., BRUGGEMEIER S. *et al.*, "Green bio-printing: fabrication of photosynthetic algae-laden hydrogel scaffolds for biotechnological and medical applications", *Engineering in Life Sciences*, vol. 15, pp. 177–183, 2015.

[LUT 09a] LUTOLF M.P., BLAU H.M., "Artificial stem cell niches", *Advanced Materials*, vol. 21, pp. 3255–3268, 2009.

[LUT 09b] LUTOLF M.P., GILBERT P.M., BLAU H.M., "Designing materials to direct stem-cell fate, *Nature*, vol. 462, pp. 433–441. 2009.

[MAD 17] MADEELI, Fabrication Additive: Enjeux et Impacts en Occitanie Pyrénées-Méditerranée, available at: http://www.madeeli.fr/wp-content/uploads/2017/01/Fabrication-Additive-en-Occitanie-Etude-2016.pdf, 2017.

[MAG 16a] MAGHDOURI-WHITE Y., BOWLIN G.L., LEMMON C.A. *et al.*, "Bioengineered silk scaffolds in 3D tissue modeling with focus on mammary tissues", *Materials Science and Engineering,* vol. 59, pp. 1168–1180, 2016.

[MAG 16b] MAGNIER M., Des chercheurs ont développé des cellules cardiaques diminuant considérablement les risques de rejets de greffes, available at: http://soocurious.com/fr/coeur-greffe-cellules-souches/?utm_source=newsletter&utm_medium=e-mail&utm_campaign=Newsletter_Journaliere_2016-03-23, 2016.

[MAH 16] MAHLA R.S., "Stem cells applications in regenerative medicine and disease therapeutics", *International Journal of Cell Biology*, vol. 2016, ID 6940283, 2016.

[MAL 15] MALAQUIN L., *Impression tridimensionnelle de tissus cellulaires et d'organes*, OMNT, Grenoble, 2015.

[MAN 13] MANNOR M.S., JIANG Z., JAMES T. *et al.,* "3D printed bionic ears", *Nano Letters*, vol. 13, pp. 2634–2639, 2013.

[MAN 16] MANDRICKY C., WANG Z., KIM K. *et al.*, "3D bio-printing for engineering complex tissues", *Biotechnology Advances*, vol. 34, pp. 422–434, 2016.

[MAR 15] MARTIN J.J., FIORE B.E., ERB R.M., "Designing bioinspired composite reinforcement architecture via 3D magnetic printing", *Nature Communications*, vol. 6, 2015.

[MAR 16] MARCHAND L., Des chercheurs américains réussissent à imprimer des tissus vivants en 3D, available at: http://www.lesechos.fr/industrie-services/pharmacie-sante/021700931522-des-chercheurs-americains-reussissent-a-imprimer-des-tissus-vivants-en-3d-1200609.php?44fmBuzArJjvPSUM.99, 2016.

[MAU 17] MAURMANN N., PEREIRA D., BURGUEZ D. *et al.*, "Mesenchymal stem cells cultivated on scaffolds formed by 3D printed PCL matrices, coated with PLGA electrospun nanofibers for use in tissue engineering", *Biomedical Physics & Engineering Express*, vol. 3, forthcoming, 2017.

[MCC 14] MCCUNE M., SHAFIEE A., FORGACS G. *et al.,* "Predictive modeling of post bio-printing structure formation", *Soft Matter*, vol. 10, pp. 1790–1800, 2014.

[MEL 12] MELCHELS F.P.W., DOMINGOS M.A.N., KLEIN T.J. *et al.*, "Additive manufacturing of tissues and organs", *Progress in Polymer Science*, vol. 37, pp. 1079–1084, 2012.

[MIC 13] MICHAEL S., SORG H., PECK C.T. *et al.*, "Tissue engineered skin substitutes created by laser-assisted bioprinting form skin-like structures in the dorsal skin fold chamber in mice", *PLoS One*, vol. 8, p. e57741, 2013.

[MIL 16] MILLER A., Bioprinted liver tissues generate organ-level responses, available at: http://3dprintingindustry.com/news/bioprinted-liver-tissue-generate-organ-level-responses-87096/, 2016.

[MIT 12] MITCHEM C., "Why digital impression", *Dental Economics*, vol. 102, pp. 52–88, 2012.

[MIT 16] MITCHELL A.C., BRIQUEZ P.S., HUBBELL J.A., "Engineering growth factors for regenerative medicine applications", *Acta Biomaterialia*, vol. 30, pp. 1–12, 2016.

[MOL 15a] MOLITCH-HOU M., New bio-printing tech to be developed by aspect biosystems with seed funding, available at: http://3dprintingindustry.com/2015/08/06/new-bio-printing-tech-to-be-developed-by-aspect-biosystems-with-seed-funding/?utm_source=3D+Printing+Industry+Update&utm_medium=email&utm_campaign=2f4d24c15b-RSS_EMAIL_CAMPAIGN&utm_term=0_695d5c73dc-2f4d24c15b-64557777, 2015.

[MOL 15b] MOLITCH-HOU M., Exclusive: bio-printing now available via your local 3D printing store, available at: http://3dprintingindustry.com/2015/09/08/bio-printing-now-available-via-your-local-3d-printing-store/?utm_source=3D+Printing+Industry+Update&utm_medium=email&utm_campaign=cd88649f0e-RSS_EMAIL_CAMPAIGN&utm_term=0_695d5c73dc-cd88649f0e-64557777, 2015.

[MOL 17] MOLDOVAN N.I., HIBINO N., NAKAYAMA K., "Principles of the Kenzan method for robotic cell spheroid-based three-dimensional bio-printing", *Tissue Engineering Part B: Reviews*, available at: http://online.liebertpub.com/doi/pdf/10.1089/ten.teb.2016.0322, forthcoming, 2017.

[MOO 14] MOONZUR R., Un pas de plus vers le "bio-printing", l'impression de tissus humains, available at: http://www.techniques-ingenieur.fr/actualite/articles/un-pas-de-plus-vers-le-bio-printing-limpression-de-tissus-humains-10706/, 2014.

[MOR 13] MORIMOTO Y., TAKEUCHI S., "3D cell culture based on microfluidic technique to mimic living tissues", *Biomatter Science*, vol. 1, pp. 257–264, 2013.

[MOS 14] MOSIEWICZ K.A., KOLB L., VAN DER VLIES A.J. *et al.*, "In situ cell manipulation through enzymatic hydrogel photopatterning", *Nature Materials*, vol. 12, pp. 1072–1078, 2014.

[MOT 15a] MOTA C., DANTI S., D'ALESSANDRO D. *et al.*, "Multiscale fabrication of biomimetic scaffolds for tympanic membrane tissue engineering", *Biofabrication*, vol. 7, no. 2, 025005, 2015.

[MOT 15b] MOTAMEDIAN S.R., HOSSEINPOUR S., ASHAIE M.G. *et al.*, "Smart scaffolds in bone tissue engineering: a systematic review of literature", *World Journal of Stem Cells*, vol. 7, pp. 657–668, 2015.

[MOU 15] MOUSSET L., L'impression 3D entre fiction et réalité, available at: http://information.tv5monde.com/info/l-impression-3d-entre-fiction-et-realite-48944, 2015.

[MUN 15] MUNJAI A., PHILIPPE J.M., MUNRO E. *et al.*, "A self-organized biomechanical network drives shape changes during tissue morphogenesis", *Nature*, vol. 524, pp. 351–355, 2015.

[MUR 14] MURPHY S.V., ATALA A., "3D bio-printing of tissues and organs", *Nature Biotechnology*, vol. 32, pp. 773–785, 2014.

[NAK 13] NAKAYAMA K., "Chapter 1. In vitro biofabrication of tissues and organs", in FORGACS G., SUN W. (eds), *Biofabrication*, pp. 1–16, Elsevier, New York, 2013.

[NAN 15] NANOSCRIBE, True 3D laser lithography, available at: https://www.youtube.com/watch?v=UoLwOYtx0TA, 2015.

[NG 16] NG W.L., WANG S., YEONG W.E. *et al.,* "Skin bio-printing: impending reality or fantasy?", *Trends in Biotechnology*, vol. 34, pp. 689–699, 2016.

[NGU 13] NGUYEN T.L., ALLEN S.J., PHEE S.J., "Exploiting 3D printing technology to develop robotic running foot for footwear testing", *Virtual and Physical Prototyping*, vol. 4, pp. 259–269, 2013.

[NGU 15] NGUYEN P., "C'est la première bio-impression qui compte", *Science et santé*, vol. 24, pp. 39–39, 2015.

[NGU 16] NGUYEN D.G., FUNK J., ROBBINS J.B. *et al.*, "Bioprinted 3D primary liver tissues allow assessment of organ-level response to clinical drug induced toxicity in vitro", *PLoS ONE*, vol. 11, no. 7, p. e0158674, 2015.

[NGU 17] NGUYEN D.G., PENTONEY JR S.L., "Bioprinted three dimensional human tissues for toxicology and disease modeling", *Drug Discovery Today: Technologies*, vol. 23, pp. 37–44, 2017.

[NIE 65] NIETZSCHE F., *Par-delà le bien et le mal*, UGE, Paris, 1965.

[NIE 15] NIEDERCORN F., Déjà utilisée en dentisterie, la fabrication additive devrait bousculer le secteur des prothèses et implants. Même les chirurgiens s'y intéressent pour mieux préparer les interventions, available at: http://www.lesechos.fr/idees-debats/sciences-prospective/021558688976-limpression-3d-simplante-dans-la-sante-1184574.php, 2015.

[NIK 17] NIKLAUS L., TANSAZ S., DAKHIL H. *et al.*, "Micropatterned down-converting coating for white bio-hybrid light-emitting diodes", *Advanced Functional Materials*, vol. 27, p. 1601792, 2017.

[NOU 15] NOURY M., La nanosanté, PhD thesis, Universités de Paris-Ouest et de Montréal, Canada, 2015.

[NOW 03] NOWOTNY H., SCOTT P., GIBBONS M., *Repenser la Science*, Belin, Paris, 2003.

[OCA 16] OCAMPO A., REDDY P., MARTINEZ-REDONDO P. *et al.*, "In vivo amelioration of age-associated hallmarks by partial reprogramming", *Cell*, vol. 167, pp. 1719–1733.e12, 2016.

[ONE 15a] O'NEILL E., 3D stem cell printing by Scottish uni, available at: http://www.scotsman.com/news/health/3d-stem-cell-printing-advanced-by-scottish-uni-1-3922232#axzz3p73bCfMM, 2015.

[ONE 15b] O'NEILL P.F., BEN AZOUZ A., VASQUEZ M. *et al.*, "Advance in 3D rapid prototyping of microfluidic devices for biological applications", *Biomicrofluidics*, vol. 8, pp. 1–11, 2015.

[OST 15] OSTEO3D, "3D printing healthcare", available at: http://www.osteo3d.com/, 2015.

[OST 16] OSTEOFAB, OsteoFab Technology, available at: http://oxfordpm.com/cmf-orthopedics/osteofab-technology, 2016.

[OUR 16] OUROBOTICS, "Introducing the revolution multimaterial 3D bioprinter", available at: https://www.weare3dbioprintinghumans.org/, 2016.

[OVS 11] OVSIANIKOV A., FARSARI M., CHICHKOV B., "Photonic and biomedical applications of the two photon polymerization technique", in BARTOLO P.J. (ed.), *Stereolithography; Materials, Processes and Applications*, Springer, New York, pp. 37–56, 2011.

[OZB 13] OZBOLAT I.T., YU Y., "Bio-printing toward organ fabrication: challenges and future trends", *IEE Transactions on Biomedical Engineering*, vol. 60, pp. 691–699, 2013.

[OZB 16a] OZBOLAT I.T., HOSPODIUK M., "Current advances and future perspectives in extrusion based bio-printing", *Biomaterials*, vol. 76, pp. 321–343, 2016.

[OZB 16b] OZBOLAT I.T., PENG W., OZBOLAT V., "Application areas of 3D bio-printing", *Drug Discovery Today*, vol. 21, pp. 1257–1271, 2016.

[PAR 65] PARETO V., *Les systèmes socialistes*, Droz, Geneva, 1965.

[PAR 15] PARK J.S., SURYAPRAKASH S., LAO Y.H. *et al.*, "Engineering mesenchymal stem cells for regenerative medicine and drug delivery", *Methods*, vol. 84, pp. 3–16, 2015.

[PAU 15] PAULSEN S.J., MILLER J.S., "Tissue vascularization through 3D printing: will technology bring us flow?", *Developmental Dynamics*, vol. 244, pp. 629–640, available at: http://onlinelibrary.wiley.com/doi/10.1002/dvdy.24254/full, 2015.

[PEN 16] PENG W., UNUTMAZ D., OZBOLAT I.T. "Bio-printing towards physiologically relevant tissue models for pharmaceutics", *Trends in Biotechnology*, vol. 34, pp. 722–732, 2016.

[PES 95] PESTRE D., "Pour une histoire sociale et culturelle des sciences – nouvelles définitions, nouveaux objets, nouvelles pratiques", *Annales – Histoire, Sciences Sociales*, vol. 3, pp. 487–522, 1995.

[PET 16] PETCH M., "3D bio-printing Pioneer Organovo Announce 2016 financial results", available at: http://3dprintingindustry.com/news/3d-bio-printing-pioneer-organovo-announce-2016-financial-results-82201/, 2016.

[POI 15] POIETIS, Qu'est-ce que la bio-impression?, available at: http://www.poietis.com/bioimpression-4d/, 2015.

[POI 16] POIETIS, L'Oréal et Poietis signent une collaboration de recherche exclusive visant la bio-impression du cheveu, available at: https://www.poietis.com/fr/post-release.php?id=9&view=0, 2016.

[POR 17] PORTER M.M., RAVIKUMAR N., BARTHELAT F. *et al.*, "3D-printing and mechanics of bio-inspired articulated and multi-material structures", *Journal of the Mechanical Behavior of Biomedical Materials*, vol. 73, pp. 114–126, 2017.

[POS 14] POST M., Dans sept ans, vous mangerez de la viande in vitro, available at: http://www.lexpress.fr/styles/saveurs/mark-post-dans-sept-ans-vous-mangerez-de-la-viande-in-vitro_1560075.html, 2014.

[PRE 17] PRENDERGAST M.E., SOLORZANO R.D., CABRERA D., "Bioinks for biofabrication: current state and future perspectives", *Journal of 3D Printing in Medicine*, vol. 1, pp. 49–62, 2017.

[QUI 15] QUINONES H., BABIARZ A., FANG L., Jetting technologies: a way of the future in dispersing, available at: http://www.nordson.com/en-us/divisions/asymtek/Documents/Papers/2002_10_Jetting_tech_way_of_future_EMAP%20Taiwan_NordsonASYMTEK.pdf, 2015.

[RAC 16] RACKHAM O.J.L., FIRAS J., FANG H. *et al.*, "A predictive computational framework for direct reprogramming between human cell types", *Nature Genetics*, vol. 48, pp. 331–335, 2016.

[RAM 17] RAMAN R., CVETKOVIC C., BASHIR R., "A modular approach to the design, fabrication, and characterization of muscle-powered biological machines", *Nature Protocols*, vol. 12, pp. 519–533, 2017.

[RAN 14] RANGA A., GOBAA S., MOSIEWICZ K.A. *et al.*, "3D niche microarrays for systems-level analyses of cell fate", *Nature Communications*, vol. 5, p. 4324, 2014.

[RAS 15] RASSKAZOVA L.A., LYTKINA D.N., SHAPOVALOVA Y.G. *et al.*, "Bioactive composites produced in situ on the basis of calcium phosphates and lactic acid oligomers", *Russian Journal of Applied Chemistry*, vol. 88, pp. 669–675, 2015.

[RAT 17] RATHEESH G., VENUGOPAL J.R., CHINAPPAN A. *et al.*, "3D fabrication of polymeric scaffolds for regenerative therapy", *ACS Biomaterials Science & Engineering*, vol. 3, no. 7, pp. 1175–1194, 2017.

[RAY 15] RAYNAL J., "Bio-impression – des machines à tisser le vivant", *Industrie et Technologies*, no. 981, pp. 38–40, 2015.

[RBT 15a] RBTH, La glande thyroïde réalisée avec une bio-imprimante 3D fonctionne, available at: http://fr.rbth.com/tech/2015/10/31/la-glande-thyroide-realisee-avec-une-bio-imprimante-3d-fonctionne_535955, 2015.

[RBT 15b] RBTH, Chérie, j'ai imprimé ta thyroïde, available at: http://fr.rbth.com/tech/2015/04/22/cherie_jai_imprime_ta_thyroide_33471, 2015.

[REN 15] RENOUARD G., L'impression 3D peut désormais créer la vie, available at: http://www.atelier.net/trends/articles/impression-3d-desormais-creer-vie_438092?banner=1, 2015.

[RIC 17] RICHARDS D., JIA J., YOST M. *et al.*, "3D bio-printing for vascularized tissue fabrication", *Annals of Biomedical Engineering*, vol. 45, no. 1, pp. 132–147. 2017.

[RIM 16] RIMANN M., BONO E., ANNAHEIM H. *et al.*, "Standardized 3D bio-printing of soft tissue models with human primary cells", *Journal of Laboratory Automation*, vol. 21, no. 4, pp. 496–509, 2016.

[ROC 13] ROCCIO M., SCHMITTER D., KNOBLOCH M. *et al.*, "Predicting stem cell fate changes by differential cell cycle progression patterns", *Development*, vol. 140, pp. 459–470, 2013.

[RU 14] RU C., LUO J., XIE S. *et al.*, "A review of non-contact micro- and nano-printing technologies", *Journal of Micromechanics and Microengineering*, vol. 24, p. 053001, 2014.

[SAN 15] SANTORO M., LAMHAMEDI-CHERRADI S.E., MENEGAZ B.A. *et al.*, "Flow perfusion effects on three-dimensional culture and drug sensitivity of Ewing sarcoma", *PNAS*, vol. 112, pp. 10304–10309, 2015.

[SAN 16] SANTÉLOG, "Bio-printing: une première preuve de concept d'organes imprimés en 3D", available at: https://www.santelog.com/news/recherche-medicale/bio-printing-une-premiere-preuve-de-concept-d-organes-imprimes-en-3d_15361_lirelasuite.htm#lirelasuite, 2016.

[SCH 14] SCHNEIDER O., EPPLE P., TEUBER E. *et al.*, "Jet printing of colloid solutions: numerical modeling and experimental verification of the influence of ink and surface parameters on droplet spreading", *Advanced Powder Technology*, vol. 22, pp. 266–270, 2014.

[SCH 16] SCHWALBE M., *Predictive Theoretical and Computational Approaches for Additive Manufacturing: Proceedings of a Workshop*, U.S. National Committee on Theoretical and Applied Mechanics, available at: http://www.nap.edu/23646, 2016.

[SCI 14] SCIENCE ET AVENIR, Quand l'imprimante façonne le vivant, available at: http://www.sciencesetavenir.fr/sante/quand-l-imprimante-faconne-le-vivant_13022, 2014.

[SEL 15] SELIMIS A., MIRONOV V., FARSANI M., "Direct laser writing: principles and materials for scaffold 3D printing", *Microelectronics Engineering* , vol. 132, pp. 83–89, 2015.

[SER 09] SERRES M., *Temps des crises*, Le Pommier, Paris, 2009.

[SEV 05] SÈVE L., "De quelle culture logico-philosophique la pensée du non-linéaire a-t-elle besoin?", in GUESPIN-MICHEL J. (ed.), *Emergence, complexité et dialectique*, Odile Jacob, Paris, 2005.

[SHA 85] SHAPIN S., SCHAFFER S., *Léviathan et la pompe à air*, La Découverte, Paris, 1985.

[SHA 14] SHAPIRA A., KIM D.H., DVIR T., "Advanced micro- and nano-fabrication technology for tissue engineering", *Biofabrication*, vol. 6, p. 020301, 2014.

[SHA 15] SHAO Y., SANG J., FU J., "On human pluripotent stem cell control: the rise of 3D bioengineering and mechanobiology", *Biomaterials*, vol. 52, pp. 26–43, 2015.

[SHE 14] SHER D., Harvard scientists inch closer to getting 3D printed organs' blood flowing, available at: http://3dprintingindustry.com/2014/02/25/harvard-3d-printed-organs-blood-flowing/, 2014.

[SHE 15a] SHER D., Bio-printing solution and dreams: an interview with 3D Bio's Dr Vladimir Mironov, available at: http://3dprintingindustry.com/2015/04/09/3d-bio-printing-solutions-dreams/, 2015.

[SHE 15b] SHER D., Hydrogels & 3D printing used to produce programmable bio-capsules, available at: http://3dprintingindustry.com/2015/06/22/hydrogels-3d-printing-used-produce-programmable-release-enzymes/?utm_source=3D+Printing+Industry+Update&utm_medium=email&utm_campaign=02cf6b2ad9-RSS_EMAIL_CAMPAIGN&utm_term=0_695d5c73dc-02cf6b2ad9-64557777, 2015.

[SHU 10] SHU X.Y., ZHANG H.H., LIN H.Y. *et al.*, "Experimental study on high viscosity fluid micro-droplet jetting system", *Science China – Technological Sciences*, vol. 53, pp. 182–187, 2010.

[SKA 15] SKARDAL A., ATALA A., "Biomaterials for integration with 3D bio-printing", *Annals of Biomedical Engineering*, vol. 43, pp. 730–746, 2015.

[SKO 14] SKOOG S.A., GOERING P.L., NARAYAN R.J., "Stereolithography in tissue engineering", *Journal of Materials Science: Materials in Medicine*, vol. 25, pp. 845–856, 2014.

[SMA 15a] SMALL LAB, Aerosol jet printing, available at: http://www.small-lab.be/aerosol-jet-printing-0, 2015.

[SMA 15b] SMARTECH, 3D printing in dentistry 2015: a ten year opportunity forecast and analysis, available at: http://smartechpublishing.com/reports/3d-printing-in-dentistry-2015-a-ten-year-opportunity-forecast-and-analysis, 2015.

[SOM 16] SOMMER P., "Les tissus humains, entre biologie et mécanique – propos d'un biologiste", *MECAMAT – CNRS*, Aussois, France, 18–22 January, 2016.

[SPU 16] SPUTNIK, Les imprimantes 3D bio russes à la pointe de la technologie médicale, available at: https://fr.sputniknews.com/sci_tech/201611171028749723-imprimante-maladie-organe-transplantation-innovation/ and https://fr.sputniknews.com/sci_tech/201611171028749723-imprimante-maladie-organe-transplantation-innovation/, 2016.

[SRE 12] SREETHARAN P.S., WHITNEY J.P., STRAUSS M.D. *et al.*, "Monolithic fabrication of millimeter scale machines", *Journal of Micromechanics and Microengineering*, vol. 22, p. 0550267, 2012.

[STA 11] STAMPFL J., LISKA R., "Polymerisable hydrogels for rapid prototyping: chemistry, photolithography and mechanical properties", in BARTOLO P.J. (ed.), *Stereolithography; Materials, Processes and Applications*, Springer, New York, pp. 161–182, 2011.

[STA 16] STACE E.T., DAKIN S.G, MOUTHUY P.A. *et al.,* "Translating regenerative biomaterials into clinical practice", *Journal of Cellular Physiology*, vol. 231, pp. 36–49, 2016.

[STR 09] STRINGER J., DERBY B., "Limits to feature size and resolution in ink jet printing", *Journal of the European Ceramic Society*, vol. 29, pp. 913–918, 2009.

[SUN 04] SUNG H.J., MEREDITH C., JOHNSON C. *et al.*, "The effect of scaffold degradation rate on three dimensional cell growth and angiogenesis", *Biomaterials*, vol. 25, pp. 5735–5742, 2004.

[SUN 16] SUNTORNNOND R., YONG SHENG TAN E., AN J. *et al.*, "A mathematical model on the resolution of extrusion bio-printing for the development of new bioinks", *Materials*, vol. 9, p. 756, 2016.

[TAK 15] TAKASATO M., ER P.X., CHIU H.S. *et al.*, "Kidney organoids from human iPS cell contain multiple lineages and model human nephrogenesis", *Nature*, vol. 526, pp. 564–568, 2015.

[TAN 14] TAN Y., RICHARDS D.J., TRUSK T.C. *et al.*, "3D printing facilitated scaffold-free tissue unit fabrication", *Biofabrication*, vol. 6, 2014.

[TAN 16] TANDON N., Lab-Grown Bones? They Could Make Painful Grafts History (Op-Ed), available at: http://www.livescience.com/52037-lab-grown-bones-could-replace-painful-grafts.html, 2016.

[TAS 13] TASOGLU S., DERMICE U., "Bio-printing for stem cell research", *Trends in Biotechnology*, vol. 31, pp. 10–19, 2013.

[THE 17] THE GRAIL PROJECT, Tissue Engineering and Regenerative Medicine, available at: http://www.thegrail-project.eu/wordpress/?page_id=11, 2017.

[THO 16] THOMAS D.J., "Could 3D bioprinted tissues offer future hope for microtia treatment?", *International Journal of Surgery*, vol. 32, pp. 43–44, 2016.

[TUR 15] TURKSEN K. (ed.), *Bio-printing in Regenerative Medicine*, Humana Press, Zurich, 2015.

[VAL 35] VALÉRY P., cited by CARRIEU-COSTA M.-J. (May 2016), "Former pour l'inconnu", *Réalités Industrielles*, pp. 5–8, 1935.

[VAN 15] VAN DUINEN V., TRIETSCH S.J., JOORE J. *et al.*, "Microfluidic 3D cell culture: from tools to tissue models", *Current Opinion in Biotechnology*, vol. 35, pp. 118–126, 2015.

[VIS 14] VISSER C.W., Fundamentals and applications of fast micro-drop impact, available at: http://doc.utwente.nl/93267/1/thesis_C_Visser.pdf, 2014.

[WAN 14] WANG J., YANG M., ZHU Y. *et al.*, "Phage nanofibers induce vascularized osteogenesis in 3D printed bone scaffolds", *Advanced Materials*, vol. 26, pp. 4961–4966, 2014.

[WAT 16] WATTS L., Denver University researchers use 3D bioprinter to create artificial body parts, available at: http://www.thedenverchannel.com/money/science-and-tech/denver-university-researchers-use-3d-bioprinter-to-create-artificial-body-parts, 2016.

[WEI 09] WEISS T., HILDEBRAND G., SCHADE R. *et al.*, "Two-photon polymerization for microfabrication of 3D scaffolds for tissue engineering application", *Engineering in Life Sciences*, vol. 9, pp. 384–390, 2009.

[WTE 15] WTEC (World Technology Evaluation Center), Global Assessment of Biological Engineering & Manufacturing, available at: http://www.wtec.org/bem/docs/BEM-Final Report-Web.pdf, 2015.

[XU 13a] XU T., BINDER K.B., ALBANNA M.Z. *et al.*, "Hybrid printing of mechanically and biologically improved constructs for cartilage tissue engineering applications", *Biofabrication*, available at: http://iopscience.iop.org/article/10.1088/1758-5082/5/1/015001/meta;jsessionid=88DF712F18E4A7C141FFF50F5176E4B8.c4.iopscience.cld.iop.org, vol. 5, p. 015001, 2013.

[XU 13b] XU T., ZHAO W., ZHU J.M. *et al.*, "Complex heterogeneous tissue constructs containing multiple cell types prepared by inkjet printing technology", *Biomaterials*, vol. 34, pp. 130–139, 2013.

[YEO 04] YEONG W.Y., CHUA C.K., LEONG K.F. *et al.*, "Rapid prototyping in tissue engineering: challenges and potential", *Trends in Biotechnology*, vol. 22, pp. 643–652, 2004.

[ZAL 15] ZALESKY A., "Meet BioBot, the 'MakerBot of biology'", available at: http://fortune.com/2015/07/13/biobot-the-makerbot-of-biology/, 2015.

[ZHO 16] ZHOU Y., "The application of ultrasound in 3D bio-printing", *Molecules*, vol. 21, p. 590, 2016.

[ZOP 13] ZOPF D.A., HOLLISTER S.J., NELSON M.E. *et al.*, "Bio-resorbable airway splint created with a 3D printer", *The New England Journal of Medicine*, vol. 368, pp. 2043–2045, 2013.

3

Some Examples of 3D Bio-printed Tissues

"We tend to see ourselves as machines. Ray Kurtzell suggests that we are forms of behavior: we are our 'circuitries.' The body and the brain are simply made up of a specific set of particles that change at a surprising speed [...]. We are a system of matter and energy that persists through time". [COC 17]

"The novelty of modern times therefore has a very particular accent. It is not just a matter of a consciousness of the time that is delineated from the past and institutes the boundary between 'yesterday' and 'today,' but of a rupture that, by rejecting the authority of models, takes away all value of exemplary nature: all antecedence is declared outdated and obsolete". [REV 12]

"What will soon seem the oldest is what initially seemed the most modern". [GID 72]

"Understanding the world, like Cervantes, as ambiguity, having to face many relative truths that contradict one another instead of a single absolute truth [...], thus possessing the wisdom of uncertainty as the only certainty, all of this demands a force that is no smaller". [KUN 86]

"It is thus best to accept the fact that a solution is never logically necessary and constraining seriously, that it never imposes itself in the absolute sense of the term – that every closure of a debate or every concensus is local by nature and can only be understood in the precise context in which it was elaborated". [PES 95]

"The author Arthur Koestler maintains that creative originality doesn't consist in creating ideas ex nihilo, *but rather in combining well-established schemata and structures through a sort of hybridization. 'The creative act is not a creation in the sense of the Old Testament. It does not create from nothing. It discovers, mixes, synthesizes facts, ideas, faculties, techniques that already existed. Everything invented will be all the more surprising as the parts are more familiar'". [CAS 17]*

"Yet, what stands out is that this world is a world of interactions. An organism's slightest activity always implies exchanges of matter, of energy, of information, of movements, of behaviors. Interactions that encourage or hinder. This may be what characterizes the ecologist: it is he/she who wishes to understand the world of the living and its interactions [...]. This is the bet made by the ecologist: as diverse and complex as the interactions of the living world can be, some are intelligible, they repeat themselves and can be anticipated – at least to a certain extent". [DEV 16]

"They end up obtaining a limitless power and dictating their own laws and rules. Technology, in return, is seeing a drastic leap and society progressively slides into a Trans-humanism rid of all complexities, in rich nations anyway". [NAT 17]

"Infobesity surrounds us from all sides and it is a serious blow to our creativity". [LEW 16]

3.1. Introduction

Klebe [KLE 88] printed biological elements for the first time using an inkjet printer: Fibronectin proteins, present in the extracellular matrix, were printed and they notably allowed the adhesion of cells to this matrix. Since that date – the previous chapter is there to illustrate it – it is possible to go much further. Why not up to the manufacture of organs? Using some examples, this is precisely what is analyzed in this part.

"3D printing technologies open [...] new pathways in terms of prostheses and organ grafts/transplants: 'a liver has been reconstructed in Israel using stem cells and a heart in Russia,' cites Uwe Diegel as examples. By 2030, bio-printed organ transplants will be possible. Research has made leaps and bounds: last November, the Russian company 3D Bioprinting Solutions managed to reproduce a thyroid and implant it into a mouse. According to the American laboratory Organovo, which is working on manufacturing human liver tissues via 3D printing, bio-printing hepatic tissues will offer an alternative to transplants and mitigate the global problem of a lack of donors" [VIN 17]. Anthony Atala, a specialist of regenerative medicine, at a TED conference said that he was capable of manufacturing a human kidney with a 3D printer and demonstrated it to the large audience. ("This pseudo-organ was in fact a sort of mold in the shape of a kidney, printed with a synthetic material, on which kidney cells were placed. An almost inert object! This misexplanation reveals the tension that has reigned in recent years in regenerative medicine, and particularly in bio-printing" [HER 14]). There is only one step between this and thinking it possible to one day "kill death". Others have a calmer scientific attitude and advance slowly, because anything else is difficult with complex subjects [FRI 17].

Indeed, since 1999, stem cells have been brought to light. They multiply and provide new cells that engender the lineage of blood cells or cells from the epidermis' basal membrane. They can transform into functional, mature differentiated cells adapting to the environment in which they are placed. Highly dependent on environmental (epigenetic) factors, they only specialize as they mature. These cells, by multiplying, have a highly promising therapeutic potential with considerable possibilities, the patient becoming the producer of his/her self-prostheses and, moreover, 3D printing can help place them in the correct location so that they act in an optimal manner. Therein lies the stake. Transplanting a group of multi-differentiated tissues and not a single or multi-tissular organ (kidney, liver, heart, lung) poses the question of mastering the immunological phenomenon, a tissue incompatibility in the heterologous transplantation [ZIN 01], a problem that normally is not present when working towards the same goal with one's own cells.

Laboratories are much more than locations where science is practiced: they are locations where, through artifacts (like that of optimizing a process), a world is created, slipping into the demands of the representations used. There is thus the presence of forms of inertia in the action and this is because, on the one hand, there is success associated with proofs of concept and, on the other hand, extraordinary attractiveness for bio-printing, in which, like a sudden crisis, an entire community starts forging ahead and getting involved in forms of experience accumulation, thinking that they are participating in the clearing of a domain that is still largely untouched.

Some failures have led to humbleness (it is somewhat necessary to give an explanation after financial support by agencies, because in the responses to calls for tender, it is necessary to make promises); then, it is possible to witness a focusing of work on "what has chances to be successful" [THO 16]. It is thus that, essentially, work is focused on printing tissues, towards "simple" organs like bones (which is not just mineral), cartilage (limited vascularization) and skin (because this seems possible and, according to Barnouin [BAR 16a], "since 2014, the American Army has been seeking to integrate bio-printing, that is, the 3D printing of living cells, into its war medicine. The goal would thus be to 3D print skin to care for the seriously wounded"). And indeed, today we are going a bit further than the proof of concept, though without having reached the stage of pure, simple transplanting.

As has been pointed out, the luck that is had is that sometimes "Nature is well made" with exemplary cellular proliferations allowing grafts to be taken to an extent that is nearly a unit. Everyone knows the ability that some animals have to partially regenerate their bodies and this information can help recreate their amputated limbs thanks to their totipotent stem cells. "However, nature offers far fewer possibilities of this to the superior vertibrates that are mammals" [KAH 05]. Thus, the works presented below relate an experimental story under construction with many short-term attempts that still hamper the technological legitimacy of bio-printing. However, can we do anything else?

The work presented below was written with the effective help of Natacha Denis at the Reactions and Process Engineering Laboratory (LRGP) in Nancy, France.

3.2. Work on cartilage

Concretely, manufacturing a biological tissue through bio-printing is performed in the following way: an initial stage involves conceptualizing the architecture of the biological tissue with a computer; the printing parameters are then programmed with an "ink" containing cells [SKA 15, MUR 14, BHA 15, PAS 16, OZB 16, RAD 17]. The "biological" tissues are then printed, layer by layer, with programmable logic controllers that reproduce the patterns conceived on the computer by depositing, for example, micro-droplets of biological inks (see, e.g. [SCH 11, STR 09]). The final step depends on the maturation of the printed tissue in a bioreactor. This is a device in which microorganisms are multiplied [GUI 14]. This stage allows the cells to organize themselves until specific biological functions emerge. Several classically published techniques exist (see [MUR 14]). Moreover, two schools of thought are developing, one that sequentially creates the support followed by cell seeding, and the other simultaneously performing these operations.

As the case of cartilage requires neither support nor vascularization, it is very interesting in terms of proof of concept in this "pre-paradigmatic" décor, because it largely avoids the pitfalls associated with other tissues. This first part of this chapter presents an analysis of the bibliography (certainly non-exhaustive) and leads to remarks on the applicability of the studies performed. Figure 3.1 illustrates the evolution of the domain in terms of the number of publications (via a search on the University of Lorraine Library's website with two keywords: cartilage and 3D Printing). The results are presented in Figure 3.1.

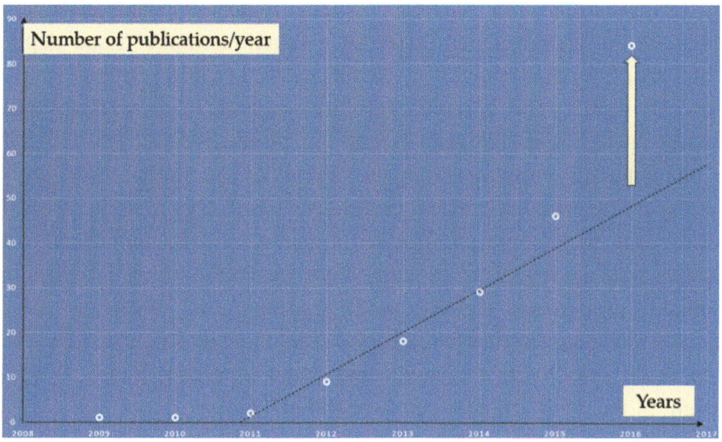

Figure 3.1. *Publications with the two keywords: cartilage and 3D printing*

As this figure shows, 2011 can be considered the "start-up" year of the dynamics on this subject with notable acceleration (in all, roughly 200 publications over the period).

3.2.1. *General remarks on cartilage*

3.2.1.1. *Definition*

Cartilage is a layer of non-vascularized and non-innervated tissue connecting the ends of bones. It is made up of collagen fibers and proteoglycans allowing cellular adhesion, mechanical support, and the transmission of mechanical and chemical signals between cells and tissues. The properties of cartilage depend on its depth in relation to the surface of the articulation. The mechanical properties of cartilage, flexible but resistant, place this organ in an intermediary position between bones and less rigid, more or less dense joint tissues like tendon or muscle. The rigidity of cartilage gives it a particularly important role in supporting the opening of different tubes that are open to the organism's air, whether this is the trachea, alar cartilage in the nostrils, or the auricle of the ear around the auditory canal. In the case of injury, their repair is slow, even nearly non-existent in adults [DOR 15]. Bio-printing thus comes as a means of attempting to perform what Nature does not do in the human.

3.2.1.1.1. Structure and composition

Three areas must be considered: superficial (10–20%), intermediate (40–60%) and deep (30–40%). The deeper the layer is, the more the cellular density is reduced and the glycosaminoglycan (GAG) content increases. The distribution and morphology of the chondrocytes differ as a function of the tissue depth. In the superficial zone, the chondrocytes are small and flat, while in the deep zone, they are larger and round. The chondrocytes are encased in an extracellular matrix (ECM) made up of water, collagen fibers and proteoglycans. The collagen fibers show a characteristic alignment in the shape of an arch. They are initially perpendicular at the articular surface in the deep zone; they change direction in the intermediate zone to become parallel at the surface in the superficial zone. This structure, also with aggregates of proteoglycans within the fibers, provides particular biomechanical properties: rigidity and resistance to shearing forces. The proteoglycans make up a veritable hydrophilic gel that takes up considerable volume in relation to its carbohydrate content, and water thus constitutes 70–75% of the liquid weight of the adult articular cartilage [DOR 15].

Proteins are present in the cartilage and their density and secretion vary according to the zone. In the superficial zone, initially clusterin, proteoglycan-4 (PRG4) and Del-1 are found. In the intermediary zone, it is "Cartilage Intermediate Layer Protein" (CILP). In the intermediate and deep zones, "Cartilage Oligomeric Matrix Proteins" (COMP) are found.

In histology, there are three types of cartilage:

– hyaline cartilage (type II collagen); the cartilage's ability to become saturated with water is limited by the collagen fiber network. The specific collagen of this cartilage is type II, made up of three twisted chains. This collagen II alone constitutes 95% of the collagen in normal cartilage (see Figure 3.2);

– elastic fiber (type II collagen and elastic fibers);

– fibrous cartilage (type I and type II collagen).

Hyaline cartilage is the most common in the body: growth cartilage, articular cartilage, the walls of the large respiratory pathways and ventral edge of the ribs. It has the following composition (Table 3.1):

Matrix	Water	70–80% of its weight
	GAG	Chondroitin S and keratan S
	Proteoglycans	Agregans
	Other proteins	Chondronectins
	Adherence proteins attaching themselves specifically to the GAG and type II collagen	
Fibers	Type II collagen	90% of the fibers
	Absence of elastic fibers	

Table 3.1. *Composition of hyaline cartilage*

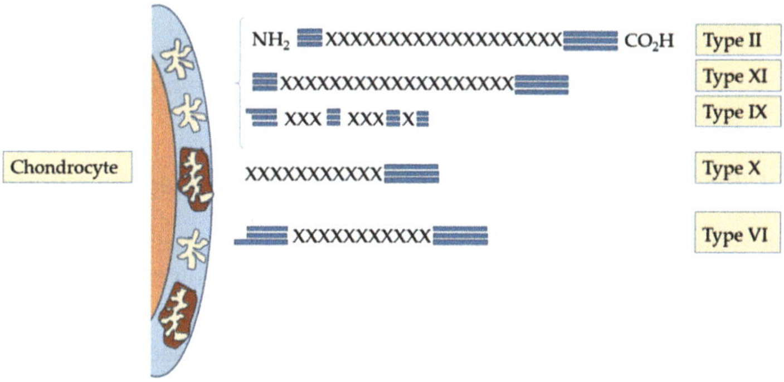

Figure 3.2. *Different types of collagen according to Bernard Mazières (CHU in Toulouse, France). For a color version of the figure, see www.iste.co.uk/andre/printing3.zip*

Beside hyaline cartilage is the perichondrium, which covers the fringes of all hyaline cartilage in the adult body except for articular cartilages. Owing to its vascularization, it plays a role in growth, nutrition and cartilaginous repair. It is made up of:

– an external fibrous layer: fibroblasts and type I collagen;

– an internal chondrogenic layer: mesenchymal cells capable of becoming chondroblasts then chondrocytes.

The nutritional elements reach the cartilage via molecular diffusion through tissues from perichondrial vessels. Concerning the nutrition of articular hyaline cartilage (absence of perichondrium), this takes place via diffusion through the adjacent articular synovial fluid. In rarer cases, this can be ensured by the vessels of a sub-chondral bone.

Figure 3.3 represents the principal elements of this cartilage, highlighting effective heterogeneity.

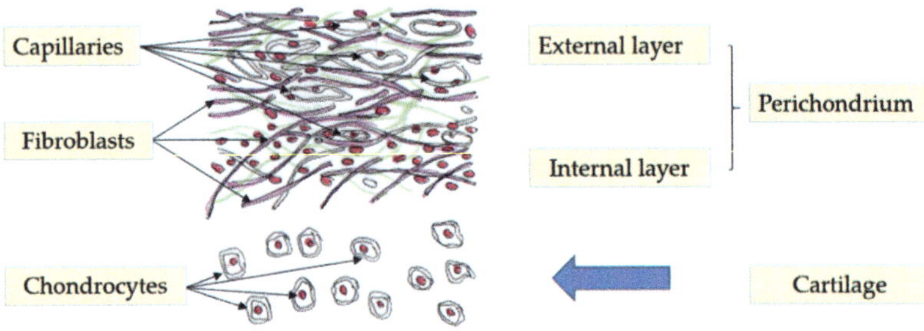

Figure 3.3. *Heterogeneities in cartilage. For a color version of the figure, see www.iste.co.uk/andre/printing3.zip*

3.2.1.2. *Difficulties to consider when conceiving a cartilaginous surface*

The advantage of cartilaginous tissue is that it presents neither vessels nor nerves and, moreover, it is made up of only one type of cell. However, by virtue of these characteristics, cartilaginous injuries cannot easily heal in a spontaneous way; the repair is made up of fibrocartilage: a cartilaginous scar tissue that does not present the properties of resistance to shearing and pressure that normal tissues can present [PER 06]. As a classical consequence, this leads to tissue degeneration and arthritis. However, cartilage, a seemingly simple tissue, thus has a very heterogeneous composition and texture that vary as a function of the tissue depth. This shows significant complexity, despite being a non-vascularized and non-innervated tissue. Furthermore, given the types of cartilage, articular cartilage presents the particularity of not being near a perichondrium to ensure its growth and nutrition. It seems more obvious to concentrate on this kind of cartilage not out of concern for complexity, but to come closer to reality and the possible simplicity of bio-printing.

From the surface to deep down, articular cartilage is morphologically divided into four layers:

– superficial or tangential layer (the collagen fibers being parallel to the surface), around 3% of the thickness;

– middle or transitional layer (≈ 5%);

– deep, radiated layer (with vertical fibers), the thickest;

– calcified layer (2–3%).

These layers correspond to variable contents of the matrix components and to a different organization of the collagen fibers in each layer. This organization leads us to think that the collagen fibers are organized in arcs whose ends are planted in the calcified layer and whose reflection is made in the middle and superficial zones, ensuring a true "framework" for the cartilaginous tissue.

3.2.2. *Cartilaginous defects and treatments*

The cartilaginous defects that can be found are osteoarthritis, age-related wear-and-tear and lesions. Indeed, a cartilaginous injury leads to a loss of the structure and function of the tissue affected by a degenerative disease. The current treatment is surgery but this is costly, invasive and complex. It leads to an increase in the morbidity rate.

3.2.2.1. *Economic considerations and constraints [DEN 17, GEL 16]*

For information on the approximate costs of orthopedic surgery in Germany, the only data accessible on the Internet (including the costs of hospitalization, surgery, anesthesia, pain management, laboratory analyses and clinical examinations) are the following:

– for a total knee prosthesis: €18,800;

– for a cartilage graft, sample, culture and transplant: € 24,400.

These costs are even more elevated when these operations do not ensure total cartilage repair. The cartilage graft slows the need for a surgical operation, but does not exclude it, hence the orientation of research towards more sustainable solutions. Thus, the 3D printing of cartilage also implying the sample, culture and transplant will obviously present prices much higher than those above, the goal being a total repair of the cartilage following a single operation. The hypothesis could be proposed that, over the patient's entire life, all of the operations would probably turn out to be less expensive via cartilage bio-printing. However, in the current phase, it is still difficult to give a robust opinion. However, these figures indicate the economic criteria to be taken into consideration, so that cartilage bio-printing can leave research laboratories.

Cartilage tissue engineering has existed for at least 20 years [PER 06]. The current strategies do not allow new cartilaginous tissue to be manufactured.

Moreover, they imply removing healthy cartilaginous tissues from around the damaged tissue in order to allow an implant. Yet, this operation leads to the degeneration of the implanted tissue (addition of necrosis). The challenge is thus being able to insert a tissue structure without removing the healthy part of the cartilage with an aim to ensure tissue regeneration, also as a function of the thickness to be implanted. This is why attention is being turned to bio-printing, which represents a very significant promise explaining the research underway.

3.2.3. *Cartilage bio-printing*

The process is the object of Figure 3.4, which revisits the 3D bio-printing methodologies presented in the preceding chapter.

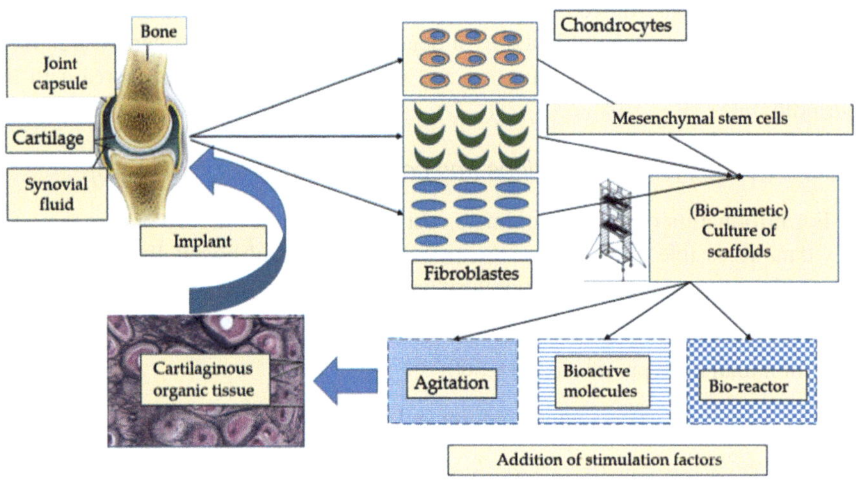

Figure 3.4. *Cartilage bio-printing (according to [BHA 15]). For a color version of the figure, see www.iste.co.uk/andre/printing3.zip*

3.2.3.1. *Bio-printing techniques for osteoarticular tissues*

Jet ink and pneumatic extrusion printers are most commonly used for printing cells contained in a hydrogel. Hydrogels are reticulated networks insoluble in water capable of inflating in it, and they provide advantages in tissue engineering as cell supports for the creation of tissue. Biocompatible hydrotissues have the ability to induce a phase change from a liquid to a semi-solid/solid through reticulation, which provides them with greater potential for 3D bio-printing (see preceding chapter; see also [ABB 16, ARM 16, CUI 12, DI 15, KLE 09, KUN 16, MAR 15, XU 09]).

3.2.3.1.1. Cell sources used

– Chondrocytes: chondrocytes are the most obvious choice for bio-printing cartilaginous tissues; the problem is that they are unstable in a single-layer culture and rare in tissues. Ideally, costal and nasal chondrocytes allow the formation of cartilage in a more significant way.

– MSC "mesenchymal stem cells": these are present in numerous tissues (bone marrow, adipose tissues) and able to differentiate thanks to TGF-ß1, for example.

– Fibroblasts.

It should be noted that in cartilage, 1 to 100 million cells are present per cubic centimeter. In cartilage engineering, when we are interested in cells, several criteria must be studied:

– the ability to have cells to print;

– the proliferative capacity;

– chondrogenesis;

– the ability to form cartilage.

We should bear in mind that the relative performance of *in vitro* cells cannot be linked to the *in vivo* performance, hence the need to study *in vivo* cellular behavior. The list of different types of cells and their principal characteristics justifying – or not – their use in cartilage engineering is presented below [DOR 15]:

– Stem cells:

- Human embryonic stem cells,

  - risk of embryo destruction during sampling;

  - risk of inducing a tumor after transplant (tumorigenicity);

  - immune rejection of the organism.

- Pluripotent stem cells,

  - ethical problem vis-à-vis gathering from a human embryo;

  - risk of tumor formation and risk of rejection.

- Mesenchymal stem cells (MSC),

  - absence of the risk of rejection because taken directly from the patient in question;

  - absence of tumor formation;

- no embryo sampling;

- ability to differentiate;

- come from human tissues (like bone marrow or adipose tissues).

– Chondrocytes:

- absence of the risk of rejection,

- better cartilage manufacturing properties than MSC,

- very limited sampling sources (limiting factor due to required *ex vivo* expansion).

3.2.3.1.2. Growth and additive factors

Table 3.2, taken from Bhardwaj, Devi, and Mandal [BHA 15], presents some growth factors that can be used to aid cellular growth. Indeed, these substances are essential for cellular proliferation. The table presents not only the growth factors that have already been tested in studies, but also the matrix in which they were inserted. This is an expert approach using substances whose effects in other biological situations are already known thanks to experimentation. The subject is thus not closed.

Growth factor	Matrix	Desired goal
TGF-b1	Poly caprolactone nanofiber	Increased chondrogenesis
TGF-b3	NiPAAm-based hydrogel	Induction of chondrogenesis
TGF-b3	Alginates	Induction of chondrogenesis
CDM P-2 and TGF-b1	Myoblastic cells	Synergies of the two factors on chondrogenesis
IGF-1 and TGF-b1	PLGA	Synergetic effect on collagen and expression of aggrecan
IGF-1, IL-4, TGF-b1, PGDF	PGA structure	Effect on cartilage production
TGF-b1, BM P-7	Chitosan-based hydrogel	Synergies of the two factors on chondrogenesis
TGF-b1, BM P-2	Stem cells from bone marrow	Regulation of chondrogenesis
IGF1, TGF-b1	PGA structure	Hydrodynamic qualities of the bio-construct

Table 3.2. *Growth factors in cartilage bio-printing*

3.2.3.1.3. Extracellular matrix

Several components of the extracellular matrix exist and they are mentioned above [BHA 15, RAD 17, SHI 16, SUN 16]. The goal is to mimic reality as much as possible. Hydrogels are interesting because they allow the presence of a large volume of water, favorable to cell growth (but other supports are also used, as there is a large quantity of polymers widely used in tissue engineering and more particularly cartilage engineering: gelatin, hyaluronic acid, collagen, chitosan, alginate, polyurethane, poly-caprolactone, etc. (see [ABB 16, KUN 16, PER 06, SHI 17, DE 15, EYR 07])). The problem of hydrogels is that their 3D structure is difficult to support, hence the search, for Yue *et al.* [YU 16], for structureless bio-printing with such a solution, the cells are extracted from animals; the process involves reating in a small tube made of alginate, a few tenths of a millimeter long, injecting cartilage cells into the tube and letting them develop for a week so that they adhere to one another. As the cells do not adhere to the alginate, the researchers obtain “strands” of cartilage. After 12 hours, the assembly structures itself and the cartilage is “raised” for 4 weeks. It is structureless bio-printing that leads to better results than those using a scaffold, but still inferior to what is done by nature. This is why synthetic materials are called upon, such as poly-glycolic acid, poly-caprolactone, methacrylates and hydroxyapatite. Hyaluronic acid is a compound naturally found in cartilage, hence used in bio-printed matrices to ensure repair, morphogenesis and matrix organization.

3.2.3.2. *Analysis concerning the different materials used and their utility*

The studies gathered in Table 3.3 tend to show that cells could modify their function and morphology as a function of the extracellular matrix into which they are integrated (see [BHA 15]). It therefore seems essential to provide the adequate structure to allow cell survival and development. The currently used materials and structures have not succeeded in perfectly reconstituting the natural cartilaginous structure, even if progress has been made. Indeed, 3D printing technologies do not yet allow spatially-oriented living materials (organized structure in the deep cartilage) to be printed and, in the assays performed on humans, this return of a random deposition to an organization has not (yet) been produced with a yield at the desired level.

Another approach is based on the use of a decellularized extracellular matrix; the cellular viability is good and the structure seems to imitate reality well. Moreover, reflection is based on the fact that there is no need to use a 3D structure: bio-printing of single cells would suffice hydrogel with a plasma, alginate and stem cell base.

Polymer	Cellular origin	Goals
Chitosan/silk fibers	Chondrocytes and mesenchymal stem cells	Effects of the chemical environment on chondrogenesis
Poly-caprolactone	Mesenchymal stem cells	Comparison between the silk fibers and cell clusters on chondrogenesis
PLGA	Bone marrow stem cells	Process of incorporating stem cells
Chitosan/polyester structures	Chondrocytes	Increase in ECM production
PLGA/collagen	Chondrocytes	Effect of the design
Hyaluronic acid	Mesenchymal stem cells	Effect of reticulation on chondrogenesis
Gelatin	Joint chondrocytes	Reticulated gel
Chitosan	Chondrocytes	Effect of the ultrasounds on chondrogenesis
Chitosan/poly-caprolactone	Chondrocytes	Effect of the mixture on differentiation
Chitosan and hyaluronic acid	Chondrocytes	Increase in cartilage ECM and incorporation of the hyaluronic acid

Table 3.3. *Matrix-living environment interactions in cartilage bio-printing (ECM: Extracellular matrix)*

3.2.3.2.1. Origin of the cells to be printed and their preparation

In order to illustrate the origin of the cells used for cartilage bio-printing and how they are cultivated and treated, the study by Kundu *et al.* [KUN 16] was used as an example. First, the human cartilage from the nasal septum (middle partition separating the nasal cavities) is surgically extracted. The cartilaginous fragment is placed in the presence of collagenase at 37°C for 12 to 18 hours. The cellular suspension obtained is filtered, centrifuged and cleaned. The collected cells are placed in suspension with antibiotics (penicillin and streptomycin) in humidified air containing 5% v/v of CO_2 at 37°C. The cells are then flattened at a rate of 10^6 cells per square centimeter on a cellular culture container. The cells will be cultivated (according to the monolayer culture method). The chondrocytes are used during the third place change, that is, 9–10 days after cultivation. For 3D printing, the chondrocytes are detached thanks to 0.05% trypsin contained in 0.53 mM of EDTA for 5 minutes at 37°C. The cells are recovered by centrifugation. The suspension is mixed with a sodium alginate solution. The deposited mixtures of cellular suspension within the alginate are reticulated with a solution of 100 mM $CaCl_2$ and 145 mM of NaCl. The constructions are then washed. Next, the cells are cultivated. Following the culture, a two-week series of treatments to once again purify and sterilize takes place. Cartilage bio-printing thus plays no role in emergency medicine applications.

3.2.4. *Primary results*

The current studies on cartilage are largely based on the study of biomaterials in order to come as close as possible to reality. But even if this is reasonable, is this the right approach? Currently, different hydrogels, cells of different origins, different types of bio-printers and different growth factors are being tested, the goal being to ensure maximized cellular viability and cartilage reformation through experimentation. According to the referenced works, it is possible to reform cartilage, but not yet in a complete way. In the studies cited, chondrocytes were used. From a clinical and medical applications standpoint, this poses certain questions concerning the rarity of these cells in the human body.

In these conditions, would it not be wiser to move towards mesenchymal stem cells that could differentiate through growth factors, allowing differentiation in chondrocytes? One of the downsides, also of chondrocytes, is their unstable phenotype. Once these cells are isolated from their matrix of origin, these cells tend to differentiate into cells presenting the characteristics of fibroblasts, which no longer produce collagen and proteoglycans, the essential and specific components of cartilage. This cellular behavior implies inserting chondrocytes into a 3D matrix similar to natural cartilage in order to maintain the native phenotype of the cells and thus the production of the right components.

Concerning the mechanical and rheological properties of the structures created, the results are heterogeneous. For example, the study based on hydrogels with a methacrylated polyHPMA-lac-PEG triblock copolymer base showed insufficient rheological properties, while the study using PEGDMA presented properties (water content, swell ratio and compressive modulus) near those found in native human cartilage. Performing an implant corresponding to reality from a rheological and mechanical standpoint is thus possible, so this is something to (absolutely) consider during studies. A distinction will be foreseeable between a pressure constraint, a shearing constraint and a wear-and-tear (fatigue) constraint. For the moment, studies are focusing only on the resistance to pressure.

A question also poses itself concerning the heterogeneous structure of the cartilage as a function of the zone: must cartilaginous tissue bio-printing be adapted according to the zone? This is a point that goes unmentioned in publications. Can a structure be adapted to the implantation area and reform itself correctly according to the depth of the cartilage? Or can the focus be turned towards cartilage with less demanding properties [HAN 16]?

Studies [MAR 15, ARM 16, SHI 17] have been interested in the effect of the porosity of the implant formed. Facing a lack of nutrient and oxygen diffusion impacting cellular growth, the idea of printing both nano- and microfibers has emerged. This would be an acceptable compromise to ensure good porosity and to guarantee the structure's mechanical strength properties.

Another study by Xu *et al.* [XU 13] focused on the combination of bio-printing with electrospinning (see Volume 2); this hybridization led to the same results on cellular growth with, in addition, improved physical properties. These two works show the importance of the process: studies on the bio-printing technique can thus also be used to improve the conception of the structure (insofar as one is used).

Finally, practically no studies cited analyzed the possible behavior of the bio-printed implant in the long term (maximum 8 weeks, except for Lian *et al.* [LIA 15], for work on animals and Haleem *et al.* [HAL 10], it seems, on humans). In the long term, could the evolution be modified and, if so, how? When it is known that cartilage is a living tissue within which there is both digestion of the matrix and renewal, it would be interesting to see if renewal over a long period of implantation can be complete. Moreover, the time that would be needed before implantation so that the bio-printed structure is viable and possibly implantable is not mentioned. The implant production time will not be the same according to a partial or total repair. Can this time be a barrier to the use of bio-printing in medicine? Is it already favorable for partial reconstruction?

According to Mok *et al.* [MOK 16], there are still a number of obstacles before healthy cartilage can be repaired for reasons mentioned in this section 3.1. Moreover, the current natural and/or synthetic materials still would not be able to mimic the complexity of the tissues and their functionalities. New investigations are therefore strongly desired to break down the present barriers, as well as to examine whether or not alternative methods can be proposed [KEL 16]. Keller *et al.* have shown that by combining stem cells and growth factors, it is possible to repair the cartilage (without bio-printing) as well as the damaged bone. "This new therapeutic approach leaves hope for a revolution in the treatment of osteoarthritis, the most widespread inflammatory disease worldwide, with more than 250 million affected. The technique updated by these researchers uses a new type of 3D implant. The first layer of this implant contains membranes with nano-reservoirs containing growth factors that will allow bone regeneration. Next comes a second layer made up of a hydrogel containing stem cells taken from the patient's bone marrow, which will regenerate the cartilage" [TRE 16]. This initial search will then have to examine the aspects of regeneration, the tissue's mechanical resistance, and its robustness over time, cellular survival and the stability of functionality.

As such, attention must be given to the media impact, progress not meaning achieving the result; the path is not marked and it is not yet known if this will allow the target to be reached. Thus, the major challenge of tissue engineering is indeed ensuring that the implanted structures "live" long enough to properly integrate to the patient's body. According to SantéLog [SAN 16b]:

– "Researchers manage to implant an ear structure and to obtain signs of vascularization one to two months after the graft;

– Printed muscular tissue implanted in a rat becomes a sufficiently solid muscle in two weeks, assuming its function and presenting vascularization allowing nerve formation to be induced".

According to Marchand [MAR 16], to test the methodology, cartilage implants have been placed under the skin of mice and rats: two months later, the ears implanted in the mice had kept their shape with the appearance of cartilaginous tissue. We were almost there.

In conclusion, experimental research on cartilage bio-printing is primarily oriented towards the study of biomaterials and their interactions. The cartilage, through its non-vascularization, is an interesting tissue for study, but it (still) raises questions that could possibly prove problematic for the total regeneration of human cartilage. However, is it not normal for a goal not to be reached when still very few works have been published and the (apparent) emergence of the field dates back to 2011. The roadmap of cartilage bio-printing is presented in the two figures below (5 and 6), so nothing can be done but to keep going!

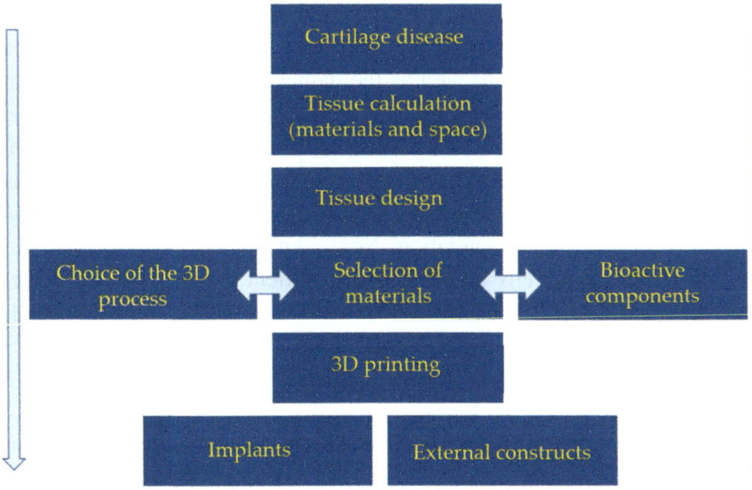

Figure 3.5. *Roadmap of cartilage bio-printing*

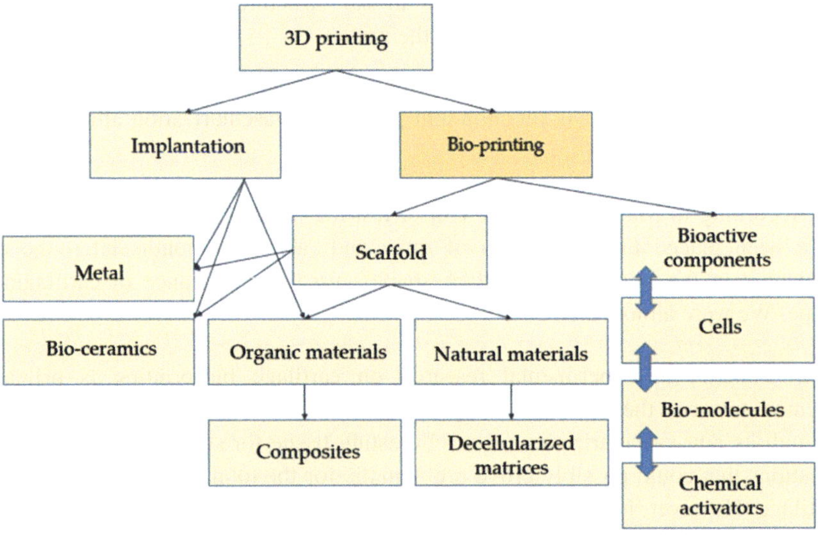

Figure 3.6. *Materials and cartilage bio-printing (according to [MOK 16]). For a color version of the figure, see www.iste.co.uk/andre/printing3.zip*

Hyaline cartilage: this corresponds to the following cartilages: cartilage of growth, articular cartilage, wall of the large respiratory tracts and ventral end of the ribs.

Chondrocytes: chondrocytes are the cells that make up the cartilage. They contribute to collagen and proteoglycan formation.

Chondroblasts: cells that will provide chondrocytes through maturation.

Electrospinning: process allowing non-woven membranes to be created thanks to the use of a polymer or melted polymer solution.

Glycosaminoglycan (GAG): long linear saccharide polymer chain present in numerous tissues in the human body.

Synovial fluid: synovial fluid is produced by the synovial membrane. It serves as a lubricant for the joints to limit friction, absorb shocks, allow nutrient diffusion for cartilage cells and eliminate waste.

Proteoglycan: combination of a protein and a GAG.

Mesenchymal stem cells (MSC): stem cells from the mesoderm present in various tissues of the adult body such as the bone marrow or adipose tissue. They are also present in the blood of the umbilical cord.

Box 3.1. *Glossary*

3.3. Skin bio-printing

"There is great complexity in the skin, it is an organ made up of several layers of tissue. With bio-printing, which consists in depositing droplets of biological ink that will form successive layers through superposition, it will be possible to constitute a three-dimensional tissue. Today, after a 15-year application development phase, we are able to print the epidermis using human cells" (Fabien Guillemot CEO of Poietis [GUI 15]). Indeed, according to Sciences et Avenir [SCI 16], if the goal, for Poietis, is "going to the clinic" to be treated for tissue repair, the company first founds its development on the enormous potential represented by tests in cosmetic and pharmaceutical research. "This is why we first worked on the skin. It is a real opportunity for development". Guillemot is not the only one getting involved in this complex path and, beyond the rigor of its production, partially brings his company to life (following the model of his National and International colleagues), in a dialectic of promise, which finds an effective reality in the application of bio-printing in cosmetic and pharmaceutical tests [GRA 12, REN 15].

According to Lipson and Kurman [LIP 14] and Lipson [LIP 13], until now, Nature has remained in the lead for performance compared with humans and computers concerning the optimal choice and possession of different stem cells to create a complex tissue! However, significant works are developing. Before dealing with what has already been published (as expressed by the number of publications in Figure 3.7), it seemed necessary to the author to recall what, in a nutshell, the skin is.

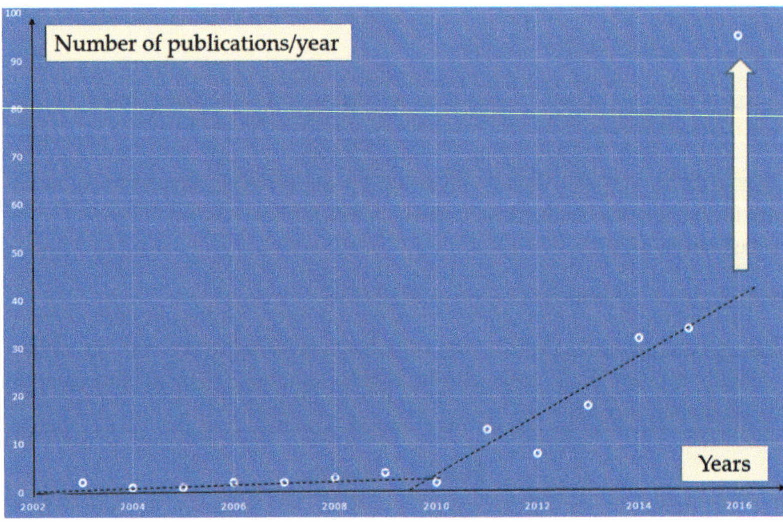

Figure 3.7. *Publications with the two keywords: Skin and 3D Printing*

As in the preceding case, the subject is very recent in terms of interest for the scientific community (linear development starting in 2010 and extreme acceleration in 2016; the subject is mushrooming).

3.3.1. *General remarks on skin*

The human skin is a complex organ divided into three layers: the epidermis, the dermis and the hypodermis. "The epidermis, in contact with the outside environment, is a multilayer epithelium; it is non-vascularized, keratinized and squamous. It is primarily composed of layers of keratinocytes, which represent 90 to 95% of the epidermic cells; it also contains melanocytes, Langerhans cells and Merkel cells. In this tissue, the extracellular compartment is reduced to the intercellular cement. The dermis, on the other hand, is made up of a small number of cells and a well-developed extracellular matrix. Fibroblasts make up the greater part of the cells in the dermis. The dermis is crossed by a ramified network of blood vessels, lymphatic vessels and nerves [a network] that connects the skin to the rest of the body. The hypodermis is the deepest part of the skin and also the thickest" [BIO 11].

These three zones defining gradients of living matter are associated with highly developed vascularization that largely exceeds the individual local nutritional needs. Blood circulation plays a role in thermoregulation, healing, immune responses, controlling the blood pressure and various exchanges: a variety of transfers in both directions (sweat and second entry of contaminants after the lung). The cutaneous nervous system is present in the three cutaneous compartments, hypodermis, dermis and epidermis (except the corneous layer).

The epidermis is made up of several cell layers (Bouschbacher [BOU 07] and Figure 3.8):

– the basal layer is made up of keratinocytes with large cylindrical or cubic nuclei, in division;

– the spinous layer is made up of 5 to 15 layers of sizeable and polygonal keratinocytes;

– the grainy layer includes 1–3 layers of flat, spindle-shaped keratinocytes deposited parallel to the cutaneous surface. They contain grains of keratohyalin (basophil granulations), that is, an assembly of proteins rich in histidine and keratin filaments;

– the corneous layer is the most superficial layer of the epidermis. It is made up of 5–10 layers of flat keratinocytes with no nucleus, called corneocytes.

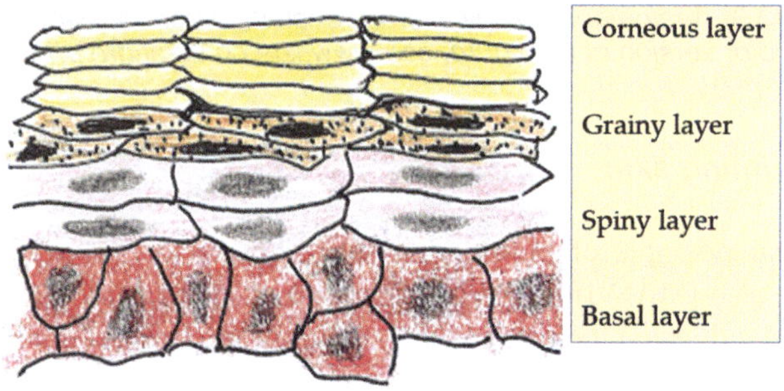

Figure 3.8. *Epidermis. For a color version of the figure, see www.iste.co.uk/andre/printing3.zip*

In case of injuries, "scarring is the result of a set of phenomena with variable intensities according to the degree and type of aggression, the degree of the tissue damage, that takes place, more or less harmoniously, in order to restitute the initial tissue integrity. Moreover, there are individual factors, some unknown and unforeseeable, that are involved in the progress of these phenomena" [VÉR 06]. As this very reduced description shows, this is a particularly complex system (with bio-printing applications with an obvious range), represented for normal skin in Figure 3.9, much more complex than only the epidermis, skin that we will have to attempt to reconstitute via bio-printing.

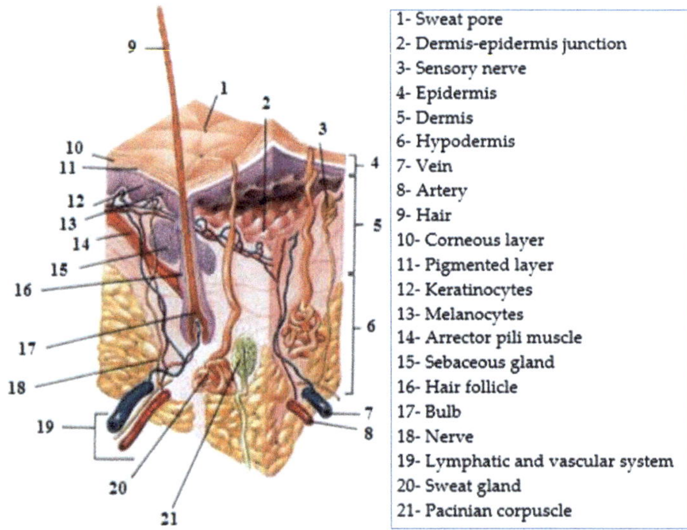

Figure 3.9. *Anatomopathological cross-section of normal skin [FUT 17]. For a color version of the figure, see www.iste.co.uk/andre/printing3.zip*

3.3.2. *Bio-printing skin*

The general principle of bio-printing skin is presented below in Figure 3.10 ([NG 16], see also [MAN 16]). The cells are extracted from the patient via biopsy, then they are cultivated before creating a bio-printed object that can later be transplanted. It is normally applicable to other issues (see preceding chapter). Figure 3.11 from Poietis presents an example of bio-printing keratinocytes on a decellularized dermis.

In a repair, it is important to understand how the stem cells that contribute to the healing of cutaneous injuries participate in this operation, migrate, proliferate and differentiate to repair a tissue after damage. "A defect in the series of cellular and molecular events that is activated to repair damage and restore the integrity of the skin can lead to improper scarring and cause acute and chronic wounds" (Blapain quoted by Huet and Hugot [HUE 17]). According to Aragona *et al.* [ARA 17], "the genetic signature of the cells that actively divide and those that migrate to repair the injury" suggest that cells protect the stem cells from infections and mechanical stress to allow a healing process", comments Sophie Dekoninck, one of the primary authors of this study (see also [BLA 10, BLA 04]). This research could demonstrate the ability of stem cells to regenerate a tissue depending on their ability to proliferate rather than their cellular origin. Considering this information, it will be useful to examine how it could be taken into consideration in bio-printing skin.

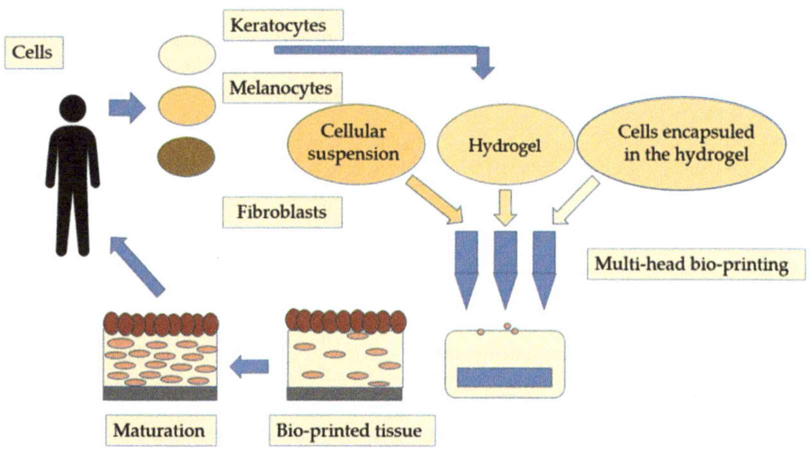

Figure 3.10. *Principle of bio-printing (applied to skin). For a color version of the figure, see www.iste.co.uk/andre/printing3.zip*

In a synthesis work, Gudapati, Dey, and Ozbolat [GUD 16] present different methods of bio-printing applied to skin (based on the use of droplets). Figure 3.12, inspired by their publication, highlights the potential of the different additive manufacturing techniques to create bio-tissues like skin.

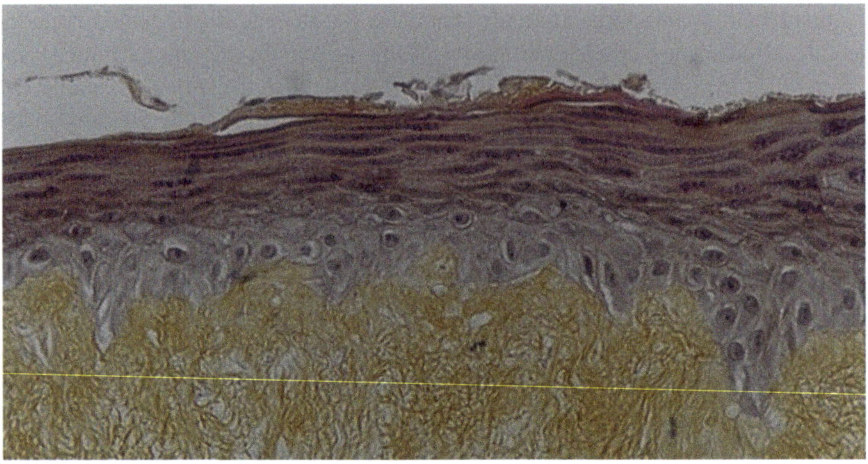

Figure 3.11. *Bio-print of keratinocytes (image generously provided by Poietis) – see Figure 3.8 for comparison with reality. For a color version of the figure, see www.iste.co.uk/andre/printing3.zip*

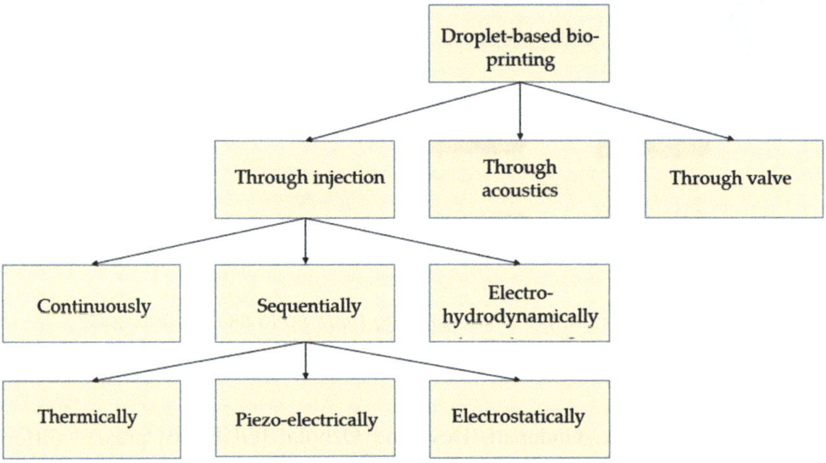

Figure 3.12. *Bio-printing methods usable for bio-printing skin*

Figure 3.13 from Kaelin [KAE 13] illustrates the desired goal: starting with cells to produce artificial skin that can be transplanted onto burn patients.

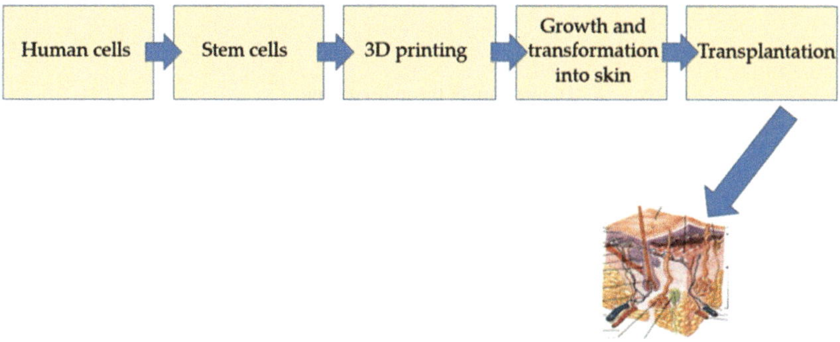

Figure 3.13. *Concept of manufacturing artificial skin. For a color version of the figure, see www.iste.co.uk/andre/printing3.zip*

For Wake Forest [WAK 15], aiming at military applications, the associated principle of 3D printing is defined by a spatial measurement of the area to be treated, cellular printing allows the burned zone to be covered and, at a final stage, an artificial epithelial cover is placed to protect the wound (see Figure 3.14). These works are in progress (see also [CHN 12]). In the case of burns, the different layers of the skin are damaged, with a loss of cutaneous cells, fibroblasts and keratinocytes. In order to allow skin to be bio-printed, one ink contains cells mixed with fibrinogen and collagen and the other contains thrombin, which promotes adhesion between layers. A digitization of the wound helps identify the shape to be printed, an object made up in the following way: fibroblasts (deeper layers of the new skin), then karatinocytes to create superficial layers of the epidermis. Similar activities exist in Japan [MIN 16].

According to Rivière [RIV 10], tomorrow, a printer placed atop a serious burn patient's wounds may then be able to deposit a fine layer of skin matrix to allow complete and rapid scarring, which will greatly minimize the risks of infection and recourse to grafts, particularly inesthetic and with the risk of rejection. However, according to Guillemot [GUI 15], "the obstacle that remains is allowing tissue vascularization. In other words, gathering cells that will form small blood vessels present in the skin. It will only be possible to proceed to the graft when the skin is successfully vascularized...".

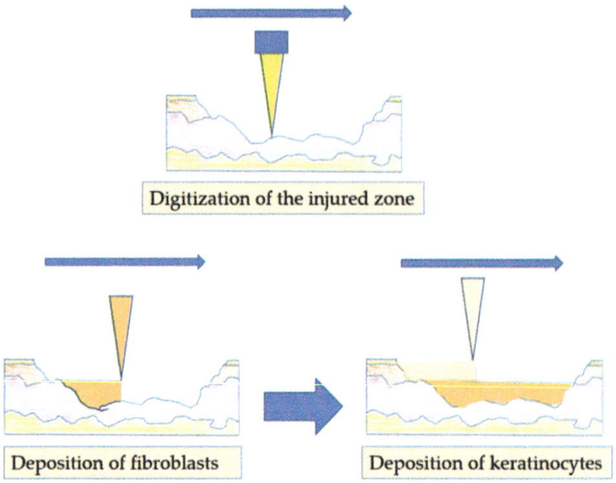

Figure 3.14. *Repairing a burn (drawing inspired by [WAK 15]). For a color version of the figure, see www.iste.co.uk/andre/printing3.zip*

In late 2015, the University of Lyon might have provided a proof of concept with the integration of viable cells and the generation of an extracellular matrix, so also possibly, before too long, 3D printed skin available for a graft. The 3D printer's ink cartridge releases a culture of living cells that, during printing, will reconstruct the different layers of the dermis. During this initial stage of the proof of concept related by Pourchet [POU 15] and Barnouin [BAR 16a]:

– cellular tissue was successfully printed and seemed to behave similarly to human skin;

– the "printed" cells survived and their growth continued after the printing process;

– the cells ended up generating a living dermal tissue;

– after 3 weeks, the dermis and epidermis seemed reconstructed, took on pigmentation and formed a tissue similar to human skin.

"These first promising results create hope for easier autologous grafts – those from a culture of the patient's cells – especially for serious burn victims, and they could confirm the potential of 3D printing in the field of regenerative medicine" [SAN 16a, ALE 16].

Very recently, Magnier [MAG 16] points out (see paper by Takagi *et al.* [TAK 16]) that the team successfully transplanted skin cultivated in a laboratory onto mice. "The researchers removed cells from the gums of a mouse and chemically transformed them into induced pluripotent stem cells (iPSC). This technique, developed in 2006, allows the iPSC to develop into nearly any organ, and thus possibly the skin. Once transformed into artificial mouse skin, the scientists grafted the created cells onto mice presenting a deficient immune system, allowing the new skin to develop more easily. The artificial skin still does not have the appearance of 'normal' skin, but was correctly developed based on natural tissue, so well that hair appeared on it and that the skin perspired normally. Above all else, this is one of the most important points during a transplant, and the new skin was able to connect to the nearby nerves and muscles. There are various applications of this new skin: replacing animals in tests conducted with cosmetics, allowing ill or burned patients to receive artificial skin grafts, etc.". These methodologies compete with classical techniques (e.g. skin banks) or with other emerging subjects like the use of fish (tilapia) skin that contains collagen proteins essential for healing [THI 17].

3.3.3. *Conclusion*

The CellMistMC and SkinGun technologies developed by RenovaCare are the fruit of nearly a decade of research aiming to find the most efficient means of accessing the regenerative properties of a patient's skin stem cells, and the best way to apply them to come to grips with the most severe wounds. Initial assays show that notable improvements appear after 4 days [EST 16]. Alternative treatment technologies are developing in the field.

Obviously, the complexity of the skin is not accessible to current bio-printing technologies in order to create a "faithful copy" of the healthy tissue that should be implanted. The question, as in the case of bio-printing, is indeed finding the right composition and shape to be put in place so that the bio-printed tissue quickly serves for repair. In any case, it is not yet foreseeable to reintegrate pigmentation or hair into the bio-construct, the priority clearly being to induce vascularization (angiogenesis; [SAR 15, RIC 16]), unless we settle for a slightly more intelligent tissue than current dressings.

3.4. Bone

Bone is among the most studied tissues because it has already been the object of prosthetists' works for years; in fact, it is the emergence of additive manufacturing processes that brought their attention back to the application with generally metallic or ceramic implants [CHA 02], as was shown in volume 1. The aspects of mechanical strength (beyond the biological part) in 3D printing are important (see, e.g. [BAR 15, QUE 16]). In section 3.3, the idea is to examine how exogenous implants can be avoided using bio-printing principles.

Bibliographic analysis, as was done for the other preceding examples, reports a higher number of publications (more than 1,000 compared with some 200 for cartilage and skin). Figure 3.15 presents the evolution of this number over time. As in the two previous cases, development takes place around 2012 after a much slower increase. In addition, as with the other two examples, the acceleration for recent years is notable.

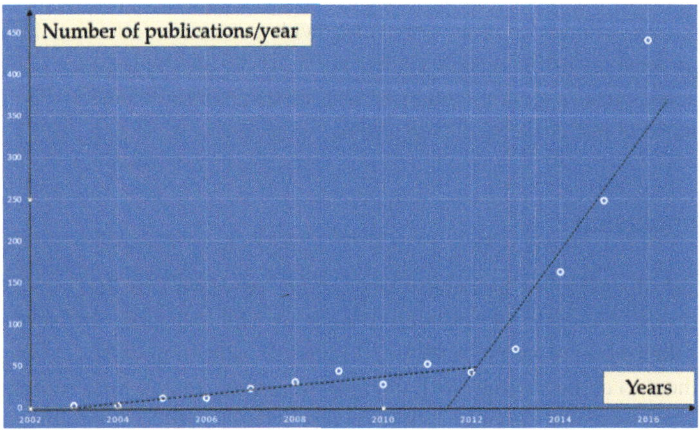

Figure 3.15. *Publications with the two keywords: bone and 3D printing*

3.4.1. *General remarks on the composition of bone*

Bones form the rigid and resistant part of the human skeleton. "A bone is made up of six different kinds of tissues:

1) the periosteum is a fibrous membrane that covers the bone, except for the articulations;

2) the compact bone, uniform and very dense, is made up of cylindrical elementary units called osteons, made up of layers called lamellae, juxtaposed like a roll of paper;

3) the spongy bone resembles a sponge with its bony lamellae defining countless cavities;

4) the articular or hyaline cartilage, which covers the extremities, looks like a rigid but still elastic jelly under a microscope;

5) the bone marrow or red marrow fills all the cavities of the spongy bone, producing 100 to 150 billion red blood cells and 1 to 30 billion white blood cells each day;

6) the yellow marrow, a fatty mass that fills the center of the diaphysis of long bones in adults" [ATL 17].

3.4.1.1. *Chemical composition*

"The living part of the bone contains 1% of proteins that form a matrix, bone cells or osteocytes, collagen fibers, and the marrow cells. It represents a third of the bone's total weight. The mineral part of the bone contains a large proportion of calcium phosphate and a bit of iron, fluoride, and mineral oligo-elements. It represents two third of the bone's weight" [ATL 17].

3.4.1.1.1. The functions of the bone

"The bone ensures four large functions:

1) Support: the skeleton serves as an anchoring point for all of the muscles and soft organs. It supports the weight of the body in every position.

2) Protection: the cranium protects the brain; the thoracic cage protects the heart and lungs; the vertebrae cover the spinal cord.

3) Storage: bone contains 99% of the body's calcium and phosphorous reserves.

4) The formation of red and white blood cells in the red marrow" [LEC 17].

According to the same source, Figure 3.16 presents a typical bone with the six kinds of tissues numbered.

Once again, this is a complex system. However, the success of different kinds of classical prostheses leads us to think that in what professionals do not (yet) know how to copy, Nature will assist, so long as a certain number of precautions are taken. It is these attempts that are presented below.

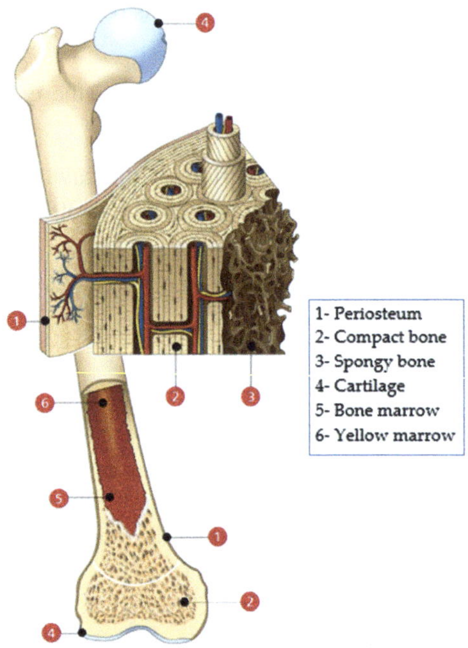

Figure 3.16. *Cross-section of a long bone. For a color version of the figure, see www.iste.co.uk/andre/printing3.zip*

3.4.2. *Bone bio-printing*

As illustrated in Figure 3.16, the bone is complex from both a morphological and composition standpoint, with living elements and chemical support structures. At the same time, the latter component can serve as a support to the bio-construct during manufacturing and for repair (so long as the mechanical quality of the structure is acceptable).

Biological ceramics play an important role among the materials used in the body to repair hard tissues; calcium phosphate ceramics in particular have a chemical composition similar to the mineral framework of bone. "Bioactive, they are considered mineral bone by the bone cells" [OSS 16]. Creating a bone poses a certain number of questions in terms of biocompatibility and mechanical properties (as close as possible to those of natural bones), being easily usable in an operating room at an acceptable cost. Various teams have dedicated themselves to this topic for years, though without yet satisfying all of these constraints at once [GOU 16].

3.4.2.1. *Inert materials and structures*

Most reconstructions are done with inert and porous materials based on hydroxyapatites with recolonizations above 70% according to D. Nimal (Osseomatrix), founder of the Assises de l'AFPR [AFP 16]. The question of the supports' porosity has been studied by Pina, Olivera, and Reis [PIN 15] and by Santoro *et al.* [SAN 15] with measured effects on intercellular communication. However, traditional materials without sufficient flexibility, a formulation based on 90% powdered hydroxyapatite, and an elastomer, poly-caprolactone, already used in the medical field, allow satisfactory mechanical qualities and good cellular recolonization (tested on animals) to be achieved [JAK 16, JAC 16, MAN 16]. Figure 3.17 presents one of the qualities of this material, its flexibility (other materials exist with similar properties: see [SEN 16]).

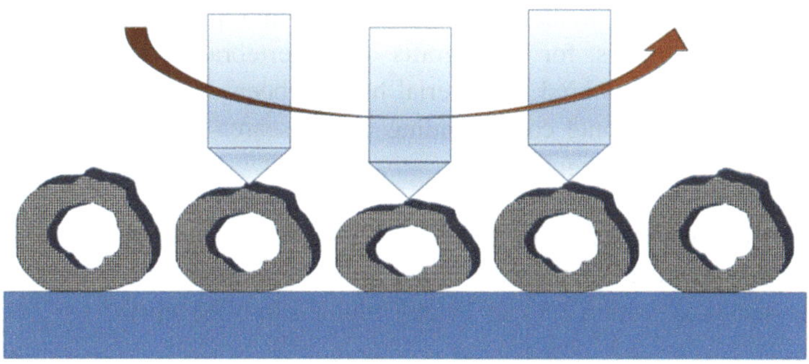

Figure 3.17. *One of the interests of the hydroxyapatite–poly-caprolactone couple: elasticity. For a color version of the figure, see www.iste.co.uk/andre/printing3.zip*

Other compositions have been proposed (see Richards *et al.* [RIC 16] for a general article): hydroxyapatites and gelatin by Wang *et al.* [WAN 16]; acrylates and other polymeric compositions by Visscher *et al.* [VIS 16] and Miao *et al.* [MIA 16]; hydroxyapatite/hydrogel couples by Bendtsen, Quinnell, and Wei [BEN 17]; silk thread reinforcements by Jackson [JAC 16a] and Compaan, Christensen, and Huang [COM 17]; calcium silicates by Pei *et al.* [PEI 16]; etc.

3.4.2.2. *Other elements*

Herberg *et al.* [HER 14] show the interest of using growth factors in the quality of the cellular recolonization. Shim *et al.* [SHI 16] introduce stem cells into the construct with significant performances (system using a hydrogel made up of atelo-collagen and supramolecular hyaluronic acid with the use of two extracellular matrices).

In short, all the possibilities have not yet been explored from an experimental standpoint and it is likely that this game of trial and error will find serendipitous solutions.

3.4.3. *Conclusion*

Llopis-Hernandez *et al.* [LLO 16] use a new technique associated with a polymer (ethyl acrylate): this polymer allows growth factors to be efficient at doses roughly 300 times lower than those classically used. The consequence is a reduced risk of side effects (but also a reduced cost of treatment). According to the authors, this polymer sublimes the action of the growth factors by facilitating a reaction with fibronectin, a protein that attaches itself to the growth factors to allow tissue regeneration, whereas the absorption of growth factors is limited in current, BMP-2-based therapies. For the moment, this material has multiple uses, for example, as a cover for hip prosthetics, for bone grafts or in vertebral column repairs. To the author's knowledge, this kind of material has not been used for bio-printing. It is thus a possible study route corresponding to the search for directly or indirectly active materials of cellular growth.

This piece of information, among others, makes bio-printing technologies competitive with other methods. For bone, it seems that the current techniques briefly presented in section 3.3 may be able to assist in bone repair.

3.5. Bio-printing and cancer

In a bibliography of a few dozen publications, it seems that bio-printing technologies that gather cells have a great potential for applications in cancer research [KNO 15]. Bio-printed cancer models actually represent a significant improvement in relation to two-dimensional models, as they mimic the complexity introduced by the third dimension by facilitating cell–cell and cell–matrix interactions, which are not taken into consideration in two dimensions. Indeed, cancer biology research has shown that tumor development is not reduced to a simple anarchical and uncontrolled multiplication of neoplastic cells, but depends on numerous interactions between these cells and the host's tissue elements. Modification of the extracellular matrix is thus an important step, necessary not only for invasion by tumor cells but also for their proliferation [CHA 02].

These new 3D models are thus capable of imitating the microenvironment of tumors, providing a support for a deeper understanding of cancer, for the search for means of cancer treatment. There are hundreds of known types of cancer and the disease is very complex, even in a single type of cancer, which makes the development of a single treatment particularly complicated. To better understand the genesis and progression of cancer, there is a need for a physiologically pertinent 3D cancer model that closely mimics the *in vivo* tumor microenvironment. Bio-printing is able to offer the possibility of forming models of highly controllable cancerous tissues, allowing better understanding of cancerological phenomena. Currently, two-dimensional models of cancerous tissues are widely used for cancer research, contributing to the basic knowledge of this disease's biology.

Moreover, as has been pointed out, there is a possibility of using a patient's cells to create a bio-printed tissue. This means that it is possible to test medications/drugs on this cell system with appreciable effects (however, over short periods). Personalized medicine consists in treating each patient individually as a function of the genetic and biological specificities of his/her tumor, but also taking into consideration the patient's environment, lifestyle, etc. In this approach allowed by the mastery of bio-printing, it is possible to create tests without an obligation to go to the molecular level. This axis should thus be developed in the future.

The expression of proteins, expression of genes, profiles of proteins, cellular markers, migration, morphology, proliferation, viability and response to medications are many factors that differ between the 2D and 3D cancerological models. Although 2D cultures already offer information concerning the pathogenesis of cancer, it is preferable to expose the cancerous cells to cell–cell and cell–matrix interactions that they would encounter *in vivo* to obtain the most pertinent results possible, physiologically speaking. Thus, studies on cancer using 3D models have already allowed more precise representations of cancerous tissues in terms of the tumor microenvironment and biological behavior with a controlled spatial distribution of cells to be obtained, which is crucial to developing early-stage diagnosis and optimal cancer treatment strategies.

3.5.1. *Examples*

A recent study [ZHA 16] was interested in a 3D printing process to construct a model of *in vitro* cervical tumors in which the HeLa cells were embedded in a gelatin, alginate and fibrinogen hydrogel. This model, created in 3D, was compared with 2D models. The comparison shows notable differences in cellular proliferation, the expression of matrix metallo-proteinase (MMP) protein, and the cells' chemoresistance between the 2D and 3D tumor models. More than 90% of the cellular viability was obtained in the bio-printed 3D model and the cells proliferated at a higher rate than in the 2D culture. The 3D HeLa cells also formed 3D cellular spheroids in contrast to the single-layer organization of the cells formed in the 2D culture. These differences may come from cell–cell and cell–matrix interactions present in 3D culture conditions. The expression of the MMP protein in the HeLa cells was also shown to be greater in the 3D printed models, very likely due to the functionality of the MMP in the degradation of the MEC. The cells in 3D printed constructions also presented greater chemoresistance with treatment using paclitaxel than in the 2D culture. The results of this study using a new 3D cell printing technique to construction *in vitro* tumor model help better characterize the formation, progression and tumor response to cancer treatments (see also [BUR 16]).

3.5.1.1. *Tumor heterogeneity*

The tumor environment presents heterogeneity and complexity of its microenvironment. These particularities can be reproduced through printing using bio-inks containing different cell types, different extracellular matrices and different biomolecules [XU 11, YAN 15]. Heterologous constructions containing tumor cells, endothelial cells and macrophages can be manufactured with a high degree of spatial control via bio-production to reproduce cell–cell interactions. In these constructions, the initial cell density can also be controlled to closely imitate the elevated cell density of a tumor and reproduce the cell–cell markings that are known to play a significant role in the behavior of cancerous cells. Moreover, 3D models printing cancerous cells and blood vessels allow real-time surveillance of the cancer metastasis process, including the invasion of tumor cells. It is thus possible to study a wide range of cancer types by printing different combinations of cancerous cells and surrounding cells to model the evolutions of cancerous metastases in a large range of tissues simply by reformulating the bio-ink. The gradients of biomolecules, which play a significant role in cancerous chemotaxis and metastases, can be generated using 3D printing methods to reveal the molecular mechanisms of biochemical sensing.

3.5.1.2. *Angiogenesis and tumor vascularization*

Huang *et al.* [HUA 14] have proposed a new method to examine cellular behavior and the effect of new drug candidates. In this research, the cancerous cells and the normal cells were sown in 3D printed biomimetic microstructures to study the differences in cellular migration between the cell types. The permeability and weak organization during the formation of vessels are distinctive characteristics of cancerous tumors. These particularities imply a different reaction when cancer medications are administered, hence these models are important considering these criteria.

3.5.1.3. *Fundamental research*

Panagiotakopoulou *et al.* [PAN 16] have studied the migration of cancerous cells through very small meshes (performed through additive manufacturing). This work contributes information about the relation between the environment's mesh and the size of the cells in relation to the interactions that they have with their near environment. This type of laboratory investigation thus participates in better knowledge of the processes of transporting cancerous cells through the body. According to Regnault [REG 17], the development of a tumor exerts a force on neighboring tissues.

In the same logic, Santoro *et al.* [SAN 15] have shown that the pores within the bones have a significant effect on the way in which the cancerous cells operate and propagate. The outside structure of the polymer synthetic bone contains artificial pores that limit the flow of liquids and apply a shearing constraint on the tumor cells. By causing the structure of these 3D printed supports and of the pores to vary, it was possible to change the flow of different liquids as well as the degree of shear stress. The combination of hydrodynamic shearing and deposition of the support creates different levels of protein production.

3.5.2. *Conclusion and perspectives*

A key advantage of the 3D microenvironment in comparison to the traditional 2D cell culture is the ability to obtain more precise and reliable data from the model. Studies using 3D *in vitro* cancer models rather than 2D models show a greater cellular viability, more physiologically pertinent proteinaceous expression profiles, a higher proliferation rate, greater chemoresistance to cancer medications and characteristics of real tumors. High output manufacturing contributes to the better characterization of tumor formation and to progression of and response to cancer treatments.

The absence of a direct correlation between the genetics and physiology of animal and human models currently limits our understanding of the way in which cancer cells behave in humans. Living microarchitectures printed using human cells are more realistic for the creation of disease models. In the interest of developing more efficient cancer treatments, a large quantity of information remains yet to be discovered and these studies can be accelerated using bio-printed cancer models. The current applications of bio-protection for cancer research allow new experimental procedures for the manufacturing of 3D cancer models to be established in order to pursue new discoveries in cancer biology or to test clinical treatments. Other studies on bio-printing will allow the high output manufacturing of 3D cancer models to shed light on the underlying mechanisms of the disease's progression, the study of the behavior of cancer cells, drug screening and the development of efficient clinical treatments.

These modeling systems have the potential to be the experimental bridges to new clinical techniques through which a tumor model specific to the patient can be created *in vitro*. Bio-inks can be generated through the proliferation of cancerous cells removed using a sample of the given tumor or a tumor bank. This would allow *in vitro* medical treatments to be tested by providing information on the most efficient type and dose of medication and by developing a personalized cancer treatment for the patient.

DEFINITION.– HeLa cells: cancerous cell line used in cellular biology and medical research.

DEFINITION.– Matrix metalloproteinases (MMP): Matrix metalloproteinases (MMP) constitute a family of proteases implied in the proteolytic degradation of numerous proteins in the extracellular matrix. These proteases play an important and complex role in several steps of cancer progression.

3.6. General Conclusion

Personalized and precision medicine, which is starting to emerge in practices, is also very promising. For example, genetic therapies could, through an analysis of the patient's genome, allow the development of customized pharmaceuticals adapted to his/her pathology. Success is becoming more and more common in publications; for example, in regenerative medicine, it was possible to recreate a human intestine using stem cells. The organ was created *in vitro* and is functional [WAT 14]. The domain is greatly developing (see, e.g. [MAO 15, HUT 15, MIT 16, QIA 12]) and it could be wise to examine how these two promising futures can be associated. It is true that it is an ambitious proposition to support interdisciplinary projects on unstable bases. However, this is already what is starting to develop in bio-printing with the addition of growth factors in bio-printed tissues.

It will be understood that, alone or supported, bio-printing represents an "immense project" that has some elementary bricks in place (still quite few in number) to show its determination to produce results with public uses in reparative medicine. Only some examples were presented, whereas we could have discussed other organs such as the heart, kidneys or liver. However, in the end, the same issues would have been shown. Even if proofs of concept exist beyond now classical applications of additive manufacturing for prostheses, research work (still) remains yet to be accomplished before seeing bio-printing technologies arrive "to a store near you".

However, as has been shown, this work is still young and it is not unusual for the current infancy, imprecisions and desire to be the first on an essential subject to be clearly visible. This is because these are pioneers that the research system is supporting. Thus, once again, let us wait a few years before passing any judgment.

In any case, the proof of the performance of bio-printing in terms of repair implies a real robustness of methods (very low failure rate, lifetime of the tissues and organs adapted to people's life expectancies), a duration of the "repair" operation adapted to the issue and an "acceptable" cost; in short, taking this position is practiced in the most diverse fields of medicine, in the running with what is being done better for a given end. The situation that aims to use bio-printing for toxicology and cancer studies is, it seems, "easier" in that the time constraints are not the same and the quantities of biological materials to be used are lower. Maybe the increased reliability of this domain should be considered with great interest (see, e.g. [NGU 17, NGU 16])?

Despite certain fears, we are still far from the emergence of groups of bio-hackers and "participatory biology" bio-printing laboratories (see Volume 1). A report quoted by Morin [MOR 14] mentions a parliamentary study on the "Stakes of synthetic biology" performed in 2012 and submitted to the government. It examines the risks: "the worries awakened by these activities result from the accumulated facilities to do synthetic biology" but "scientific and financial obstacles present garage biologists from performing works that could have marvelous ends" (see also [DEL 13]). It is thus necessary to wait. Nevertheless, beyond the aspect of bio-printing technologies' attractiveness, an ethical approach is proposed in the next chapter.

If the emergence of this domain must be allowed and if we want it to be able to fulfill its promises, it certainly must be helped along; however, as Figure 3.18 from Irvine and Venkatraman [IRV 16] reminds us, it is necessary to consider a very large set of parameters to successfully achieve its goal (by initially disregarding interdependences, some of which are known to be critical). This poses questions of scientific knowledge, but also of a successful heuristic approach for the exploration of the complexity (which will be the object of Chapter 5).

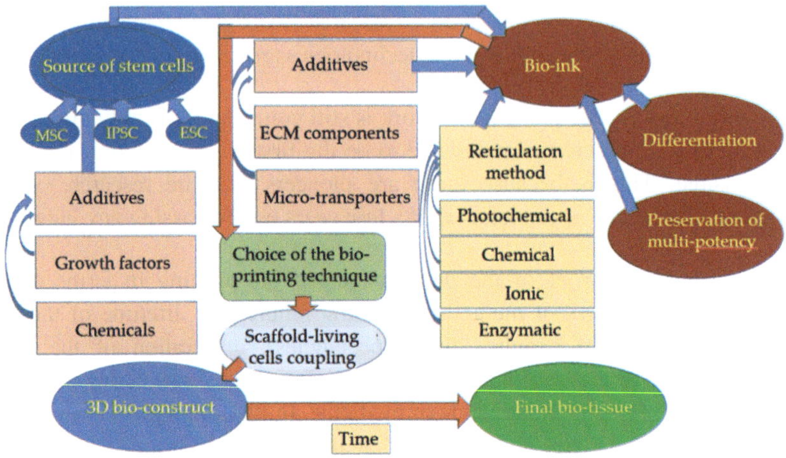

Figure 3.18. *Some elements to consider for bio-printing. For a color version of the figure, see www.iste.co.uk/andre/printing3.zip*

3.7. Bibliography

[ABB 16] ABBADESSA A., MOUSER V.H.M., BLOKZIJL M.M. *et al.*, "A synthetic thermosensitive hydrogel for cartilage bio-printing and its bio-functionalization with polysaccharides", *Biomacromolecules*, vol. 17, no. 6, 2016.

[AFP 16] AFPR, "Assises Européennes de la Fabrication additive", available at: http://www.afpr.asso.fr/les-assises-2016/, Paris, 2016.

[ARA 17] ARAGONA M., DEKONINCK S., RULANDS S. *et al.*, "Defining stem cell dynamics and migration during wound healing in mouse skin epidermis", *Nature Communications*, vol. 8, p. 14684, 2017.

[ARM 16] ARMSTRONG J.P.K., BURKE M., CARTER B.M. *et al.*, "3D bioprinting using a templated porous bioink", *Advanced Healthcare Materials*, vol. 5, pp. 1724–1730, 2016.

[BAR 15] BARRAUD A., Evaluation des nouvelles technologies de fabrication additive dans le développement d'implants orthopédiques, Thesis, University of Quebec, available at: http://espace.etsmtl.ca/1526/1/BARRAUD_Anthony.pdf, 2015.

[BAR 16a] BARNOUIN V., "TOP 10 des applications de l'impression 3D dans le domaine militaire", available at: http://www.3dnatives.com/top-10-armee-impression-3d-05102016/, 2016.

[BAR 16b] BARNOUIN V., "TOP 5 des vidéos de la semaine: Impression 3D et réalité virtuelle, la bio-impression sur France 2…", available at: http://www.3dnatives.com/top-5-videos-impression-3d-06112016/, 2016.

[BEN 17] BENDTSEN S.T., QUINNELL S.P., WEI M., "Development of a novel alginate-polyvinyl alcohol-hydroxyapatite hydrogel for 3D bio-printing bone tissue engineered scaffolds", *Journal of Biomedical Materials Research A*, vol. 105, no. 5, pp. 1457–1468, 2017.

[BHA 15] BHARDWAJ N., DEVI D., MANDAL B.B., "Tissue-engineered cartilage: the crossroads of biomaterials, cells and stimulating factors", *Macromolecular Bioscience*, vol. 15, pp. 153–182, 2015.

[BIO 11] BIOLOGIE DE LA PEAU, "La peau humaine normale", available at: http://biologiedelapeau.fr/spip.php?article9, 2011.

[BLA 04] BLANPAIN C., LOWRY W.E., GEOGHEGAN A. *et al.*, "Self-renewal, multipotency, and the existence of two cell populations within an epithelial stem cell niche", *Cell*, vol. 118, pp. 635–648, 2004.

[BLA 10] BLANPAIN C., "Stem cells: skin regeneration and repair", *Nature*, vol. 464, pp. 686–687, 2010.

[BOU 07] BOUSCHBACHER M., "L'hypoderme, le derme, l'épiderme", available at: https://tel.archives-ouvertes.fr/tel-00285597/file/Verso_mars08.pdf, 2007.

[BUR 16] BURKS H.E., PHAMDUY T.B., AZIMI M.S. *et al.*, "Laser direct-write onto live tissues: a novel model for studying cancer cell migration", *Journal of Cellular Physiology*, vol. 231, pp. 2333–2338, 2016.

[CAS 17] CASTILLO A., "La créativité ou l'art d'exhumer des analogies cachées", available at: https://www.letemps.ch/economie/2017/02/16/creativite-lart-dexhumer-analogies-cachees, 2017.

[CHA 02a] CHANTRAIN C., DE CLERCK Y.A., "Les métallo-protéases matricielles et leurs inhibiteurs synthétiques dans la progression tumorale", *Médecine/Sciences*, vol. 18, pp. 565–75, 2002.

[CHA 02b] CHARTIER T., CHAPUT C., DOREAU F. *et al.*, "Stereolithography of structural complex ceramic parts", *Journal of Materials Science*, vol. 37, pp. 3141–3147, 2002.

[CHN 12] CHNEIWEISS H., *L'Homme réparé; espoirs, limites et enjeux de la médecine régénérative*, Plon, Paris, 2012.

[COC 17] COCKER C., "Le soldat du futur", in ENTRAYGUES O. (ed.), *La rupture stratégique*, available at: http://www.defense.gouv.fr/content/download/495587/8447516/file/Etude_IRSEM_n48.pdf, 2017.

[CUI 12] CUI X., BREITENKAMP K., FINN M.G. *et al.*, "Direct human cartilage repair using 3D bioprinting technology", *Tissue Engineering*, vol. 18, pp. 1304–1312, 2012.

[COM 17] COMPAAN A.M., CHRISTENSEN K., HUANG Y., "Inkjet bioprinting of 3D silk fibroin cellular constructs using sacrificial alginate", *ACS Biomaterials Science & Engineering*, vol. 3, no. 8, 2017.

[DEL 13] DELFANTI A., *Biohackers; the Politics of Open Science*, Pluto Press, London, 2013.

[DEN 17] DENIS N., Private communication concerning bioprinting, 2017.

[DEV 16] DEVICTOR V., "Qu'est-ce que l'écologie scientifique?", available at: http://sciences-critiques.fr/quest-ce-que-lecologie-scientifique/, 2016.

[DEW 15] DE WINDT T.S., VONK L.A., BUSKERMOLEN J.K. *et al.*, "Arthroscopic airbrush assisted cell implantation for cartilage repair in the knee: a controlled laboratory and human cadaveric study", *Osteoarthritis & Cartilage*, vol. 23, pp. 143–150, 2015.

[DIB 15] DI BELLA C., FOSANG A., DONATI D.M. *et al.*, "3D Bioprinting of cartilage for orthopedic surgeons: reading between the lines", *Frontiers in Surgery*, vol. 2, p. 39, 2015.

[DOR 15] DORAN P.M. (ed.), *Cartilage Tissue Engineering. Methods in Molecular Biology*, Humana Press, New York, 2015.

[EST 16] ESTEBAN-VIVES R., YOUNG M.T., ZHU T. *et al.*, "Calculations for reproducible autologous skin cell-spray grafting", *Burns*, vol. 42, pp. 1756–1765, 2016.

[EYR 07] EYRICH D., WIESE H., MAIER G. *et al.*, "In vitro and in vivo cartilage engineering using a combination of chondrocyte-seeded long-term stable fibrin gels and polycaprolactone-based polyurethane scaffolds", *Tissue Engineering*, vol. 13, pp. 2207–2218, 2007.

[FRI 17] FRICAIN J.C., DE OLIVERA H., DEVILLARD R. *et al.*, "Impression 3D en médecine régénératrice et ingénierie tissulaire", *Medecine Sciences*, vol. 33, pp. 52–59. 2017.

[FUT 17] FUTURA SCIENCES, "La peau", available at: http://www.futura-sciences.com/sante/definitions/biologie-peau-7189/, 2017.

[GEL 16] GELENK-KLINIK, "Tarif chirurgie du genou", available at: https://clinique-orthopedique-arthrose.com/tarifs-chirurgie-genou.html, 2016.

[GID 72] GIDE A., *Les faux monnayeurs*, Poche, Paris, 1972.

[GOU 16] GOUBET F., "L'impression 3D révolutionne les greffes osseuses", available at: https://www.letemps.ch/sciences/2016/09/28/limpression-3d-revolutionne-greffes-osseuses, 2016.

[GRA 12] GRAVITZ L., "Printing muscle", *Technology Revolution*, vol. 115, pp. 80–82, 2012.

[GUD 16] GUDAPATI H., DEY M., OZBOLAT I., "A comprehensive review on droplet-based bioprinting: past, present and future", *Biomaterials*, vol. 102, pp. 20–42, 2016.

[GUI 14] GUILLEMOT F., "Bio-printing: imprimer du vivant ne relève plus de la science-fiction", available at: http://lesclesdedemain.lemonde.fr/sante/bio-printing-imprimer-du-vivant-ne-releve-plus-de-la-science-fiction_a-11-3707.html, 2014.

[GUI 15] GUILLEMOT F., FABRE M., "L'impression de peau en 3D inventée par un breton", available at: http://www.ouest-france.fr/leditiondusoir/data/627/reader/reader.html?t=1448557688780#!preferred/1/package/627/pub/628/page/9, 2015.

[HAL 10] HALEEM A.M., EL SINGERGY A.A., SABRY D. *et al.*, "The clinical use of human culture–expanded autologous bone marrow mesenchymal stem cells transplanted on platelet-rich fibrin glue in the treatment of articular cartilage defects: a pilot study and preliminary results", *Cartilage*, vol. 1, pp. 253–261, 2010.

[HAN 16] HAN S.E., OH K.S., "Satisfactory surgical option for cartilage graft absorption in microtia reconstruction", *Journal of Cranio-Maxillo-Facial Surgery*, vol. 44, pp. 471–478, 2016.

[HER 14] HERTEL O., "Quand l'imprimante façonne le vivant", http://www.sciencesetavenir.fr/sante/quand-l-imprimante-faconne-le-vivant_13022, 2014.

[HER 14] HERBERG S., KONDRIKOVA G., PERIYASAMY-THANDAVAN S. *et al.*, "Inkjet-based biopatterning of SDF-1β augments BMP-2-induced repair of critical size calvarial bone defects in mice", *Bone*, vol. 67, pp. 95–103, 2014.

[HUA 14] HUANG T.Q., QU X., LIU J., CHENA S., "3D printing of biomimetic microstructures for cancer cell migration", *Biomed. Microdevices*, vol. 16, pp. 127–132, 2014.

[HUE 17] HUET J., HUGOT V., "Découverte de la dynamique des cellules souches à l'origine de la cicatrisation des plaies", available at: http://www.diplomatie.gouv.fr/fr/politique-etrangere-de-la-france/diplomatie-scientifique/veille-scientifique-et-technologique/belgique/article/decouverte-de-la-dynamique-des-cellules-souches-a-l-orginie-de-la-cicatrisation, 2017.

[HUT 15] HUTMACHER D.W., HOLZAPFEL B.M., DE-JUAN-PARDO E.M. *et al.*, "Convergence of regenerative medicine and synthetic biology to develop standardized and validated models of human diseases with clinical relevance", *Current Opinion in Biotechnology*, vol. 35, pp. 127–132, 2015.

[IRV 16] IRVINE S.A., VENKATRAMAN S.S., "Bio-printing and Differentiation of Stem Cells", *Molecules*, vol. 21, p. 1188, 2016.

[JAC 16a] JACKSON B., "Researchers print hyper-elastic bones", available at: https://3dprintingindustry.com/news/researchers-print-hyperelastic-bones-96457/, 2016.

[JAC 16b] JACKSON B., "Bone fractures fixed by 3D bio-printed silk", available at: https://3dprintingindustry.com/news/bone-fractures-fixed-3d-bio-printed-silk-97115/, 2016.

[JAK 16] JAKUS A.E., RUTZ A.L., JORDAN S.W. *et al.*, "Hyperelastic "bone": a highly versatile, growth factor–free, osteoregenerative, scalable, and surgically friendly biomaterial", *Science Translational Medicine*, vol. 8, p. 358127, 2016.

[KAE 13] KAELIN B., "Bioprinted human skin on the horizon", available at: http://www.3dprinterworld.com/article/bioprinted-human-skin-horizon, 2013.

[KAN 05] KAHN A., PAPILLON F., *Le secret de la salamandre; la médecine en quête d'immortalité*, Nil Editions, Paris, 2005.

[KEL 16] KELLER L., SCHWINTÉ P., GOMEZ-BARRENA E. *et al.*, "Smart implants as a novel strategy to regenerate well-founded cartilage", *Trends in Biotechnology*, vol. 35, no. 1, 2016.

[KLE 88] KLEBE R.J., "Cytoscribing: a method for micro-positioning cells and the construction of two- and three-dimensional synthetic tissues", *Experimental Cell Research*, vol. 179, pp. 362–373, 1988.

[KLE 09] KLEIN T.J., MALDA J., SAH R.L. *et al.*, "Tissue engineering of articular cartilage with biomimetic zones", *Tissue Engineering, B-Reviews*, vol. 15, pp. 143–157, 2009.

[KNO 15] KNOWLTON S., ONAL S., YU C.H. *et al.*, "Bioprinting for cancer research", *Trends in Biotechnology*, vol. 33, pp. 504–513, 2015.

[KUN 86] KUNDERA M., *L'art du roman*, Gallimard, Paris, 1986.

[KUN 16] KUNDU J., SHIM J.H., JANG J. *et al.*, "An additive manufacturing-based PCL-alginate-chondrocyte bioprinted scaffold for cartilage tissue engineering", *Journal of Tissue Engineering and Regenerative Medicine*, vol. 9, pp. 1286–1297, 2016.

[LEC 17] LE CORPS HUMAIN, "Les secrets du corps humain: les os", available at: http://www.lecorpshumain.fr/corpshumain/1-os.html, 2017.

[LEW 16] LEWIS C., *Too Fast to Think: How to Reclaim Your Creativity in a Hyper-connected Work Culture*, Kogan Page, Paris, 2016.

[LIP 13] LIPSON H., "New world of 3D printing offers "completely new ways of thinking", *IEEE Journal*, vol. 4, pp. 12–14, 2013.

[LIA 15] LIAN Q., CHEN C., UWAYEZU M.C., "Biphasic mechanical properties of in vivo repaired cartilage", *Journal of Bionic Engineering*, vol. 12, pp. 473–482, 2015.

[LIP 14] LIPSON H., KURMAN M., *Impression 3D: la prochaine révolution*, First Interactive, Paris, 2014.

[LLO 16] LLOPIS-HERNANDEZ V., CANTINI M., GONZALES-GARCIA C. *et al.*, "Material-driven fibronectin assembly for high-efficiency presentation of growth factors", *Science Advances*, vol. 2, p. e1600188, 2016.

[MAC 17] MACIEL E., Tilapia Skin Burn Bandage, available at: http://www.rexresearch.com/macieltilapia/macieltilapia.html, 2017.

[MAG 16] MAGNIER M., "Créée à partir de cellules souches, cette peau artificielle est un véritable espoir pour soigner les grands brûlés", available at: http://soocurious.com/fr/peau-artificielle-science-greffe/?utm_source=newsletter&utm_medium=e-mail&utm_campaign=Newsletter_Journaliere_2016-04-06, 2016.

[MAN 16] MANCHUELLE J., "Des chercheurs viennent d'élaborer une technique qui va révolutionner le soin des fractures", available at: http://dailygeekshow.com/os-synthetique-3d/, 2016.

[MAO 15] MAO A.S., MOONEY D.J., "Regenerative medicine: current therapies and future directions", *PNAS*, vol. 112, pp. 14452–14459, 2015.

[MAR 16] MARTEL A., "3d.FAB, un laboratoire français dédié à l'impression 3D du vivant", available at: http://www.3dnatives.com/3dfab-francais-impression-3d-vivant-19022016/, 2016.

[MAR 16] MARCHAND L., "Des chercheurs américains réussissent à imprimer des tissus vivants en 3D", available at: http://www.lesechos.fr/industrie-services/pharmacie-sante/021700931522-des-chercheurs-americains-reussissent-a-imprimer-des-tissus-vivants-en-3d-1200609.php?44fmBuzArJjvPSUM.99, 2016.

[MAR 15] MARKSTEDT K., MANTAS A., TOURNIER I. *et al.*, "3D bioprinting human chondrocytes with nanocellulose – alginate bioink for cartilage tissue engineering applications", *Biomacromolecules*, vol. 16, pp. 1489–1496, 2015.

[MIA 16] MIAO S., ZHU W., CASTRO N.J. *et al.*, "4D printing smart biomedical scaffolds with novel soybean oil epoxidized acrylate", *Scientific Reports*, vol. 6, 2016.

[MIN 16] MINAMI H., "New skin tissue from iPS cell could advance burn treatment", available at: http://www.asahi.com/ajw/articles/AJ201604160016.html, 2016.

[MIT 16] MITCHELL A.C., BRIQUEZ P.S., HUBBELL J.A. *et al.*, "Engineering growth factors for regenerative medicine applications", *Acta Biomaterialia*, vol. 30, pp. 1–12, 2016.

[MOK 16] MOK S.W., NIZAK R., FU S.C. *et al.*, "From the printer: Potential of three-dimensional printing for orthopaedic applications", *Journal of Orthopaedic Translation*, vol. 6, pp. 42–49, 2016.

[MOR 14] MORIN L., "Biohackers, l'internationale des savants fous", available at: http://www.liberation.fr/futurs/2014/11/09/l-internationale-des-savants-fous_1139780, 2014.

[MUN 15] MUNJAI A., PHILIPPE J.M., MUNRO E., "A self-organized biomechanical network drives shape changes during tissue morphogenesis", *Nature*, vol. 524, pp. 351–355, 2015.

[MUR 14] MURPHY S.V., ATALA A., "3D bio-printing of tissues and organs", *Nature Biotech.*, vol. 32, pp. 773–785, 2014.

[NAT 17] NATIONAL INTELLIGENCE COUNCIL, Global trends – Paradox of progress, available at: https://www.dni.gov/files/images/globalTrends/documents/GT-Full-Report.pdf, 2017.

[NG 16] NG W.L., WANG S., YEONG W.E. *et al.*, "Skin bioprinting: impending reality or fantasy?", *Trends in Biotechnology*, vol. 34, pp. 689–699, 2016.

[NGU 16] NGUYEN D.G., FUNK J., ROBBINS J.B. *et al.*, "Bioprinted 3D primary liver tissues allow assessment of organ-level response to clinical drug induced toxicity in vitro", *PLoS ONE*, vol. 11, 2016.

[NGU 17] NGUYEN N.G., PENTONEY JR S.L., "Bioprinted three dimensional human tissues for toxicology and disease modeling", *Drug Discovery Today: Technologies*, vol. 3, pp. 37–44, 2017.

[OSS 16] OSSEOMATRIX, "Céramiques biologiques: pour des os de synthèse sur mesure", available at: http://www.osseomatrix.com/blog/2015/03/01/ceramiques-biologiques-pour-des-os-de-synthese-sur-mesure/, 2016.

[OZB 16] OZBOLAT I.T., PENG W., OZBOLAT V., "Application areas of 3D bioprinting", *Drug Discovery Today*, vol. 21, pp. 1257–1271, 2016.

[PAN 16] PANAGIOTAKOPOULOU M., BERGERT M., TAUBENBERGER A. *et al.*, "A nano-printed model of interstitial cancer migration reveals a link between cell deformability and proliferation", *ACS Nano*, vol. 10, pp. 6437–6448, 2016.

[PAS 16] PASHUCK E.T., STEVENS M., "From clinical imaging to implantation of 3D printed tissues", *Nature Biotechnology*, vol. 34, pp. 295–296, 2016.

[PEI 16] PEI P., QI X., DU X. *et al.*, "Three-dimensional printing of tricalcium silicate/ mesoporous bioactive glass cement scaffolds for bone regeneration", *Journal of Materials Chemistry B*, vol. 4, pp. 7452–7463, 2016.

[PER 06] PERETTI G.M., XU J.W., BONASSAR L.J. *et al.*, "Review of injectable cartilage engineering using fibrin gel in mice and swine models", *Tissue Engineering*, vol. 12, pp. 1151–1168, 2006.

[PES 95] PESTRE D., "Pour une histoire sociale et culturelle des sciences; nouvelles définitions, nouveaux objets, nouvelles pratiques", *Annales; Histoire, Sciences Sociales*, vol. 3, pp. 487–522, 1995.

[PIN 15] PINA S., OIVERA J.M., REIS R.I., "Natural-based nanocomposites for bone tissue engineering and regenerative medicine: a review", *Advanced Materials*, vol. 27, pp. 1143–1169, 2015.

[POU 15] POURCHET L., "Skin equivalent through bio-printing", available at: http://www.3dbioprintingconference.com/3d-bioprinting-conference/lea-pourchet-universite-de-lyon-presents-skin-equivalent-through-bioprinting/, 2015.

[QIA 12] QIAN L., HUANG Y., SPENCER C.I. *et al.*, "In vivo reprogramming of murine cardiac fibroblasts into induced cardiomyocytes", *Nature*, vol. 485, pp. 593–598, 2012.

[QUE 16] QUEVEDO-GONZALEZ F.J., Computational design of functionally graded hip implants by means of additively manufactured porous materials, PhD Thesis, University of Quebec, available at: http://espace.etsmtl.ca/1756/1/QUEVEDO_GONZALEZ_Fernando_Jos%C3%A9.pdf, 2016.

[RAD 17] RADHAKRISHNAN J., SUBRAMANIAN A., KRISHNAN U.M. *et al.*, "Injectable and 3D bioprinted polysaccharide hydrogels: from cartilage to osteochondral tissue engineering", *Biomacromolecules,* vol. 18, pp. 1–26, 2017.

[REG 17] REGNAULT M., "Interdisciplinarité: la mécanique pour mieux comprendre le cancer et l'embryogenèse", available at: http://curie.fr/actualites/interdisciplinarite-mecanique-mieux-comprendre-cancer-l%E2%80%99embryogenese-007652, 2017.

[REN 15] RENOUARD G., "L'impression 3D peut désormais créer la vie", available at: http://www.atelier.net/trends/articles/impression-3d-desormais-creer-vie_438092?banner=1, 2015.

[REV 12] REVAULT D'ALLONNES M., *La crise sans fin – essai sur l'expérience moderne du temps*, Le Seuil, Paris, 2012.

[RIC 16] RICHARDS D., JIA J., YOST M. *et al.*, "3D Bio-printing for vascularized tissue fabrication", *Annals of Biomedical Engineering*, vol. 45, no. 2, 2016.

[RIV 10] RIVIÈRE J.P., "De la peau fabriquée par une imprimante jet d'encre modifiée!", available at: http://www.doctissimo.fr/medicaments/news/de-la-peau-fabriquee-par-une-imprimante-jet-d-encre-modifiee, 2010.

[SAN 15] SANTORO M., LAMHAMEDI-CHERRADI S.E., MENEGAZ B.A. *et al.*, "Flow perfusion effects on three-dimensional culture and drug sensitivity of Ewing sarcoma", *PNAS*, vol. 112, pp. 10304–10309, 2015.

[SAN 16a] SANTÉLOG, "Soin des Brûlures: Le bio-printing ou de la peau imprimée en 3D", available at: https://www.santelog.com/news/infection-plaies-cicatrisation/soin-des-brilures-le-bio-printing-ou-de-la-peau-imprimee-en-3d_15251_lirelasuite.htm#lirelasuite, 2016.

[SAN 16b] SANTÉLOG, "Bio-printing: Une première preuve de concept d'organes imprimés en 3D", available at: https://www.santelog.com/news/recherche-medicale/bio-printing-une-premiere-preuve-de-concept-d-organes-imprimes-en-3d_15361_lirelasuite.htm#lirelasuite, 2016.

[SAR 15] SARKER M., CHEN X.B., SCHREYER D.J., "Experimental approaches to vascularization within tissue engineering constructs", *Journal of Biomaterials Science*, vol. 26, p. 12, 2015.

[SCH 11] SCHNEIDER O., EPPLE P., TEUBER E. *et al.*, "Jet printing of colloidl solutions: numerical modeling and experimental verification of the influence of ink and surface parameters on droplet spreading", *Advanced Powder Technology*, vol. 22, pp. 266–270, 2011.

[SCI 16] SCIENCES ET AVENIR, "Produire des tissus humains par imprimante 3D: le défi de Poietis", available at: http://www.sciencesetavenir.fr/sante/20160209.OBS4272/produire-des-tissus-humains-par-imprimante-3d-le-defi-de-poietis.html, 2016.

[SEN 16] SENATOV F.S., NIAZA K.V., ZADOROZHNYY M.Y. *et al.*, "Mechanical properties and shape memory effect of 3D-printed PLA-based porous scaffolds", *Journal of the Mechanical Behavior of Biomedical Materials*, vol. 57, pp. 139–148, 2016.

[SHI 16] SHIM J.H., JANG K.M., HAHN S.K. *et al.*, "Three-dimensional bioprinting of multilayered constructs containing human mesenchymal stromal cells for osteochondral tissue regeneration in the rabbit knee joint", *Biofabrication*, available at: http://www.annualreviews.org/doi/abs/10.1146/annurev-bioeng-071812-152428?journalCode=bioeng; https://www.researchgate.net/profile/Dong-woo_Cho/publication/292989545_Three-dimensional_bioprinting_of_multilayered_constructs_containing_human_mesenchymal_stromal_cells_for_osteochondral_tissue_regeneration_in_the_rabbit_knee_joint/links/56c16fe608aee5caccf68517.pdf, vol. 8, p. 014102, 2016.

[SHI 17] SHIE M.Y., CHANG W.C., WEI L.Y. *et al.*, "3D printing of cytocompatible water-based light-cured polyurethane with hyaluronic acid for cartilage tissue engineering applications", *Materials*, vol. 10, p. 136, 2017.

[SKA 15] SKARDAL A., ATALA A., "Biomaterials for integration with 3D bioprinting", *Annals of Biomedical Engineering*, vol. 43, pp. 730–746, 2015.

[STR 09] STRINGER J., DERBY B., "Limits to feature size and resolution in ink jet printing", *Journal of the European Ceramic Society*, vol. 29, pp. 913–918, 2009.

[SUN 16] SUN K., LIB R., JIANG W. *et al.*, "Comparison of three-dimensional printing and vacuum freeze-dried techniques for fabricating composite scaffolds", *Biochemical and Biophysical Research Communications*, vol. 477, pp. 1085–1091, 2016.

[TAK 16] TAKAGI R., ISHIMARU J., SUGAWARA A. *et al.*, "Bioengineering a 3D integumentary organ system from iPS cells using an in vivo transplantation model", *Science Advances*, vol. 2, 2016.

[THI 17] THIEBERT C., "De la peau de poisson pour soigner les grands brûlés", available at: http://sante.lefigaro.fr/article/de-la-peau-de-poisson-pour-soigner-les-grands-brules, 2017.

[THO 16] THOMAS D.J., "Could 3D bioprinted tissues offer future hope for microtia treatment?", *International Journal of Surgery*, vol. 32, p. 43, 2016.

[TRE 16] TREGOUËT R., "Les cellules-souches ouvrent la voie vers une nouvelle médecine", available at: http://www.rtflash.fr/, 2016.

[VER 06] VEROLA O., "Aspects anatomo-pathologiques de la cicatrisation", available at: http://www.cicatrisation.info/notions-de-base/histologie-de-la-cicatrisation.html, 2006.

[VIN 17] VINCENT M., "L'homme de 150 ans est déjà là, Uwe Diegel", available at: https://www.maddyness.com/innovation/2017/01/26/maddy-keynote-tuer-la-mort/, 2017.

[VIS 16] VISSCHER D.O., FARRÉ-GUASCH E., HELDER M.N. *et al.*, "Advances in bioprinting technologies for cranio-facial reconstruction", *Trends in Biotechnology*, available at: http://dx.doi.org/10.1016/j.tibtech.2016; https://www.researchgate.net/profile/Dafydd_Visscher/publication/301586894_Advances_in_Bioprinting_Technologies_for_Craniofacial_Reconstruction/links/571b4dd108aee3ddc569d8d6.pdf, 2016.

[WAK 15] WAKE FOREST, "Military applications", available at: http://www.wakehealth.edu/Research/WFIRM/Research/Military-Applications/Printing-Skin-Cells-On-Burn-Wounds.htm, 2015.

[WAN 16] WANG X.F., SONG Y., LIU Y.S. *et al.*, "Ostogenic differentiation of 3 dimensional bi-oprinted constructs consisting of Human adipose-derived stem cells in vitro and in vivo", *PLoSo ONE*, vol. 11, 2016.

[WAT 14] WATSON C.L., MAHE M.M., MÚNERA J. *et al.*, "An in vivo model of human small intestine using pluripotent stem cells", *Nature Medicine*, vol. 20, pp. 1310–1314, 2014.

[XU 09] XU T., BAICU C., AHO M. *et al.*, "Fabrication and characterization of bio-engineered cardiac pseudo tissues", *Biofabrication*, vol. 1, p. 035001, 2009.

[XU 11] XU F., CELLI J., RIZVI I. *et al.*, "A three-dimensional in vitro ovarian cancer co-culture model using a high-throughput cell patterning platform", *Biotechnology Journal*, vol. 6, pp. 204–212, 2011.

[XU 13] XU T., BINDER K.W., ALBANNA M.Z. *et al.*, "Hybrid printing of mechanically and biologically improved constructs for cartilage tissue engineering applications", *Biofabrication*, vol. 5, p. 015001, 2013.

[YAN 15] YANG W., YU H., WEI F. *et al.*, "Selective pattern of cancer cell accumulation and growth using UV modulating printing of hydrogels", *Biomed Micro-devices*, vol. 17, p. 104, 2015.

[YU 16] YU Y., MONCAL K.K., LI J. *et al.*, "Three-dimensional bio-printing using self-assembling scalable scaffold-free "tissue strands" as a bio-ink", *Scientific Reports*, vol. 6, no. 28714, 2016.

[ZIN 01] ZINK S., Heart transplant's uneasy alliance: Medicine, technology and corporate health care, PhD Thesis, Temple University, 2001.

[ZHA 16] ZHAO Y., YAO R., OUYANG L. *et al.*, "Three-dimensional printing of HeLa cells for cervical tumor model in vitro", *Biofabrication*, vol. 6, p. 035001, 2016.

4

Ethical Issues and Responsible Parties

"In this technical and epistemic context, which makes it credible, the hypothesis of convergence imposes itself all the more as the biotechnologies and nanotechnologies refer to a single model of technical action, which is manufacturing". [LAR 17]

"I think now it will only take a few words to update the point of resemblance between watchmaking movement and a body. It is quite simply that the latter also has a solid support – the aperiodic crystal constituted by the hereditary substance – largely removed from the disorder of thermic agitation". [SCH 44]

"If our cells are free, their organization must obey the principles of economy or ecology that structure societies or ecosystems. In an ecosystem like the forest as a whole, there are general regulation phenomena that control the temperature or composition of the ground. However, there is no genome of the forest that, programming each of the inhabitants, ensures the realization of global phenomena. Homeostatis, that is, the constancy of the inner milieu of the forest, is not written anywhere. There is no nature police". [SON 00]

"The culture of impudence is also the culture of irreverence, of demystification, and of the devalorization of ideals". [WUR 77]

"The engineer works with respect towards dignity and for the performance of services for his/her neighbor, without distinction of origin, social position, and worldview". (Manifesto of the VDI (Germany) in 1950, quoted by Didier, [DID 08])

"The pretention of science to dissolve in the anonymity of the mechanical, physical, and chemical environment, those centers of organization, adaptation, and invention that are living beings... Hence the insufficiency of all biology that [...] would like to eliminate any consideration of meaning". [CAN 65]

"Doesn't modern civilization [...] run, by the very fact of scientific progress, a mortal risk? In the future, it will be necessary to fight against scientific pride and control its realizations: the question is knowing if the educated have the right to be disinterested in the consequences that can be deduced from their research and discoveries". (Isaac [ISA 23] quoted by Kaspi, [KAS 02])

"In fact, the mechanisms that increase or reduce the risks remain relatively opaque and have not been mastered". [DEC 01]

"We then count on technical progress to later escape the problems that we encounter and that we know are real – practices and attitude we have not escaped". [LER 11]

"Doesn't risk finally appear to be the diminished and reductive manner in which Man in techno-scientific societies, no longer managing to give meaning to his unhappiness, notices what is happening to him". [DUP 02]

"'The next battle for the freedom of the Internet could deal with 3D printing' proclaimed a headline on the renowned technology blog, TedCrunch, in August 2012". [MOR 14]

4.1. Introduction

In Volume 1, the desirability aspects of additive manufacturing were the object of the author's interest, but then reflection was essentially focused on the technological dimension of the technology. Its social acceptance, with the publicity of uncontestable scientific success (see, e.g. the recent publication by Zhu *et al.* [ZHU 17] on the manufacturing of blood vessels implanted in monkeys), has become an inescapable element for realizing projects affecting Man, which leads to exploring the feedback of a small part of society that has responded to a survey on bio-printing [ARC 16]. When the living is introduced, these worries, which are revisited in this chapter, are expanded to an ethical reflection on the possible applications of manufacturing tissues, which could aim at an enhancement of Man, even, why not, a search for immortality.

These two sections are completed by a third part concerning the methods of functioning (more) responsibly in research; this element furthermore poses the question of the difficulty exploring convergence situations that bio-printing cannot escape.

4.2. Reflection on the acceptance of bio-printing

NOTE.– The survey presented below was conducted with Natacha Denis (engineering student at ENSIC in Nancy and holder of a pharmacist's diploma).

To obtain responses, nothing is more natural than asking questions. The survey through questionnaires is, in this regard, a practical means of quickly collecting information and an efficient decision-making tool. Even if there is no miracle recipe to conduct a good survey and to obtain pertinent results every time, there are incontestable rules to respect at each stage and to avoid disrupting the respondent [PAP 13]. According to Survey Magazine [SUR 16], the 10 commandments for a quality survey would be the following:

– establishing clear, precise and operational goals for the survey in writing and then having them validated by those concerned;

– precisely identifying the goal of the survey (base population) and choosing a representative sample (the method of surveying by Internet, quick but biased, does not allow this method of selecting a population);

– choosing a restricted number of quotas and having recourse to simple quotas rather than crossed quotas;

– concentrating the questions asked on the single objective of the survey and not adding useless questions, even if they are otherwise interesting (upon fact-checking, this is what was attempted);

– organizing the questionnaire into clear parts by going from the general to the specific and from neutral questions to engaging questions. (The choice of mixing neutral and engaging questions has been restricted to have a permanent challenging of the questions by respondents; the order of the questions can play an important role, because one question can influence another through the "contamination effect". This must be verified during a pre-test or the order of the questions must be varied.)

– not multiplying open-ended questions that contribute much less information than well-formulated, closed questions (there were no open-ended questions);

– using clear, simple language understandable by anyone (the questionnaire was tested before transmitting it);

– paying attention to the presentation of the questionnaire and clearly indicating the instructions and information necessary for the researchers and/or respondents (idem);

– insisting on the need for great rigor (not expressed, as it is important to have a return) with all those involved in general and particularly in researchers and data entry personnel;

– being prudent in the interpretation and restitution of the results by being well aware of the margins of error (this is the goal of the discussion of this survey work).

What is the maximum length of an online questionnaire?

According to Dussaix [DUS 09], it has been noted that, generally, the shorter a questionnaire is, the more efficient it is. There is always a great temptation to add questions. "There is no need to hesitate to delete questions that are not directly related to the study, even if they are interesting. It is also necessary to ensure that the questions are not doubled (unless there is a need to verify the validity of the response). The length of the questionnaire largely determines the number of people who will accept to respond to it. The longer the questionnaire is, the greater the fatigue felt and the less precise the responses are. When the questionnaire is too long, those surveyed tend to do a rush job or give any response". Quick, inaccurate responses are given to the last questions when the survey is too long!

A good survey generally has 20 to 25 questions. Beyond this, the term "expert survey" is used. They are not highly compatible with the desired relation with the general public. A series of 10 to 12 extra closed-end questions make the survey boring. One trick thus involves breaking the rhythm of the survey then by adding an open-ended question in the middle or an image that distracts the respondent. Neither the questions themselves nor the questionnaire as a whole should be overly long. The time necessary to respond must be calculated so as not to allow boredom. Overly long questionnaires bore the respondent and negatively influence the response rate and the quality/reliability of the responses received!

Box 4.1. *Framework of a survey*

The survey concerning bio-printing essentially aims at the perception by the public. The response to each question can be significant when deciding to launch certain research and ethical reflections. It also allows examination of the research subjects that could be developed. The survey conducted online in April/May 2016 exploits the questioning methods of works conducted by professional investigators, questions having been introduced in the text in a slightly surreptitious manner so that the respondents' attention is not focused on certain questions central to this work.

4.2.1. *Raw survey data*

We initially proposed a series of 21 questions to students and university personnel (teacher-researchers, researchers, doctoral students, technical and administrative personnel); in addition, a certain number of people in our/their surroundings agreed to respond to this questionnaire. This allowed a rather significant sample to be established (n = 320 respondents), even if this cannot be considered representative of the French population. Furthermore, the greater part of those surveyed were familiar with science, which should definitely accentuate the imbalance in responses compared with future national studies. It is in the comparison of the results and the cross-fertilization of the information obtained that the interest of this finally original study lies. So as to be able to compile the responses as fast as possible and have a clear and easy-to-use interface, we conducted the questionnaire through a Google Documents form. This allowed the evolution of the responses to be followed in real time.

According to the way in which they are formulated, the questions can provoke effects and reactions related to psychological or psychosocial phenomena in the respondent. These are the subject's defense mechanisms. Indeed as a classical trend, every individual will try to defend, even quite unconsciously, a certain image of himself/herself, or a certain integrity-unity of his/her person in the face of this questioning of confronting a questionnaire, especially when it is related to one's health and privacy.

According to Giezendanner [GIE 12], the tendency to acquiesce is a systematic response error that is magnified by tendentious wording, so we obtain answers that are totally suggested by the question. For formulations like, "Stricken with a serious illness, only bio-printing can save you, but with only a fifty percent chance of survival, would you be willing to test its use? Yes – No", the expected/desired result seems rather clear. To counter this tendency, we have:

– Reduced the number/avoided binary closed-end questions (Yes/No, True/False, etc.), switching from time to time when asking an individiual's personal opinion.

– Had recourse to the principle of alternation by varying the wording of questions so that the response is favorable or unfavorable to the same opinion.

– Formulated questions in the positive/negative.

– Alternated positive and negative propositions at an array of levels.

– Alternated the questionnaire's favorable and unfavorable wording.

– Reduced the number of questions like, "Do you agree with…" "Do you admit that…"

– Asked the same question twice by reversing the question, that is, making 2 counterbalanced formulations after one another.

There was an attempt to consider these remarks in the survey itself. The results thus obtained are gathered below. Color versions of the graphs are available at www.iste.co.uk/andre/printing3.zip.

Question 1

Have you heard of "traditional" 3D printers?

Yes:	307	97.2%
No:	9	2.8%
Don't know:	0	0%

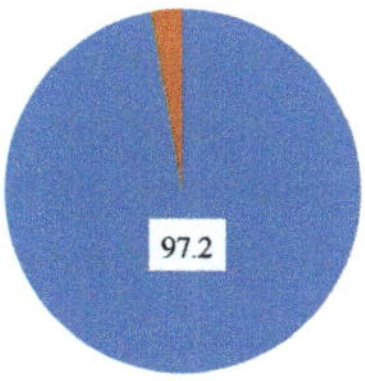

Question 2

Have you heard of bio-printing?

Yes:	174	55.1%
No:	141	44.6%
Don't know:	1	0.3%

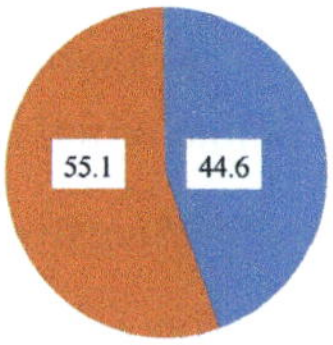

Question 3

If yes, do you remember where?

Internet:	75	35.9%
In person:	48	23%
Literature (newspapers, magazines, publications, etc.)	76	36.4%
Television	61	29.2%
Other	20	9.6%
Don't know	28	13.4%

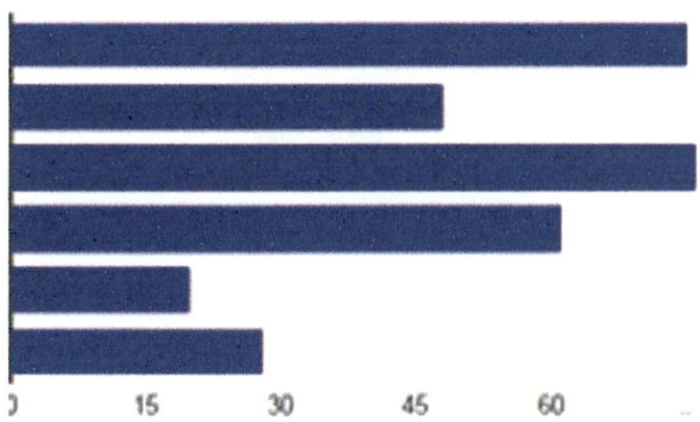

Question 4

In general, would you like to know more about the evolution of this subject and its potential applications?

Yes:	255	81.2%
No:	19	6.1%
Don't know:	40	12.7%

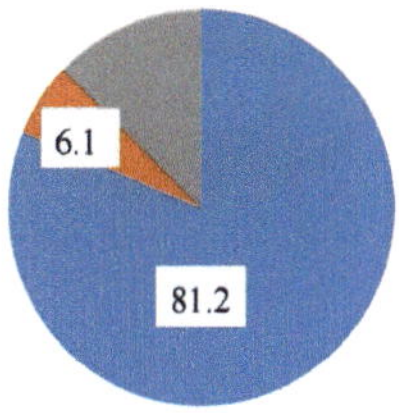

Question 5

Will this technology improve health and quality of life?

Yes:	232	73.7%
No:	8	2.5%
Don't know:	75	23.8%

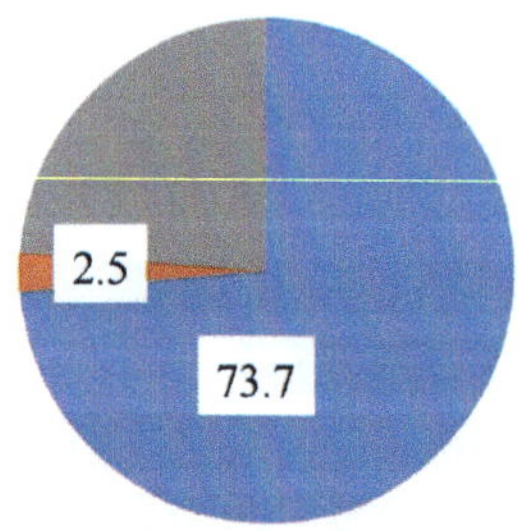

Question 6

Is printing living material still science fiction or close to reality?

Science fiction:	74	23.6%
Reality:	211	67.2%
Don't know:	29	9.2%

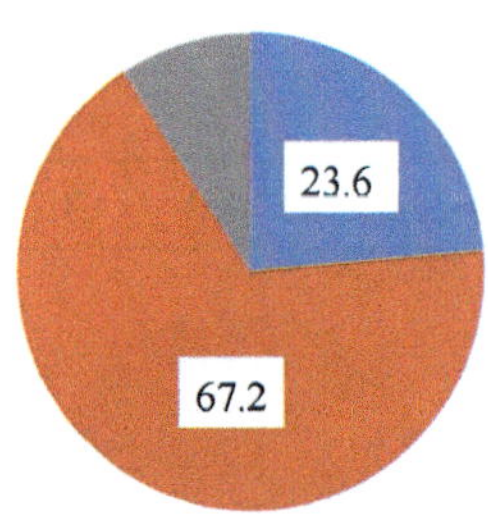

Question 7

For you, how far can/should bio-printing go?

Tissues (skin, bone, etc.):	237	75.2%
Organs:	226	71.7%
Cloning:	12	3.8%
Don't know:	21	6.7%

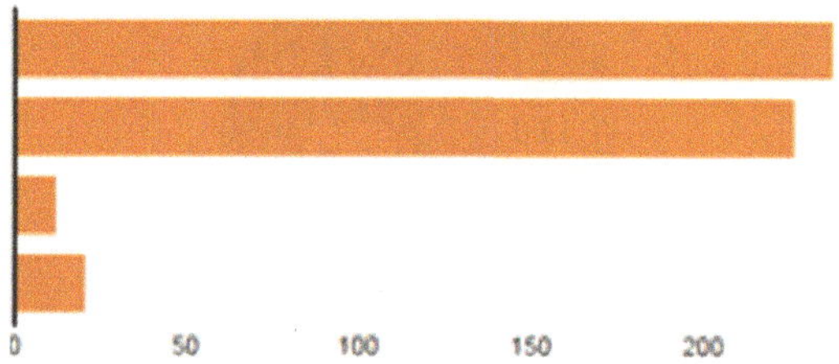

Question 8

Stricken with a serious illness, only bio-printing can save you, but with only a fifty percent chance of survival, would you be willing to test its use?

Yes:	228	72.2%
No:	32	10.1%
Don't know:	56	17.7%

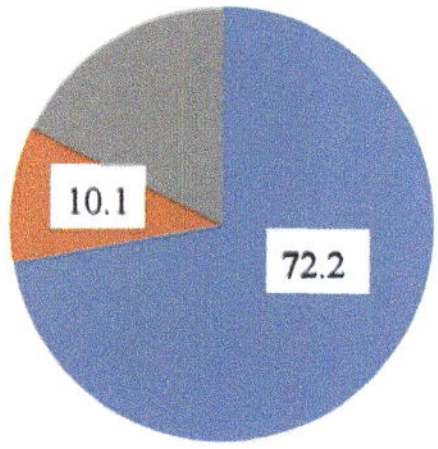

Question 9

In the future, would you be willing to test bio-printing if the situation allowed?

Yes:	218	69.2%
No:	41	13.0%
Don't know:	56	17.8%

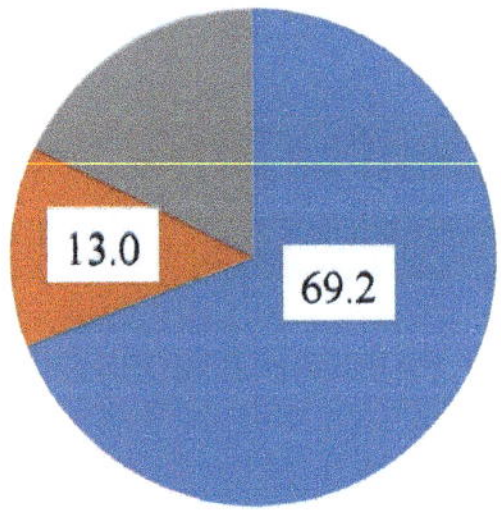

Question 10

Do you hope that bio-printing will one day be used routinely in medicine?

Yes:	242	77.1%
No:	27	8.6%
Don't know:	45	14.3%

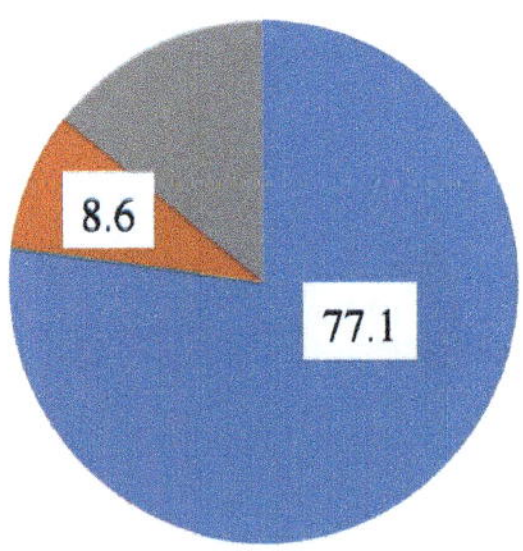

Question 11

If so, by when?

A few years:	56	19.5%
Ten years:	113	39.4%
Several decades:	70	24.4%
Don't know:	43	15%

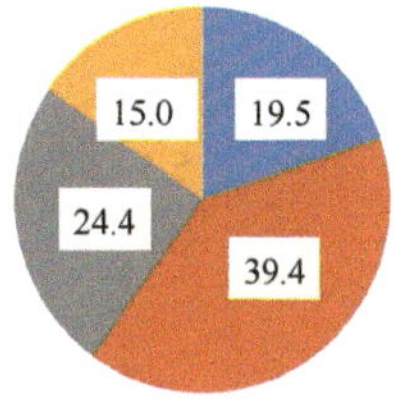

Question 12

What could be the derived uses of this technology?

Sports:	149	47.2%
Military:	172	54.4%
Cosmetic surgery:	174	55.1%
Rejuvenation:	155	49.1%
Cloning:	190	60.1%
Other:	35	11.1%
Don't know:	30	9.5%

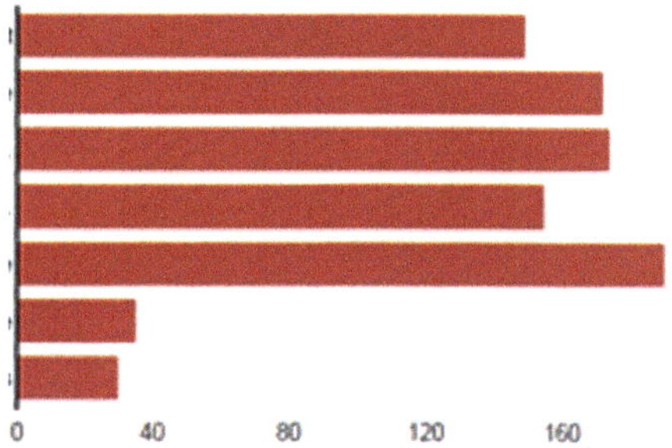

Question 13

Imagine you are too short; would you be willing to pay to have your legs extended thanks to bio-printing through the conception of new bone tissue?

Yes:	17	5.4%
No:	274	87%
Don't know:	24	7.6%

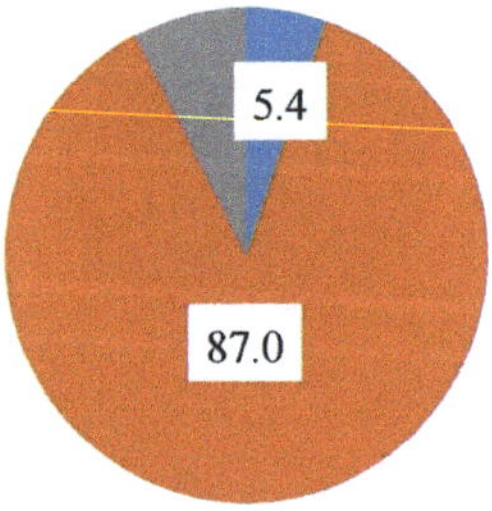

Question 14

If yes or maybe, how much would you be willing to pay?

1000€:	6	3.4%
10,000€:	12	6.7%
100,000€:	0	0%
Don't know:	161	89.9%

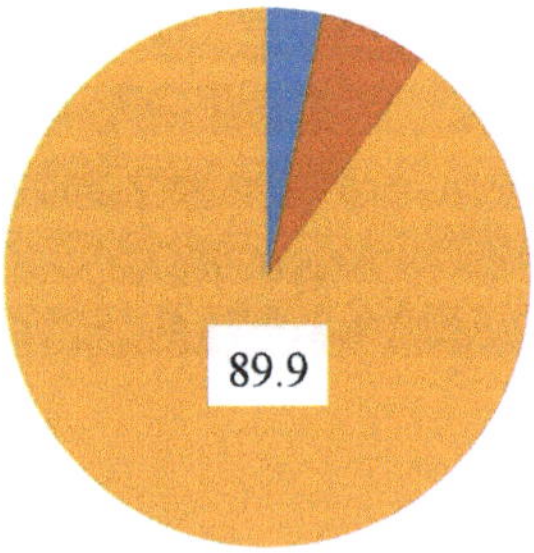

Question 15

If it were one day possible to develop structures such as a heart, a finger, etc. does this pose a moral issue?

Yes:	65	20.6%
No:	238	75.3%
Don't know:	13	4.1%

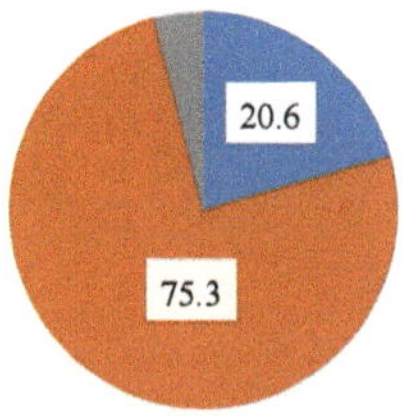

Question 16

Among the following uses, which would increase your trust in this new technology?

Regulation:	242	76.8%
Reliability:	223	70.8%
Medical control:	239	75.9%
Ethics committee:	223	70.8%
Reimbursement by social security:	68	21.6%
Preliminary testing on animals:	120	38.1%
Other:	16	5.1%
Don't know:	9	2.9%

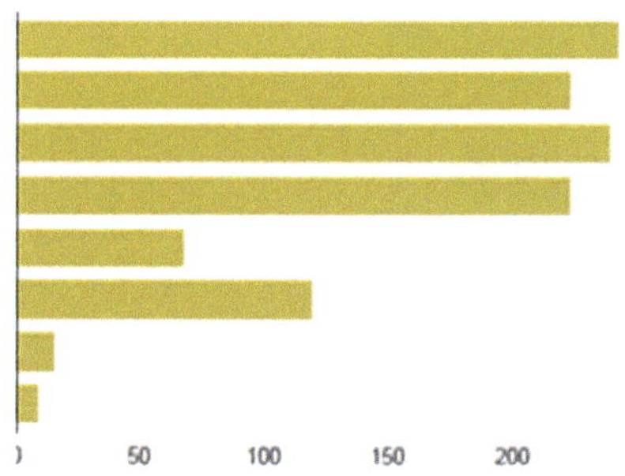

Question 17

With more and more affordable machines (a few thousand €), will the possibly resulting market be controllable?

Yes:	79	25.1%
No:	172	54.6%
Don't know:	64	20.3%

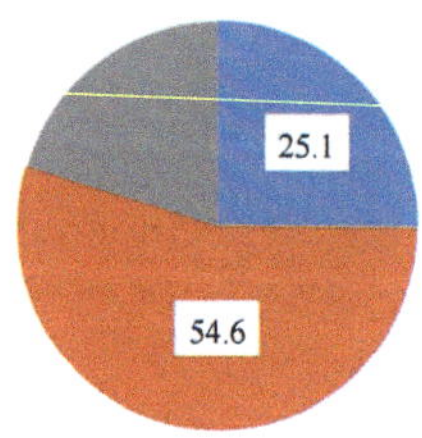

Question 18

By whom?

The State:	129	47.1%
Medical professionals:	159	58.0%
Companies using bio-printing:	73	26.6%
Other:	24	8.8%
Don't know:	57	20.8%

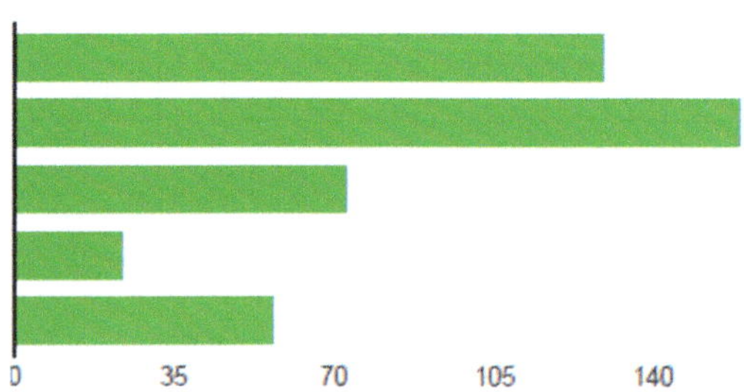

Question 19

To this end, should there be a legal framework to control the use and evolution of bio-printing?

Yes:	301	95.6%
No:	6	1.9%
Don't know:	8	2.5%

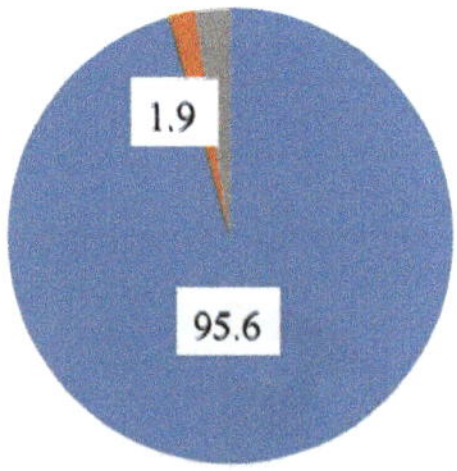

Question 20

How old are you?

Under 18:	4	1.3%
18-24:	59	18.7%
25-34:	57	18.0%
35-49:	92	29.1%
50-64:	65	20.6%
Over 65:	39	12.3%

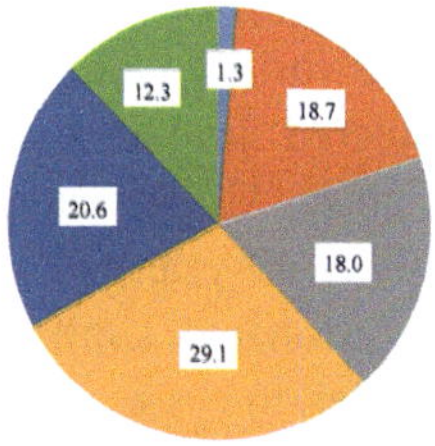

Question 21

What is your occupation?

Student:	73	23.2%
Higher education and research personnel:	144	45.7%
Farmer:	0	0%
Craftsman, retailer, company director:	1	0.3%
Liberal profession, senior executive:	45	14.3%
Intermediate profession (technician, supervisor, etc.):	10	3.2%
Employee, worker:	8	2.5%
No occupation or retired:	34	10.8%

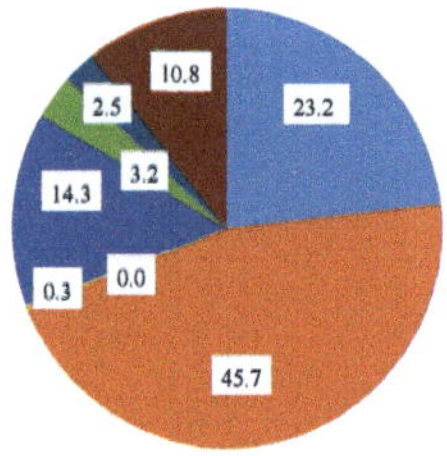

4.2.1.1. *Critical analysis of the raw data*

First of all, the foundations of social agreement on the development of bio-printing as a desired means of caring for and repairing damaged tissues, even more, are confirmed by this survey, with the same trends toward an individualistic approach. In view of the respondents' occupations, the numeric significance of people working in higher education and research can be highlighted. It represents 45.7% of the population of participants. It was also noted that a student population represents one-fourth of the people who responded and around one-tenth for liberal professions, senior executives, and for the category with no occupation or retired. These categories will be considered in the analysis. The division of the age categories is relatively homogeneous. All age categories are represented, however, with one leading category: that of 35–49-year-olds at 29.1%. The categories for ages 18–24, 25–34 and 35–49 are proportionately similar, around 20%. The category over 65 is slightly smaller. Those under the age of 18 can be neglected. It will be interesting to later study the responses by age group. However, with a still very modest worldwide volume of scientific publications on bio-printing (roughly 1,000–2,000 in peer-reviewed journals at the end of 2016), the university population that participated in the survey is probably no better informed than the general population; the respondents have been informed by different media that have sensitized them to the emergence of bio-printing, maybe even the survey itself. In this context, the essentially university population acts, in the end, as the general population (but probably with different pretences), which corresponds to an important and significant result, considering the limited number of respondents (320).

Now, if we deal with the different questions, what seems interesting to point out is that the respondents are already familiar with 3D printing (either additive manufacturing or 3D printing) since 97% of the people know this recent technology. What was not expected is that more than half had already heard of bio-printing (55.1% vs. 44.6%). This half of the population was familiar with bio-printing thanks to four means of communication: the Internet, literature, television and interpersonal communication (this number must be compared with the 50,000-100,000 publications on 3D printing, which is of the order of 50-100 times greater than that for bio-printing). This statement stems from either an over-mediatization of the emerging matter or the fact that certain respondents invented knowledge that seemed possible to them Let's not forget that, in a recent work by Bayon and André [BAY 15], it had been shown that in a framework far-removed from this subject, that of wind turbines, university respondents had a tendency to respond in a positive way much more often than the general population, even when they had no credible quantitative elements.

Thus, in the selected case of wind turbines, it was asked that they propose the quantity of energy necessary to manufacture, install and maintain an onshore wind turbine relative to what it can produce during its average lifetime (20 years). The calculation made using the data available to us reported an "energy cost" relative to the order of 10%. Yet, with an essentially scientific cohort (with roughly 30% of respondents believing not to know), the average value (representing the perception of those surveyed) is of the order of 30-35% for this energy cost. This means that it is believed that a third of what the wind turbine will produce during its average lifetime will be consumed by the establishment of the electricity-producing device. This kind of result limits the pertinence of surveys, because it poses the question of the credibility of the views expressed by certain groups.

This kind of result probably indicates:

– a general lack of knowledge on the matter on the part of the intellectual world;

– more asserted knowledge that integrates this kind of renewable energy into the more general context of electricity production, exploiting gas-fired thermal power stations to compensate for wind intermittency. Yet, in this case, it is clear that, costs being added, the average figure proposed is above the 10% reality; or

– more simply, from a cultural evolution standpoint, an agreement on the development of renewable energy with modest energy profitability in the framework of sustainable development. Could the virtuous formatting undertaken in the past few decades pay off?

However, the subject of bio-printing is recent and has not undergone a "media hype", and, furthermore, for this kind of subject, there are not large differences between "university" populations and the general population. People really seem to have an interest for 3D printing of the living; indeed, 81.2% would like more information on this field, but how do they see the technology today: operational or simply almost mature? The possible game of the respondents to perhaps become engaged in the possible development of supposed myths of modernity, anchored in an ever-positive vision of technological progress, a (still) unwavering faith in this scientific progress and a vision of the Promethean conception of Man. The origin, in terms of the respondents' training, could possibly give some indications on this matter, which would ultimately be very useful.

In the present survey, a very important point deserves to be underlined: for question 6 concerning knowing whether bio-printing comes from science fiction or not, 100% of respondents provide a view (deep or shallow as it may be), while for question 2, only 55% of respondents claim that they have heard of the bio-printing process.

The survey by Baker [BAK 16] conducted with 1,576 scientists indicates that more than 70% of researchers have failed to reproduce the experiments of another scientist, and more than half have not succeeded in reproducing their own experiments. The data also stem from sometimes-contradictory attitudes towards reproducibility. Although 52% of the people interviewed agree that there is a "crisis" of reproducibility, less than 31% think that the fact of failing to reproduce published results means that these results are probably false. Nevertheless, most say that they still trust the published literature. More than 60% of respondents said that the pressure that pushes them to publish research results and the selective communication of these results always or often contribute to problems of reproducibility. Among the other factors, more than half emphasized an insufficient number of replications or low statistical power. Should we think that approximation is a requirement in the university world?

Independent of this aspect, among the response biases generally mentioned by the literature, there is a strong presence of the bias of social desirability and artifacts of questioning [HER 07, BUT 10]. The former results from the respondent's will to show himself/herself in his/her best light [CRO 60]. The latter designates "the measurement error resulting from the adoption of a specific response behavior by the respondent in a situation of questioning if he/she believes he/she has discerned, even partially, the objective of this questioning" [HER 07]. In this research, it is possible to re-question the existence of these two types of bias in light of the opportunities presented by the spread of Internet technologies. Numerous researchers have emphasized that online studies allowed time and effort to be saved in collecting quantitative data [ARA 00, COU 00, GAL 00, COB 01, GAN 04]. However, their ability to reduce the bias of social desirability and the artifacts of questioning are not all known. Finally, the question of the value and utility of uncertainty in discourse and the media begs to be asked, with the consequences of erroneous responses to surveys [BES 07]. According to McGoey [MCG 09, MCG 12], uncertainty creates a demand for solutions to the ambiguity that it perpetuates by encouraging debate on a world to be constructed, and in which the culture of the respondents is expressed. The techniques are, to varying degrees, dependent on one another with certain coherence among them, which an educated public can feel. The novelty disrupting this previous order can then be seen subjectively in negative, explaining the results; "our desire for certainty almost always leads us to sacrifice reality to an abstraction that is translated by postulates" [FUS 75].

In a second part of the questionnaire, the participants were asked for their view on the potential applications of bio-printing today and those foreseen in the near future; 73.7% think that this can improve health and quality of life. For 67.2%, it is thus a reality (while only assemblies of a few tens to thousands of cells are created for research and tests in cosmetics or medications). It is nevertheless to be noted that 23.6% consider printing the living to be science fiction (this is indeed the question dealt with by André and Larger [AND 16] on epistemological research concerning bio-printing). For the participants in the survey, bio-printing can primarily extend to the printing of tissues and organs. Few, apparently, think of applications in the field of cloning (see the question of biases).

The trend towards a normative representation, called a bias of social confirmism, may lie at the origin of this result: through his/her responses, the person surveyed translates a social and moral ideal, the desire to conform to the social norm. He/She gives responses in line with social norms, expected responses. In these conditions, the respondent may tend to idealize himself/herself, to present prestige reactions and to wish to project a positive image, valorizing his/her person. There is thus a bit of lying, exaggeration or feigning of socially well-viewed habits or opinions. He/She will naturally tend to respond according to what is valued by society, what he/she considers to be the social norm.

In an Internet survey, it is less likely for the people who respond to the survey to adopt a façade behavior in the sense that there is "saving place" corresponding to the fear of being negatively judged through their responses. Someone surveyed in front of a physical person may wish to make a good impression and give a valorizing response (as a function of what he/she judges to be socially desirable and socially valuable), which could contribute to giving himself/herself a certain prestige, avoiding being negatively judged. This behavior is also translated by the minimization of opinions, defensive simulation and refuge towards stereotypes (i.e. socially allowed clichés). But before a computer screen? (Unless he/she believes he/she can be identified?)

The trust that respondents have in this emerging technology is noticeable: a non-negligible part of the population would be willing to test bio-printing during a serious illness with no other alternatives (72.2%), if the situation allowed it (69.2%), and routinely from a medical standpoint (77.1%). Surprisingly, many people think of the routine use of bio-printing rather than in serious situations with no therapeutic alternative. Respondents see the possible use of bio-printing in the medical world by different deadlines. The views are split: 40% project the application in around ten years, 20% in a few years and the rest several decades (25%). The opinion for the use in the scope of esthetics and increased performances is clearly negative at 87% (see above concerning possible biases associated with the "socially correct").

Concerning the ethical part, the conception of organs poses no moral problem for three-fourth of the participants, although 20.6% stated that this is problematic for them. Furthermore, for 54.6%, the market that could lead to bio-printing machines will be controllable; 25.1% think the opposite (see reflection expressed in the preceding chapter on bio-hackers). The population emphasizes two sectors that could control this market: medical professionals and the State. To a lesser degree, companies producing the machines could also control bio-printing (!). To increase their trust in this new technology and its access to the market, the participants want control through regulation, good reliability of the technology, medical control, and ethical committees. Preliminary testing and reimbursement by the French social security system are relegated to the second tier (which may explain why financial aspects are set aside in these responses). A legal framework to control the use and evolution of bio-printing seems like evidence for people having responded to the questionnaire, but, at the same time, these same people would accept human experimentation if they were strongly affected.

4.2.1.2. *Crossed analyses*

Two types of analyses were performed, one linked to age and another allowing us to examine whether there are differences between populations (the general population and that of the university environment). Table 4.1, presenting the responses as a function of the respondents' age, does not allow the existence of a significant intergenerational disparity to be shown (close to the precision of the measurements).

	Total population	Under 35	35 – 49	Over 50
Knowledge of 3D printing	97%	96%	97%	99%
Knowledge of bio-printing	55%	52%	57%	57%
Desire for more information on the subject	81%	86%	79%	79%
Improvement of health and quality of life	74%	77%	75%	68%
Bio-printing as reality	67%	63%	68%	70%
Bio-printing as science fiction	24%	29%	29%	13%
Bio-printing test if seriously ill	72%	65%	71%	65%
Bio-printing test if the situation allowed	69%	73%	72%	64%
Bio-printing test routinely	77%	77%	83%	72%
Timeframe for the arrival of bio-printing				
A few years	20%	11%	22%	13%
Ten years	40%	38%	39%	37%
Several decades	25%	30%	23%	16%
Absence of a moral problem	75%	68%	86%	74%
Moral problem	21%	28%	11%	19%
Controllable market	25%	27%	27%	20%
Uncontrollable market	55%	47%	54%	64%

Table 4.1. *Significant absence of intergenerational disparity in responses*

However, some results emerge (or seem to emerge) that would probably need to be looked at in greater depth: older respondents have a greater belief in bio-printing technology, but do not think that the nascent technology will improve the comfort of their lives as often.

Table 4.2 corresponds to the analysis of the results concerning the qualification of the university population relative to the "general" population, as well as to the same population minus the university population.

	Total population	Total population minus the "university" population	People working in higher education and research
Knowledge of 3D printing	97%	97%	98%
Knowledge of bio-printing	55%	48%	63%
Desire for more information on the subject	81%	79%	84%
Improvement of health and quality of life	74%	72%	77%
Bio-printing as reality	67%	65%	72%
Bio-printing as science fiction	24%	27%	17%
Bio-printing test if seriously ill	72%	73%	70%
Bio-printing test if the situation allowed	69%	69%	69%
Bio-printing test routinely	77%	75%	80%
Timeframe for the arrival of bio-printing			
A few years	20%	20%	19%
Ten years	40%	41%	38%
Several decades	25%	26%	22%
Absence of a moral problem	75%	71%	83%
Moral problem	21%	25%	13%
Controllable market	25%	23%	28%
Uncontrollable market	55%	56%	53%

Table 4.2. *Difference between the "general" population and the university environment in the responses to the survey*

If we assume that the university environment has greater knowledge of bio-printing (maybe because they responded to the survey), that they are slightly more inclined to consider bio-printing a reality, the greater absence of a moral problem relative to the general population could be underlined. In any case, there is no greatly significant difference in the responses to the survey in the different populations selected.

4.2.2. *General discussion: whom to trust?*

Paraphrasing Badiou [BAD 09], it is possible to show that the definition of a scientific-technological concept and/or activity, as associated with a complex debate with the public, cannot escape the production of its effects, the multiplicity of its attributes, such that it can also indeed be said that only the concept exists or that only its attributes exist. This remark thus imposes an analysis of the "bio-printing" object in connection with its consequences, within interdisciplinary expertise. Invisible damage or simply bad choices in the present can show themselves to be irreversible years later. The inability to anticipate and foresee futures thus imposes a sharing of the risks and desirabilities between deciders, researchers, and society [BER 10]. Yet, in a provocative way, but leading to reflection, Brune [BRU 85] writes, "There is no equivalence between the functional individual and the high-level technocrat. The former functions in the service of the latter. One serves the society of consumption production in the measure of its total depolitization; the other manipulates the technocratic ideology to exercise real power, under the alibi of the complexity of managing modern economies. Functional normality reinforces the function of normalizers, those who rise up as specialists of power to better dissuade the citizen from looking, understanding, and responding". Do we not also reproach experts, hence the temptation to support ourselves on another "portmanteau" concept, the expert citizen [AND 02]. It is indeed on such bases that the principles of Socially Responsible Research (SRR) were put in place.

The (initially undesired) example of the questionnaire on the wind turbines [BAY 15] cannot challenge these principles. However, the methods of relating science-technology-society should probably be revisited to consider a certain number of elements:

– It is necessary for the researcher not to confuse belief, truth and proof. This is doubtless a delicate and significant job that originates in current research's method of functioning; in fact, it is necessary to take things into consideration between verifiable elements (which was initially performed before launching the survey) and conjectures.

– It is important to analyze how the stakes of society, strongly articulated by the media, can lead (or not) to obscurantist forms of manipulation. In this framework, this cannot be the "common" culture, borrowed from the socially correct, shared beliefs that must impose their law on the scientist.

– The obligation on the researcher's part to analyze, if it exists, how a false opinion defended by a minority of people (but who seem convinced that it is true) can spread [GAL 03, GAL 08].

– The use of the notion of probability to define the reliability degree of a diagnosis on a scientific and technological project (e.g. from a survey) can lead to disadvantageous mistakes for society.

– Trust in the public's views must be analyzed to explain its role controlling research.

– Putting in place scientific committees to validate projects (beyond those that grant financing) may not be the solution to the question: criteria of innovation, social and economic impact, morality, etc., stem from a moment of civilization. The risks of conservative reactions and a lack of deepening must be foreseen.

It is not, however, a question of coming to the end of the essential question of seeking harmonious relationships between those who create and those who use (and who suffer when risks exist), nor between those who stop scientists from researching in this article. An example that would need to be confirmed is not enough for an ensured conclusion. Nevertheless, it poses the question of the operationalization of the SRR principles that still need to be refined.

4.2.3. *Preliminary conclusion*

In any case, it is difficult to think that objective knowledge, media storytelling, etc., are obvious advantages to uniformly change the collective behaviors of the different layers of society, even if, generally, the entire public is rather favorable to the development of safe medical practices as a solution, though certainly not the only one, for maintaining good health. Then the question is posed of how to calmly interact between the science that is being built and the public with obvious ethical aspects that would have to be taken into consideration (see section 4.2 below).

Anticipation on the side of science and technology must allow wait-and-see attitudes and reactions in emergencies to be surpassed by ensuring beforehand behavior analyses, the availability of tools and points of reference to face them, to honestly encourage the desired (and desirable) future, which implies that the possessors of scientific and technical knowledge accept leaving their ivory tower by getting involved in a responsible relationship with the society [NWO 14, AND 13].

However, the survey conducted using the best possible knowledge on the diversity of the representations (though reduced to the possibility of receiving responses) does not allow the estimation of the possibility of having a clarification of the role/interests of the different actors with their identity facets, to valorize diversity, and the possibility of more in-depth reflection on particular axes, nor to search for other partners allowing, as much as possible, the debate to be balanced. In this sharing of an important share of ignorance, it could have been possible to report the uncertainties of scientific and technical knowledge, approximations of knowledge, abuses of interpretation and limits of competence, to measure at least the extent of the unresolved questions and the question marks. This desired approach could have authorized taking distance from all (?) the ideologies, overly reassuring proposals and abusive simplifications. This aspect could thus be considered more in-depth in another context.

For the "lay" public, going off its perception, assimilating this to the project to study its relationships supposes that the project's guidelines can be understood in themselves, beyond all intense mental activity, short of language [BRE 90]. Yet, this preliminary maturation work for the debate on bio-printing has not, for obvious reasons, really been engaged. It is actually very important to take an interest in the way in which the complexity of the systems can be explored; "how the links of dependence and interdependence are woven, opening new possibilities, potentially developing a network of supplementary constraints, modeling the horizon on which desires, hopes, and fears are projected, determines the concrete framework in which the others exist" [BRE 90]. Learning to assess, at the level of the "basic" citizen, is a means of introducing the notion of honest compromise, learning from the problems of others, in short, returning to a socialization of the sharing. Is it not true that Hans Jonas [JON 90], in his approach of "fear heuristics", reminds us that it is necessary to cultivate a disinterested fear, likely to detect the hazards of the technology? This is a work to be carried out in another context; in any case, for the researcher engaged in developing new technological solutions directly or indirectly affecting Man, it can be difficult to draw closer to the society, all the more as, in the chosen example (which would require confirmation on other textbook cases), some of those who are more likely to reflect on the theme provide elements distant from the expected responses [SCI 14]. Let us then try to invalidate this reflection by Walter [WAL 08]: "the prophets of unhappiness today are scientists"!

Bio-printing

Today, everyone has heard of 3D printing. Thanks to this technology, in the medical world, it has been possible to create hip prostheses, dentures, etc. Today, ever more ambitious science is investing in 3D printing of the living. Indeed, why not apply this concept to printing organs: skin, cartilage, bone tissues to reconstruct a bone, muscles, liver, etc. And in the future, for its replacement or to increase its performances.

1) Have you heard of "traditional" 3D printers?

– Yes

– No

– Don't know

2) Have you heard of bio-printing?

– Yes

– No

– Don't know

3) If yes, do you remember where?

– Internet

– In person

– Literature: newspapers, magazines, publications, etc.

– Television

– Other

– Don't know

4) In general, would you like to know more about the evolution of this subject and its potential applications?

– Yes

– No

– Don't know

5) Will this technology improve health and quality of life?

– Yes

– No

– Don't know

The goal of bio-printing is to print an "ink" containing specific cels from a patient in three dimensions on an adequate support. The cells are then allowed to develop to form a tissue. There are multiple interests of this technology:

– Replacing the use of test animals with human tissues. There would no longer be a need to test new medications and cosmetics on animals; they could instead be tested on representative organs;

– Creating living organs or tissues for grafts, allowing their healthy life to be extended;

– No longer testing medications directly on patients: tests on tissues from the patient's own cells: Individualized medicine (the right treatment for the right patient).

Thanks to bio-printing, Man could be "repaired": changing tissues and organs, but also increased performances (enhanced Man), which could pose ethical questions.

6) Is printing living material still science fiction or close to reality?

– Science fiction

– Reality

– Don't know

7) For you, how far can/should bio-printing go?

– Tissues (skin, bone, etc.)

– Organs

– Cloning

– Don't know

8) Stricken with a serious illness, only bio-printing can save you, but with only a fifty percent chance of survival, would you be willing to test its use?

– Yes

– No

– Don't know

9) In the future, would you be willing to test bio-printing if the situation allowed?

– Yes

– No

– Don't know

10) Do you hope that bio-printing will one day be used routinely in medicine?

– Yes

– No

– Don't know

11) If so, by when?

– A few years

– Ten years

– Several decades

– Don't know

12) What could be the derived uses of this technology?

– Sports

– Military

– Cosmetic surgery

– Rejuvenation

– Cloning

– Other

– Don't know

13) Imagine you are too short; would you be willing to pay to have your legs extended thanks to bio-printing through the conception of new bone tissue?

– Yes

– No

– Don't know

14) If yes or maybe, how much would you be willing to pay?

– 1,000€

– 10,000€

– 100,000€

– Don't know

15) If it were one day possible to develop structures such as a heart, a finger, etc., does this pose a moral issue?

– Yes

– No

– Don't know

16) Among the following uses, which would increase your trust in this new technology?

– Regulation

– Reliability

– Medical control

– Ethics committee

– Reimbursement by social security

– Preliminary testing on animals

– Other

– Don't know

17) With more and more affordable machines (a few thousand €), will the possibly resulting market be controllable?

– Yes

– No

– Don't know

18) By whom?

– The State

– Medical professionals

– Companies using bio-printing

– Other

– Don't know

19) To this end, should there be a legal framework to control the use and evolution of bio-printing?

– Yes

– No

– Don't know

For our information

20) How old are you?

– Under 18

– 18-24

– 25-34

– 35-49

– 50-64

– Over 65

21) What is your occupation?

– Student

– Higher education and research personnel

– Farmer

– Craftsman, retailer, company director

– Liberal profession, senior executive

– Intermediate profession (technician, supervisor, etc.)

– Employee, worker

– No occupation or retired

Box 4.2. *Bio-printing Survey (without figures)*

4.3. Ethics and bio-printing

What if all of humanity's problems were caused not only by outside factors, but were inherent to our biological and cognitive limits? For millennia, we have sought to transcend our most fundamental limits. Today, with the advent of new technological possibilities, some calmly foresee rationally organizing their transformation towards immortality [MIN 07] and/or the augmentation of their performances [SUS 12]. In short, the emergence of 3D technologies not only disturbs society and production, but also poses ethical questions [NEE 15, NEE 16].

According to Stelarc [LEB 07], the post-human body would be a homogenous and synthetic machine: a cybernetic, which is a robotic being that would no longer be subject to the dysfunctions and biological flaws abounding in the human body. In all its performances, Stelarc's processes are attached to recreating and refounding the human being, with the concern of examining and modifying the body. Here, we see values from modernity being surpassed, for David Le Breton [LEB 90]: "The cardinal values of modernity, those that are set forth publicity, are those of health, youth, seduction, flexibility, and hygiene. These are the cornerstones of the modern story on the matter and its obliged relationship to the body".

By inverting the modern matter, Stelarc would no longer have only one concern, that of the body/object in interaction with the technological object. He is pursuing this modern process of disembodiment and idealization of the body by pushing it relentlessly not to become only an object and image. This state of urgency is revealed in all of its performances: all the body's motor and sensory actions are due to be amplified and prolonged through the incorporation of prostheses and interfaces. It constitutes an ideal subject for bio-printing allowing (in principle) technological exploits applied to Man. Technology, formerly considered a device of technocratic oppression, becomes a tool of emancipation in this context. From this nascent post-humanity, Stelarc would be the incarnation of the cyborg, the bionic man of the post-industrial era, with the tools of his own metamorphosis at his disposal.

Trans-humanism is hidden behind bio-printing. Today, this is supported by computer science specialists much more than by researchers from traditional domains, for example, biologists. Actually, behind trans-humanist propositions, we may see the appearance of the fundamental debate that will shake science in the coming years: on the one hand, the followers of "all information" who think that living beings, maybe even the entire universe, are above all other information supports that can be transferred to other materials, even compressed or formulated differently; on the other hand, researchers who, to the contrary, think that the material support maintains a fundamental importance [SUS 12]. Yet, the epistemological debate engaged on this subject must be capable of explaining this important question in the field of bio-printing, which is ultimately rather narrow (relative to all possible fields).

Steiner [STE 10] draws attention to the place of the commodification of the body in business between human beings concerning organ transplants. This brings out the complex process of creating a new social resource: the transplantable organ. In most cases, the organ is only available after the death of a person: more than generosity, death is the operator thanks to which societies proceed to separating the resource and the person. "Societies have not undergone this redefinition of death passively. New forms of social solidarity have appeared to accompany the optimal use of death, this way of acting through which contemporary societies force themselves to extract everything possible from it to care for the living". The market normally does not allow the rich to count on transplantable medical resources allowing their life to be saved in regulated markets like those in France. "Everything would chance once the bio-market took on an international dimension, with the trade of transplantation. Calling on the commercial device would be the euphemized form of social cynicism on a worldwide scale" [STE 10]. However, by using bio-printing performances, using one's own cells and one's own tissues, this sort of deviation would no longer take place (especially if this is one day possible to do in one's own garage with the help of bio-hackers).

In a fully evolving world, bio-printing multiplies the offers of innovation for a society avid for the best technology, which incites Man to treat his body as an object that can always be made more perfect. It is in this "Promethean disposition of Man" that this technology allows a new consideration of the ill or handicapped body, as it allows hope for access to effective prosthetic devices. However, the hopes/promises linked to organ repairs can involve people in forms of dependence associated with the representation of their image, of their "agism". Moreover, behind this proposition of the "repaired" Man full of hope, another vision of the human is profiled in the background, that of the "enhanced" Man in his physical capacities, and beyond the field of bio-printing, of his intellectual capacities. These hybrid connections of prostheses directly integrated into the body are part of the imagination of an "ideal" human; this is a promise that has not yet been fulfilled by bio-printing.

Yet, at the same time, it is possible, with a vision of personalized medicine, to use this technology to study cancer, to test medications on living systems that do not come from animals (tests on animals are more and more limited in the Western world), to repair humans (particularly bone, but maybe soon skin and other tissues or organs, etc.). This technology opens new doors that must be considered.

Thus, upon reading about the elements concerning bio-printing, some ethical questions emerge, even if the applications are still far off, even hazardous or science-fiction, but for the author, it seemed necessary to open the debate. Indeed, must we adopt, why do so, the engineer's underlying vision? Do we wish to create an "artifactualized" nature? Will there be the domination of Man by those who master bio-printing due to extraordinary promises of indefinite repair (killing death!)

and enhancement? What could be the effects on the population if only a financial elite accesses this technology's contribution (maybe still happily become one)? Do the improvements linked to printing organs constitute improvements of our lifestyle and thriving?

"Diversity, complexity, imperfection, vulnerability, here is strength of Ulysses, the force of Man. Ulysses is not seeking to steal the divinity from the gods; he opposes them with his humanity, which is the key of his freedom. Before the progress of science, it is not morality that must limit technology, it is reason. The world that is built to be more human, but most importantly to survive, must be inspired by the teachings of Homer" [LÉO 10].

Chneiweiss [CHN 12] reminds us of article 3 of the European Charter of Fundamental Rights: "Everyone has the right to respect for his or her physical and mental integrity. In the fields of medicine and biology, the following must be respected in particular: the free and informed consent of the person concerned, according to the procedures laid down by law...". For Paul Ricœur [RIC 90], "ethics is the very movement of freedom that seeks a good life with concern for others and proper use of social institutions". But in 1926, Bernanos, in the prologue of "Under the Sun of Satan", wrote a very hard phrase, but one that clearly illustrates the subject of ethics for a postmodern society involved in forms of "unhope" [GUI 12] linked to liberalism today and the unbridled freedom of markets: "Realism is the good conscience of bastards". Reflections must be made on than just recipes whose application has the advantage of perpetuating the new established order to advance. That is somewhat the stake of this reflection in section 2 of this chapter capable of serving bio-printing and particularly its evolution. The promises of an ever more prehensible living are part of a broader context. In the era of digitization, Man could have made his third historic breakthrough, after the emergence of language, then writing. Thus, the experience of an individual is more and more transmitted to him/her and less and less experienced live. The body has never seemed so obsolete in the face of the "true life" that is spoken, written and computerized, and with the emergence of new technologies like bio-printing. "Here, we find Plato's idea of simulacrum-body, to the point of asking the question: could Man be tempted to free himself from his body? To do away with this finitude? Unless the materiality of bodies remains essential? Wouldn't understanding/taking the body be a means of improving ourselves?" [VAN 15].

4.3.1. *Framing elements*

Wiener wrote these prophetic lines in 1971: "The machine, just like the living body, can be considered a device that seems, locally and temporarily, to resist the general trend of growing entropy. Through its capacity to make decisions, it can produce an organization zone around itself in a world whose general tendency is to disorganize". This is indeed the final stake of bio-printing, being able to order living matter in view of a certain functional determinism. The passions associated with this project due to the hopes it awakens are very significant in an individualistic society that has made a determining element of social recognition out of the utopia of the perfect and eternally young body [SFE 95, MOR 92, KUN 90, LEB 13]. Yet, "the ethical question *par excellence* is the question of people who act. The action has the particularity of placing the actor before an undetermined situation and this is precisely the greatness of the actor, to choose the orientation to impress upon events" [FEL 14]. It is difficult not to have a guide, but this allows responsibility and somewhat forced reflection. Indeed, the existence of unknown risks is inherent to a deficit of knowledge, but also of coherent and mastered concepts in relation to the ambitions of bio-printing [POU 06]. It only remains to revisit a rational use of the precautionary principle.

"Here, we differentiate the external (supported) and internal (implants, hybridity) systems undertaken. If the former only pose a problem in matters of resisting usage, training users, or an adapted technological environment, the latter are confrontd by physical problems of immunity, psychopathology of bodily force, technological obsolescence, and mastery of energy, each assuming an important ethical aspect" [CLA 13] According to R&D Taxsavers [R&D 15], in the USA, since 1981, there has been a "research tax" (unlike the French system) that allows the support of research:

– on new technological products or processes;

– avoiding uncertainties (improved robustness);

– on experimental processes.

The eligible costs include salaries, assays, research contracts and patenting costs. It is a question of whether the operation will be extended in the future.

Bio-printing, which is included as a means in the second category, depends on notions from the sciences of complexity, biology, information and cybernetics. At the same time, it is an application of computer science, based on reflections on the nature of information and communication with its associated notions like self-organization [KAU 93], self-replication, retroaction, coding and digital compression

[SUS 09, GUI 11, FRE 99, FRE 03]. This is how it can be the object of a "research tax" credit in the USA.

Acting in the field of bio-printing allows, in principle, interaction to be directly induced in the bricks, thus the coupling of scientific disciplines (biology) and technologies like materials science, the microelectronics of computers and cellular engineering [AGO 14, CCC 12]. Neuroscience will also profit from bio-printing, even from micro-fluidics: miniaturization can facilitate the implantation into the brain (and the body in general) of electronic stimulation, data-storage, actuator, transducer and imaging systems; coupling with the cognitive sciences would thus allow an influence on education and learning, the formation of ideas, formatting citizens, even decision-making. What does man then become in the unfolding of this comprehensive process? Who will think for whom? (see, e.g. the fearful cry of PMO [PMO 04] in France in another context).

"It is good to question the norms that were believed to be fair, the truths held to be inarguable. It is through these rectifications, modifications, new contributions, that humanity has progressed and continues to progress, through a desire, a will to always go higher, always go further, towards the best. However, the very widespread error is believing that the new is always better and that there is some sort of dishonor in not contributing and constantly bringing triumph to previously unseen principles" [LEV 03]. Actually, a society's demand for complete (!) well-being cannot avoid passions and new risks [LIP 06]; hence, this small reflection on ethics adapted to bio-printing.

Each individual must probably think that there is no problem surviving an accident or a serious illness, in the best possible conditions, thanks to all the medical technology like bio-printing, which could be a new element among the panoply used by medical professionals (see section 1). It may involve considering the body as a customly reparable artifact, forming the hypothesis that it is possible to modify the foundations to complete it or make it conform to the idea held of it [LE 13a]. In this context, Hunyadi [HUN 15] believes that the use of this type of medical technology is not universally neutral "given that it gives rise to the idea of humanity detached from the inertia of its own life, which is usury and degradation". By freeing ourselves from certain bodily constraints, he considers this liberation to "dialectically turn into servitude". True or false, the generally asked question is translated by the search for a certain social consensus temporarily placing the cursor "in the right place". The risk is then that habituation, as in the frog syndrome (that is, when a change takes place in a sufficiently slow manner, it escapes consciousness and, most of the time, arouses no reaction, no opposition and no revolt), may be translated by forgetting and an inflation of the fields of application; for example, going from repair to enhancement, as trans-humanists would like.

Independent of this aspect, "two types of discoveries can be made in science, some foreseen by theory, implying two met conditions: very advanced science, for example physics, and the simplicity of phenomena; the other are unforeseen: they arise unexpectedly during experimentation, not as corollaries of theory and specific to confirming it, but always beyond it, and consequently contrary to it" [BER 55/56]. Considerable scientific openness authorized by additive manufacturing processes allows all sorts of audacity, particularly in strongly interdisciplinary fields. On these disruptive bases, some scientists believe that bio-printing will thus considerably magnify the strength of current technologies, on the one hand, and give birth to new technologies, on the other hand, whose enormous possibilities are still not well measured (for, beyond the applications that are currently imagined, they will likely lead to unsuspected applications if they succeed).

Even before applications, fundamental research exploiting this subject of convergent technologies hopes to bring about new properties of living matter, unknown phenomena that will surprise us: the researcher will then have succeeded when he/she is surprised by the result (serendipity); this makes philosophers say that there is no longer a search, as with Descartes, for the mastery of matter, but the non-mastery; it is hoped that the whole will be more than its parts, even built from the elements that constitute it, whereas science, by necessity, traditionally reduces the whole to its components in order to be able to study them, because the whole is too complex. The convergence of technologies (see section 3) would allow it to be measured, to be mastered, with the complexity of reality, even to be recreated or created anew [EGL 14].

Moreover, what is more complex than life? One of the ultimate dreams of this exploration of complexity could be seeing the appearance of life as a property emerging from the elementary bricks (e.g. stem cells), even molecular constituents. Although remaining a young discipline and momentarily lacking tangible results, biological research on the profound (genetic, metabolic, molecular) causes of aging are seeing a booming growth. Numerous researchers agree today in saying that the life expectancy could rather easily be pushed up to 120 to 150 years for Man in the near future given current progress in the area [LOU 15]. Bio-printing, in principle, can participate in this goal. Thus, in the era of modern biotechnologies, it is not only capable of modifying the living, but even of manufacturing “pieces” of the artificial living. Magnien [MAG 13] poses some questions stemming from ethics:

– Risk/benefit relation: for example, “bio-security” and “bio-safety”, notably with problems of the robustness of prints.

– Ethics of risks: responsibility/precautionary principles.

– The pursued goals: example of the difference between the “repaired Man” and “enhanced Man” thanks to bio-printing, with all questions on the meaning of Man that this poses. Here, we find the usual and redoubtable question of the limit, to trans-humanist goals. It is the “power to be oneself” that Habermas [HAB 02] will discuss in terms of the impact of biotechnologies.

– The relationship to the living and life: bio-printing may challenge our points of reference between the natural and the artificial, the living and the inanimate, and interrogate our responsibilities in the potential generation of new biological beings.

– The difference between life and the functions of the living (from the living to the lived): The functionalities of the living must be distinguished (even if they are connected to it) from the exercising of these functions in the lived, partly conscientized for Man.

Let us not forget that ethics is not there to slow the development of science and technology, but to help discern what is going towards the human and thus making modern technologies perform better. A proposition by Sicard [SIC 06] explains the idea: “Ethics is not the vigil that captures, for its own benefit, the paths that are opening to the imaginary, intelligence, and human creativity; ethics is not the high priest of immobilism”. The challenge of our era is the foundation of a necessary balance between the progress of different scientific, technical, ethical, cultural, social, economic and political spheres. It is a matter of proposing an educational stance that brings “benevolence and ethical vigilance” together concerning today’s bio-printing, all while pursuing the coupling of research and ethical questioning.

On this basis, in the educated public, closed supports are found for the development of what some call “the enhanced Man” because they allow profound evolutions in humanity (well-being, longevity, enhanced performances, immortality, etc.) [KUR 06, SIL 97, KAM 15]; others are formally opposed [FUK 04]. Then naturally the question is asked of the position of research in this seductive and alarming context. Are there rules to frame it or ethical foundations that authorize works? The question of the links between body and repair/enhancement via bio-printing is not a given. Rather broadly, the very idea of the “technicized body” refers to multiple registers. The coupling “body and technology” produces multiple representations on Man and society that, for some, go towards well-being, security, performance and a healthy life, while others can more easily point their finger at obscure facets in a legitimate way.

After having tried to return to the very context of the concept of ethics (see Figure 4.1), it seemed interesting to examine in some general cases how, in the current situation of a research world strongly supported by contracts, including many with companies, it is possible to advance, but attempting to escape a "Pontius Pilate" position to take part in a responsible approach. This is the path proposed below. In a "Jonassian" logic, Fressoz and Pestre [FRE 13] remind us, "[The risks] are no longer side effects of progress; they constitute the primary challenge of our societies. The diverse knowledge that allow invisible but universal knowledge to be known, the kind that produces technoscience, is thus found at the heart of the political and must become our guides". The task is thus quite intensive.

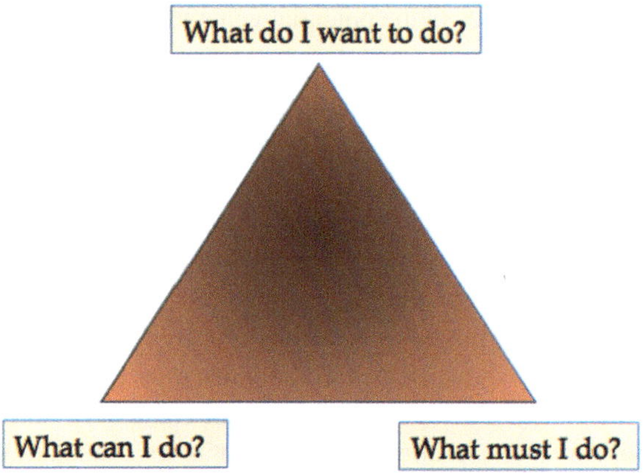

Figure 4.1. *The fields of ethics*

4.3.2. *Return on the concept of ethics*

Ethics comes from words meaning "moral science", "place of life; habit; mores; character; state of the soul; mental disposition" and "morals". It corresponds to a practical philosophical discipline (action) and a normative one (rules) in a natural and human environment. It aims to indicate how humans must behave, act and be, among themselves and towards their surroundings. The rapid evolution of the world (with a double role of spectator and actor) poses problems that no human concerned with progress (which is already an ethical attitude) can ignore because the application of bio-printing on humans can transform them, for better and for worse.

It can be believed that it is humanity's conscience, and it alone, that must indicate and give meaning to these transformations. But how do we define good and bad? How is the cursor placed between these two concepts? How to differentiate between truth and veracity [KLE 08]? Who decides? Here we have "loose ideas" of the questions asked in research that is extolling because it is open to the unknown.

"The notion of free will in the choice the individual makes concerning their body is one of the essential principles of the laws of bioethics. Autonomy of decision and the free consent to a physical performance improvement technique require each individual to have the knowledge to understand, judge the possible consequences on his/her health, and express a completely voluntary choice, with no pressure" [MAL 16].

In an attempt to illustrate the difficulty of the ethical question, Nordon [NOR 90] wrote, "The certainties common to all Men in a single society are so intimate to them that none of them perceive them. Invisible, they are self-evident like a healthy organ, until, potentially, elapsed time makes them ill, replaces them with others, and gives the necessary space to explain them *a posteriori*". These few words highlight some very important properties for the exploration of the matter: belonging to a community of values, the notion of emergence, emerged problem and the possible reinvention of history in the name of new classifying criteria. Indeed, everything coexisting requires a minimum of agreements and rules on the "good" and the just, but also on the integration of less commonly accepted interpretations (the aforementioned "cursor" problem) [FRE 02].

However, in the facts according to Ayache [AYA 16a], "universal" ethics does not exist and the world is envisioned through fragments with specific targets that are distinguished through their degree of generality (applied ethics (extending to specific concepts like business ethics), for example, lack the degree of generality of general ethics): bioethics, medical ethics, handicap ethics, end-of-life ethics, environmental ethics, animal ethics, research ethics and even commercial and economic ethics. They are also distinguished by their object (such as bioethics, environmental ethics, etc.), or by their cultural foundation (which may be the habitat, the religion, the culture of a society specific to one country, a social group or an ideological system).

In every case, ethics aims to respond to the question "How best to act?" Ethics has both feet strongly planted in the real: it is not just a set of abstract concepts. This notion is always marked by nuances: nothing is black or white. It is thus necessary to know how to nuance the shades of gray with the "optimal" search for the cursor's position. But if this notion can be confused, how, and in the name of which ethics must a researcher be stopped from trying to understand? "Such is the path that the ethics of our times has chosen to withdraw from the world: parcellizing it, fragmenting it, atomizing it, pulverizing it – literally: analyzing it" [HUN 16]. Ethics has become analytical and for Hunyadi [HUN 16], it is the surest means to subtract the world itself from all fundamental questioning and to "allow the systematic forces that govern it to develop without constraints".

Ethics has thus become, in some way, the accomplice of the economic system based on the Rights of Man, verifying that the liberal ethics of individual rights is respected, thus that wrongs are not inflicted, and that discrimination is avoided and private life preserved. Normative ethics exploiting various committees, but which, if well reflected upon, serves to validate what they are henceforth unable to criticize. Mark Hunyadi [HUN 16] takes the example of the precautionary principle, which he has long studied: "In risk management, the precautionary principle is brandished to avoid excessive or unreasonable waste, but it is to help the technoscientific device better fit into its whole. The precautionary principle does not serve to criticize the technological influence on the world, but to make this influence more fluid".

Independent of these aspects, Ogien and Tappolet [OGI 09] think that it is practically impossible to define very precise criteria to delimit what could be good (or bad); highly dissimilar elements can be qualified as good by some whose judgment also depends on very different considerations. The devil does indeed lie in the details! To drive this home, when speaking of ethics, we are also speaking about morality, deontology, even for the most educated of consequentialism and rational choice theory! According to the authors quoted above, the following definitions can be set forth:

– The ethics of virtue: an agent must carry out an action if, and only if, it is the one that a virtuous agent would carry out in those circumstances.

– Deontology: an agent must carry out an action if, and only if, this action is required by absolute moral principles that are applied regardless of the consequences.

– Consequentialism: an agent must carry out an action if, and only if, this action promotes good (with equal consideration vis-à-vis everyone).

– Rational choice theory: an agent must carry out an action if, and only if, he/she is maximizing its expected utility (if a goal is considered good and the agent has the choice between several options to achieve it, he/she must choose the one that allows him/her to do it in the best way).

These tautologically coherent definitions use slightly commonplace words such as good, values, virtue and morality that can claim to explain extremely complex things in a simple way. For example, is the list of virtues that characterize character traits to be valued to achieve "happiness" universal? If not, if some large principles can be agreed on (the 10 Commandments, for example, for those who respect the Old Testament), from one region to another, from one era to another, there is a risk of getting involved in concepts of peaceful cohabitation that differs greatly in their foundation and expression (some examples are presented in Table 4.3, illustrating the concept).

Period	Names	Characteristic elements
Greeks	Aristotle	The golden rule of ethics is seeking the final cause, which will find its realization through the right measure, the search for the proper average, in view of happiness. The ethics of virtue or moral greatness seeks actions that make Man good (Aristotle).
	Epictetus	For stoics, ethics is the normative knowledge of human behavior whose end is upright knowledge and action. Epictetus believes that it is useless for an isolated individual to want to influence events that we do not master, and that we must only feel responsible for events on which we can have an influence.
Christian ethics		Do unto others as you would have them do unto you [LEM 14]
17th Century	Spinoza	"Good" is defined here as "what we know with certainty to be useful to us", rational and intuitive knowledge primarily being this insofar as they indicate to us how to involve our individual and collective action into the love of freedom for everyone.
	Kant	If we had to base morality on facts, it would quickly be ruined by them. One single issue: building duty, far from being a reality, is a norm of reason, valuable for all reasonable beings.
19th Century	Nietzsche and Bergson	We must escape imposed constraints and erect a geneology of values by asking ourselves what forces support it. At this stage, the free spirit founds the obligations to which it wishes to submit itself, beyond all institutionalized social constraint, in all intelligence and reason [SAR 04]. "Thought, by following the common thread of causality, can reach to the furthest abyss of being and that it is capable, not only of knowing itself to be, but even of correcting it" [NIE 89].

	Francis Galton	"Its goal [eugenics] is to regulate human unions so as to obtain the greatest number of individuals capable of composing a society considered to be the best" [DRO 95].
19th Century	William James	He emphasizes utilitarianism. Ethics is perceived as stemming from the utility for society. This standpoint often appears the most pragmatic.
20th Century	Hans Jonas	He places ethics in the domain of responsibility concerning universal risks that technoscience can run in relation to society and future generations.
	Jean Bernard	"With the progress of knowledge [...], the need to encourage its happy effects, to limit its perverse effects" [BER 90].
21 Century	Jean-Pierre Dupuy [DUP 02]	"Enlightened catastrophism" that imagines a worst-case scenario, catastrophic enough that we all agree not to want it, and credible enough not to bring us to do everything so that it does not come about.
	Trans-humanism	Genetic design, modification of cognitive and emotional functions, increased life expectancy, improvement of athletic performances [KUR 03] Genetic lottery versus genetic choice Society must accept great diversity of what is the good life and admitting that each individual has the right to defend his/her own conception of personal fulfillment [MIS 09]. Humanity+, formerly the World Transhumanist Association (WTA), is an international organization campaigning for the ethical use of new technologies to proceed to the improvement of Man [WIK 15f]

Table 4.3. *Some options concerning "peaceful coexistence"*

However, the emergence of an ethical issue is not a natural and spontaneous phenomenon; it requires questioning, generally supported by a pre-alert, a premonition, reasoning through analogy (e.g. a parallel between asbestos and carbon nanotube in the scope of a health risk). Imaginaries then constitute a powerful action matrix of technological development, insofar as they draw the horizon of the thinkable [CAU 06, CHO 06] that orients the strategy of the actors involved in the field and naturally the desirability of the technology, even more simply of its "acceptance". On this subject, this is because bio-printing could aspire to a new form of control over living matter, because it could instrumentalize the living that it obliges to renounce the positivist illusion that scientific statements are taken from all metaphysics, to accept questioning on the representations and values underlying research programs. The fear of sorcerers' apprentices flushes out the ethicist because bio-printing can spread out on a foundation of going beyond fundamental cultural values, particularly distinctions between nature and artifice as well as between nature and culture.

Let us be clear: it is not a matter of refusing the transgression of borders between nature and artifice on principle, because all technology sooner or later goes over this border. A reflection on values does not necessarily aim to sanctify nature or the human, to install ethics of respect. It is rather a matter of becoming aware that, by manipulating objects, we are manipulating values, and of questioning oneself on the type of relations that we hope to install between these three fundamental poles of our civilization, which are: nature, technology and culture. Moreover, "substances like betablockers, melatonin, and caffeine allow healthy subjects to simply feel 'good,' as they always should have been, some of them say. Are they improved/enhanced? Are they drugged? The debate is not scientific, because this gray zone in the continuum between the normal and the pathological obviously refers to the WHO's definition of health and to the means a society accepts to have" [CCN 13]. But, as Benvin's provactive Figure 4.2 suggests [BEN 15], how far is too far?

Figure 4.2. *Apocalyptic vision of one application of bio-printing to "enhance" humans (according to Marithé http://marithe.over-blog.com/article-28754085.html)*

Concerning the ends, it is certainly legitimate to state that science has its end in itself, rather than in applications, but this stops neither the researcher nor the public from wondering about the ends of current bio-printing research, the legitimacy of agreed-upon investments for the exploration of this world that is being built in laboratories. These questions of meaning are mostly common sense questions that each researcher, like every citizen, should ask himself/herself: Why this research? Why this rather than any other? To what needs for the good of society does it respond? Whom does it profit? Does it justify the agreed-upon investments? Who will assume the responsibility if this goes badly? (see [THE 13]).

In order to allow the research actors to exercise their responsibility as citizens, it is indispensable to guarantee their right to report their questions and to be understood (but beforehand, it would be best to inform them to make them into "enlightened citizens"), the ethical committee of CNRS [COM 06] and the Charter of Socially Responsible Research (SRR) of the INSIS-CNRS [AND 13]. Several questions are associated with this need, the consideration of a field of possible unexpected events that lead to (or will lead to) a debate by a non-instrumentalized social body and its clarification due to the process of euphemization, banalization and lateralization of the imaginary associated with the attractiveness of the technology in question [TOP 13, GRA 03]. This is indeed what is covered by the SRR aiming to unleash the possible impacts of research on humanity (directly or indirectly) and on his environment, which must situate these impacts in the technoscientific dynamic, but with a cultural field still mostly untouched by a number of "hard" scientists. With reculturation, "neutral" information and active and informed monitoring are thus necessary to serve as a counterbalance to certain economic tactics and strategies obsessed with immediate gain or certain ideologies such as trans-humanism.

Nevertheless, even if science does not yet authorize it, it may be necessary to consider this reflection by Liogier [LIO 10], who writes, "Contrary to the ethical process that puts the cart before the horse, if you will, by hoping to provoke adherence, profound belief in humanity, through simply intellectual adhesion, at the very level of the most deeply rooted desires". Are we sure that philosophers and ethicists are not (too) ahead of the facts, the possible "manufacturing" of a Man taken from the fatality of his original limits [MUN 13]? But does wanting to keep Man as he is stem from another ethical question or not? Still if there is a desire to compensate for physical and mental deficits, why and on what bases they forbid improving a healthy body? To achieve what goal? That of successfully creating the technological utopia with its social transformations [MUS 13]? Incidentally, the CBHS in the USA [CBH 15] warns of possible detours, recalling the aspects of health risks (authorization to release on the market, called AMM or "authorisation de mise sur le marché" in France) and ethics aiming at human enhancement. If the AMM can reasonably be foreseen at the moment with biocompatible inert materials (e.g. case of titanium), it is not currently capable of bio-printing associated with cell therapies (see also Dodds, [DOD 15]).

In this slightly disconcerting décor, would it not be necessary, as proposed by the CPP – the French Ethical Committee for the protection of the citizen – [CPP 10], to recommend research in the field of:

– promoting an in-depth evaluation of the available options for managing risks, whatever shape and nature they may assume;

– taking responsibility for the wealth of formulations of one question, of the diversity of the representations that are associated with it, and of the interdependencies with other questions (see Le Marec, [LEM 02]);

– adopting a process that is revisable in time as a function of new knowledge and its added possibilities;

– associating the decision-makers to do this so as to reinforce action decisions (see also NCB, [NCB 16]).

"The general objective of these recommendations is less to develop ethically correct research through a series of norms or prohibitions to respect in a practice entirely focused on the means and positive results of research than to develop ethical vigilance through a series of measures meant to encourage reflection on the values and ends of research. This is a profound change of mentality to be undertaken in research, where ignorance, even reticence, remains overwhelming with regards to ethics. This change demands awareness on the part of researchers from every discipline. Awakening ethical reflection on science and technology cannot be conceived as a temporary loan that could be contributed by specialists in ethics. This awareness will take time and the measures must be sufficiently significant to allow long-term work" [COM 06].

Box 4.3. *Recommendations by the Ethical Committee COMETS – CNRS – France [COM 06]*

4.3.3. *What can be foreseen?*

"Imbeciles find this world reasonsable because it is wise, whereas life shows us every day that it is among the perfectly unreasonable wise, that science does not necessarily grant common sense or virtue. The modern world that brags of the excellence of its techniques is actually a world delivered to instinct, I mean its appetites. This is why it is oriented towards experiments that only seem daring because they are in no way proposed by reason, but inspired by instinct. It draws its vanity from the fact that these experiments are new, without great concern for knowing if they can be performed or not, because it prides itself on being able to conquer every difficulty through its techniques If such experiments cannot be performed, technology could not, however, allow them to be pursued to the end, but they are likely already capable of taking them far enough to make them irreversible, that is, to involve our species in dead-end streets" [BER 53].

If "no human problem can be resolved without humanity" [LEV 03], several kinds of considerations can be foreseen at the stages of research, ethical expertise or applications.

4.3.3.1. *Research: towards a responsible approach*

"We believe that at all times, we could, given a desire, master everything through anticipation" [WEB 59]. The emergence of interdisciplinary works associating legal experts, philosophers, economists, sociologists, specialists in risk management and vulnerabilities, etc., with the best scientists (in a still limited number) working towards the technological development of bio-printing is indispensable for the systems of governance in question to be better known and for their pertinence to be periodically re-evaluated. On these bases, the normative ranks should be possible to anticipate in time and thus allow the benefits of this emerging technology not to be radically challenged by the appearance of unexpected risks or arguments that are harmful because they are unfounded. The studies on the implication of publics in emerging technologies, the epistemology to construct for bio-printing, the training of future generations with new knowledge, the economic and structural organization of research and its development, the legal frameworks concerned by the field, etc., are so many guarantees to be granted to the responsible development of this sector (if only so that it can develop harmoniously) [LAR 10].

However, visibly, there are as many moralities as political parties, social classes, schools of thought, religions, nations and homelands, and all these moralities, some of which are contradictory, could honestly be opposed to "universal" ethics. There is no reason to do away with it, the inexistence of morality because those who develop this theoretically nihilist attitude would not only approve of others taking what belongs to them, disregarding their promise, abusing their trust, etc. There are thus some invariants to be researched to determine what, at a given moment, can be considered true and fair by a large community. Ethics is actually born of the confrontation of local values (recognized by a group) and the real that is being built. "Pure" science can be led to deny the former, without being able to get involved with the latter due to "mutilating" and reductive disciplinary research (all while supporting it, because the search for the truth does not give it meaning).

By exploring the direction of the application, particularly biomedical, it can no longer justify the slightest distancing from an ethical question [CHR 91], "even if scientists react like most humans: they claim it is not their fault" [LEL 11]. However, putting ethical reflection into practice implies a sharp mind and a true will to think of the future, insofar as it is necessary to understand the processes that could be implemented, to adapt to the possible reality, to be "in harmony" with society, whereas there are numerous specific micro-research projects mixed with a high number of decisions and interventions before the application can see the light of day [RIC 94]. Therefore, not everything is accessible to the researcher "who is fully focused on" or involved in writing up a project or a publication.

The development of scientific education normally allows a growing minority made up of people belonging to the most diverse trends imaginable to gain greater and more responsible understanding of the problems posed to all of humanity. It is largely the job of this educated part of society to reflect first on the belief of the fatality of progress or development concerned with the society as a whole. However, the sought-after democratic dialogue likely cannot foresee the direction of the progress expected from habitual practices with an essentially technical and financial (top-down) origin. The authority of scientific, medical and technological experts, supported by the State and industrial lobbies, can quickly diminish any debate and consequently limit democratic operations by imposing visions from the powers in place (even if some are disjointed) to the public that then no longer has access to the decision-making process, conferred on the national representation.

Yet, the questions of opinion(s) are not those of engineers and scientists; they have a cognitive origin, and are even speculative and globalizing. The responses of these "authorities" cannot be solidarized from "citizen" questions. It would be necessary to find a place of exchange to appreciate the knowledge of opinion for what it is, on condition of informing them, taking the time to listen to them, and wanting to take them seriously. However, if it is necessary, obviously, to be responsible, this does not mean that creativity and innovation must be curbed by a fundamentalist ideology, because it will always be necessary to fight against all conservatism and respect the constraints of financing and disciplinary evaluation agencies [MIR 13, GRI 15].

To come closer to this context, INSIS, one of the 10 institutes of the CNRS, has proposed a Charter of Socially Responsible Research (SRR) [AND 13] to avoid remaining in a tropism overly oriented towards certainty that could lead researchers to sacrifice reality to abstraction (calculations often hide ignorance, according to Bouleau, [BOU 13]). Indeed, research associated with the development of technology is an element that is really starting to be discussed by intellectuals, but which requires escaping "strictly scientific" discussion to also invest aspects of subjective perception [CHA 13]. If some hope for a prudent attitude considering the social dimensions of the technological innovations allowed by science before expecting to be enlightened by public agitation (induced by other intellectuals), it should be noted that technoscientific development underhandedly transforms the meaning of what is human (improved well-being, but plausible risks for health, the environment, diverse alienations, etc.).

On this basis, there is a desire to make scientists' disciplines and behaviors evolve so as to integrate the awareness of their social responsibilities into their research activities. Thus, the foundations of SRR, elements of the sustainable development charts, are:

– SRR covers social and environmental matters closely connected to the future axes and researchers of a team in the team's activity;

– SRR is not and should not be separated from the Research Unit's action strategy because it is a matter of integrating the social and environmental concerns into the activities;

– SRR is a voluntary concept;

– an important aspect of SRR is the way in which laboratories interact with their internal and external decision-makers ("employees", clients, close environment, administration, partners, etc.).

A non-exhaustive list of actions at this stage of reflection is presented below:

– traceability: this investment is essential to reinforce trust between partners, particularly with companies (quality management, good practices, etc.);

– responsibility: respect for regulations relative to research operations: protection of researchers and the environment;

– respect for the law, particularly concerning ethics (human experimentation, animal experimentation, computer science, freedom, etc.);

– originality/novelty: reflection on launching research operations whose result is not foreseeable with the knowledge of the "former art"; this means a better creation of effectively new knowledge through scientific in-depth research or the exploitation of interdisciplinary projects;

– conscious analysis of the uses of research results for the society (this is a useful but difficult reflection due to uses of the same concept for different ends); in-depth reflection on the uses of artifacts in the society: use and associated disturbances, recycling, risks of irreversibility, sustainable development, effects on the environment in the short, medium and long term, relations with humanity: risks, justice, social equity, development of the human personality, etc.;

– periodically revisiting the different items as a function of the advancement of scientific knowledge and its effects;

– mission to alert the hierarchy (administration) if new risk situations seem to emerge.

Beyond the respect for the regulations in effect, the researcher is responsible for his/her own ethics; he/she is, as expressed by Abel [ABE 94], "abandoned to his/her own responsibility" because there is not ready response, but this forces the researcher to define a "code" of conduct that he/she must attempt to uphold. In the pursuit of this quest, it would be necessary, as mentioned previously, to privilege the plurality of standpoints, that is, to have a more deliberative approach towards science and medicine, more controversial, in short, to practice new forms of deliberative rationalism due to their collective nature: learning to be able to create an object and a scientific problem applied to the medical world (with real patients!) with others demonstrating expertise. The organization of debates, already conducted by the research unit, may be a means of expanding reflection and better connecting science, technology and Society. But this is only a starting point.

If more controversial research is practiced, it is necessary to call even more strongly on the scientific and/or medical researcher's assumption of responsibility. At the same time, all exceptional situations, State reasons or emergencies invoked can create lawless situations. Thus, "escaping ethics" must not be equated to the fact of escaping law, which leads to legally conditioning the "unethical". Let us recall that there is doubled social responsibility for the researcher, precisely because he/she is a scientist, "savant", and thus, in principle, authorized to be "knowing" and an expert. Then, let us try to avoid this comment from Taleb [TAL 07], who writes, "The researcher is in the position of an airplane pilot so absorbed in his/her on-board instruments that he/she no longer knows how to turn his/her head to the window to see that the engine is in flames".

4.3.3.2. *Ethical committees*

The previous comments illustrate the fact that it is difficult to advance on one's own, even by debating in a research laboratory, to define the "good" for an entire community, but they also lead to questioning the "optimal" choice of the participants in ethical reflection, which must essentially cover a largely interdisciplinary field of investigation, and a field of awareness and opinions as large as possible, with a desire for objectivity on the part of every participant! Furthermore, without wanting to add to it, it is best to have a minimum number of people so that the exchange is productive. A wonderfully paradoxical injunction! The problems raised are so broad and major that they require the cooperation of several disciplines and demand new methods of investigation. Research can certainly be locally disciplinary: the "hard" sciences, law, sociology, etc., must stop there. However, no specific discipline can conveniently solve them on its own. Bioethics, whether it is applied to bio-printing or not, is necessarily interdisciplinary [RES 73].

In fact, consulting different partners is expressed through a diversity of schools revealing the dimension, complexity and tensions of the emerging domains. It results from the difficulties of "cooperation" between strongly heterogeneous actors. Indeed, there are already ratios of power between technological, disciplinary scientific, knowledge-transfer, environmental, public health, etc., ends for scientists. These separate characteristics are often based on an appearance of a shared definition. In fact, the new domain, from a scientific or technological standpoint, is probably defined at least as much by ends, "systems of meaning", as by a field of questioning or effective problems or a list of industrial results and medical exploits, etc.

The existence of certain fuzziness may be due to an absence of individual clarification, but also to the possibility of some exploiting this unstabilized framework to act in an engaged way. In this case, dysfunctions, even ruptures of dialogue between stakeholders, can be expected in the long term. From this fact proven in cases where uncertainty reigns (e.g. managing nuclear waste, nanotechnologies, GMOs), knowledge of the diversity of representations is a prerequisite necessary for positive exchange action and honesty. It is indeed a matter of finding a position that will decide between the positions of technophiles and technophobes, often on heuristic and veracity standpoints (often without there being robust enough information to beyond one's view). On this matter, Klein [KLE 08] writes, "On the one hand, enthusiastic technophiles who think that new technology is safe so long as it has not been shown that it could be dangerous; on the other, radical technophobes who believe that no innovation is inoffensive so long as it has not been successfully proven that it is free of hazard".

Several considerations/interests may be at work:

– disciplinary considerations: new knowledge, innovation, creation of wealth, technical progress, care and medicine, technical well-being, risks, toxicity and resource management;

– ethical and moral considerations: prevention, ethics of life, preservation of life, intergenerational solidarity, sustainable development and long-term effects;

– personal conditions targeting the imagination of improved performances (from demand and expectation to addictive forms);

– scope of daily life: nuisances, pollution, hygiene, security, health, stress, living conditions and comfort;

– political and social framework: prior to employment, social control of decisions, civic-mindedness, democracy, neighborhood solidarity (e.g. waste management), "police" control and the associated security, and terrorism.

Using the best possible knowledge on the diversity of representations (which would need to be quantified, which is certainly difficult in the case of bio-printing, if only by affecting some points of each response to the proposed criteria), there is a chance to have clarification of the role/interests of the different actors, to valorize diversity and to reflect more in-depth on particular axes, but also to seek other partners allowing, insofar as it is possible, the debate to be balanced. In this sharing of a great deal of ignorance, it must be possible to report on the uncertainties of scientific and technical knowledge, approximations of knowledge, abuses of interpretation, limits of competency, and to measure at least the extent of the unresolved questions. This approach allows distances to be taken from every (?) ideology, overly reassuring proposals and abusive simplifications. But to do this, there must be a sufficiently calm or partially "neutral" framework to advance. But the neutral aspect does not mean a neutral decision, based on the search for a soft consensus [AND 13]. The construction of expertise must thus bring out the behaviors of each component concerned in the social body, the most intelligible possible and the most explicit possible. "By conforming to expectations and prescriptions, [the expert] makes his/her behavior possible to anticipate, thereby facilitating interaction with other individuals and integration into the group" [BEN 13].

Normally, but with a very modest citizen representation, it is on these kinds of foundations and values that ethical committees are created, whether it is the COMETS (Ethical Committee of CNRS), the CCNE (French National Ethics Advisory Committee) for the French State, the EGE (European Group of Ethics) for the European Union, etc. It is a matter of realizing pluralist expertise between disciplinary experts. But “peaceful cohabitation” demands a minimum of understanding on the good, the fair and the reasonable, on the means of combining contradictory interactions from every representative of the social body. In the end, the interactions with the society must, with acuity, pose questions of how to think of ethics in an advanced democracy and how to articulate it with scientific and technical progress, with differences between the approach based on an organization listening to every citizen and/or a more reductive vision but, in principle, more immediately operational. It is best, in forms of interaction that largely remains to be invented, to foresee the emergence of specificities, the progress expected by society and the constraints, and also the lacking information, in a case-by-case manner (to escape sterile generalizations). It is actually necessary to be able to respond to questions that are raised as being vital concerning later political decisions on the applications of bio-printing progress.

Today, scientific experts are no longer considered harbingers of uncontestable truths, being able to express themselves on complex subjects with a single voice. Moreover, often financed by certain companies, within “New Public Management”, they can, sometimes understandably, be suspected of partiality. What is generally observed with an educated population in a social context that has greatly changed since the “Trente Glorieuses” after the last World War (and that is leaving the present reflection) is that the old bipolar schema between experts and laymen is disappearing. It is giving way to a strongly heterogeneous pluralist schema in which multiple groups of interests get involved to face one another or sometimes, on a particular aspect, to cooperate.

Furthermore, Bensaude-Vincent [BEN 13], in another framework, poses the important question of formatting questioning, focused on benefit/risk aspects (see Figure 4.3), without other, less reductive aspects emerging in bio-printing research for medicine, every researcher being “obsessed” with worldwide scientific competition. She believes that this closedness potentially encourages extremists and confrontations. In fact, the public could thus be forced to get involved, with a manager’s rationality, in the dichotomous exploration of this benefit/risk ratio. “Balancing the risks and benefits like an accountant gives the impression that everything is under control, that the situation is being managed” (situation in which very heterogeneous elements (e.g. the possible problems emerging in different timeframes) are shamelessly associated with immeasurable aspects (e.g. peaceful cohabitation and the image connected to the use of additive manufacturing technologies on humans, etc.)).

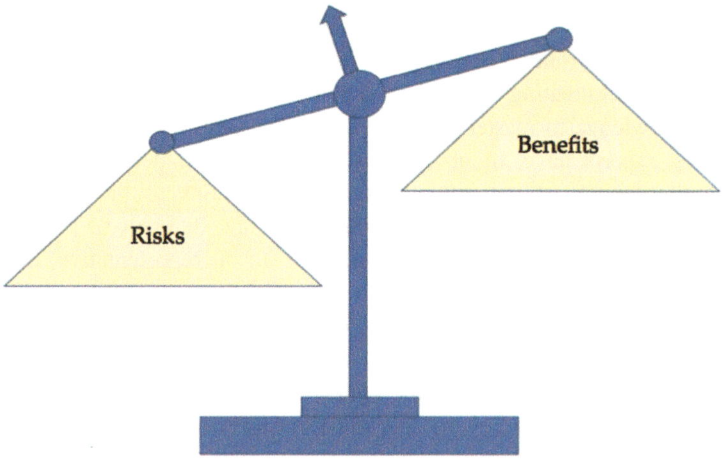

Figure 4.3. *Reductionism "benefits/risks"*

In the real world, part of the complexity stems from the irrationality of the actors, including that of the political actors (and their decisions, even their lack of decision), as well as from the multitude of impacts once an open system is considered open, as can be the case of bio-printing. Specific difficulties appear in the technical world: identifying the entities that will play a role in the system's evolution, their definitions, their roles, the rules that humans apply to them and authentification processes, mastering specific associated risks [MOR 90]. However, in the exploration of this complexity, it is essential to master the effects of one parameter on others. When technology and the living are brought together, things become even more complex.

In this so-called systematic framework, the concept of complex interdependency refers to the idea that every actor or parameter is sensitive and vulnerable to the behaviors of the other actors and vice versa. How then to justify a reductive model that only considers the preponderant values coming from former knowledge, without seeking to verify the potential importance of factors considered to be secondary and their effects on the complex interdependencies that may be the source of significant problems? It is indeed in light of this reflection that the difficulty of "good" ethical expertise appears, without it being easy to know if new research actions must be undertaken or if it is necessary (or not) to go more in-depth with others.

These few comments show the challenge for experts to master all of the ethical risks, especially those located at the interface between normally disjointed scientific and/or technological disciplines. This situation, where the field of involvement is strongly concentrated on scientists' knowledge, does not consider (or may not consider) expert and lay knowledge from other disciplines or other forms of knowledge. It risks creating distrust in the ability that may be shown by the medical applier of knowledge to master a complex situation, still not implemented. All the same, these experts have an obvious useful role in the definition of norms to authorize the launch of reparative medical applications of bio-printing on the market, human experimentation (Huriet-Sérusclat law in France), playing on the development of a regulatory system adapted to a representative democracy. However, in the uncertain fields that are the object of discussion, it is still difficult to find a serious scientific justification to be content with a certain level of trust, which incidentally is not even calculated using mathematical models (not very robust in this case). This is another form of the ethical and/or political question that cannot stem from the scientific framework that must foresee what may be an "acceptable" risk [LEM 11].

Therefore, without wishing to oppose the honest aspects of a society, which has ambiguous positions vis-à-vis technical progress in medicine, against that of experts, there is a need to reflect on the best possible interactions with citizens or patients. Comblin [COM 03], however, thinks that institutions seek rather to standardize people, leading for some, who wish to escape the rule, to feelings of guilt or frustration. Transgression is not far off. To drive this home, Fuster [FUS 75] reminds us, "Being virtuous is a good process for getting off the hook. Intolerance is always on the side of virtue [...]. It is in the name of virtue, of the great principles of human promotion and purity, that victims are piled on the scaffolds and in crematoria".

In short, the cursor's position between "good" and "bad" is not always easy to define! It may then be useful to remember that a tolerant analysis of the facts and their prolongations involves not disdaining the opinion of others, avoiding a good conscience and respecting the ambient dogma. This statement means that such an approach must bring all the forces face to face, explore the foundations of opinions and behaviors, avoid superficiality, and not select only the pertinent data (that responding to the goal) from others [CIT 13, LAG 12], all while avoiding formatting individuals. How to avoid autistic forms and study "the blind points of our discourse, which often fairly point to those undergoing progress, those refusing some of its forms, those dreaming of other possibilities" [PES 10].

However, the difficulty is being able to anticipate problems that could emerge in the long term with major aspects of irreversibility and interdependence [DUP 02, BEC 09]. How to be sure of having made the right decision? "The action takes place in a context where all large-scale employment of an ability unleashes, despite the agents' good intentions, a series of effects closely linked to the beneficial, immediate, and intended effects, a series that, over a cumulative process, leads to harmful consequences sometimes exceeding the desired goal" [JON 97]. Certainly, the precautionary principle [AUV 13] must be used, but will it be enough to maintain trust?

4.3.3.3. *Applications*

By taking the close but much more widely mediatized example of nanotechnologies, if citizens are questioned (see, e.g. the French public debate in 2008-2009), they are generally opposed to their use. In this case, it was possible to bring about a greatly dissymmetric schema opposing experts and the public; its intentions then transmitted by the media. However, speaking of the public in general is forgetting the existence of a very heterogeneous pluralism with specific interests. In this respect, if a survey concerning nanomedicine is conducted [AND 15], the positions of the audience who responded are less opposed, even favorable to the use of nanotechnologies for their health. One of the questions was, "Research work being heavy and requiring time and consequent means, should the results of research be awaited or should care be given taking every precaution possible?" The response is clear, since more than 60% of respondents are very favorable or favorable (see Figure 4.4), which complements the information provided in section 1. From a sociological standpoint, it would be interesting to examine how, over several decades, the transition towards growing individualism has taken place, possibly explaining this result. Can we not envision a certain correspondence between nanotechnologies and bio-printing?

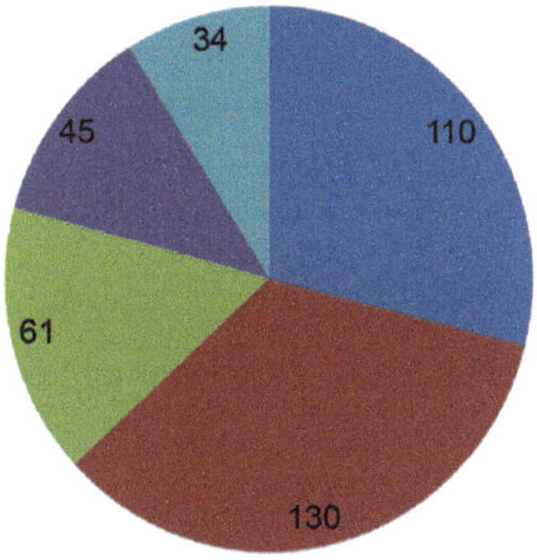

Figure 4.4. *Responses to the survey (dark blue: very favorable; red: favorable; green: rather unfavorable; purple: unfavorable; light blue: no response). For a color version of the figure, see www.iste.co.uk/andre/printing3.zip*

In the economy of knowledge explored today, bio-printing technologies propose radical transformations in unique offers desired by users. The question asked by Michèle Debonneuil [DEB 07] is knowing if we are engaged in what is called the "quaternary revolution", towards qualitative aspects corresponding to the market launch of new products, services, or systems that satisfy a reinforcement of a "well-being demand", for which it is acceptable to pay a certain price (notion of attractiveness) and not the lowest price or faster production. There are then new challenges to overcome. Among the structuring and important factors, it is necessary to take into consideration the place of "baby boomers" in our fully evolving society. Indeed, the baby boomers, great in number, have the time and money, even if their skill to master certain innovations is sometimes considered modest. They have a high level of education and are aware of the new stakes affecting their health. Yet, it is now recognized that, in general, these are not the marketing exercises that awaken an interest in individuals for a product; it is people's intrinsic needs that create effective demand, which then can spread to a broader public. Then, the profile of the population has a crucial importance for choices made, leading the other components of the social body into the dynamics of production or service created. The social agreement comes at this price!

However, in the realm of facts, current bio-printing micro-firms aiming at applications in medicine in France, following in the footsteps of biotechnology firms dedicated to the life sciences, see (as with "classical" additive manufacturing) constraints limiting their developmental potential, even threatening their difficult durability. Among these constraints, it is especially worth mentioning the structural difficulties proving the existence of a real, financed market, reaching a critical size, the lack of adequation of public aids and supports with the never-ending wealth of research and R&D costs, the reluctance of financial markets to accompany technological projects perceived as an overly uncertain investment in terms of industrial, medical and commercial applications, and manufacturers' willingness to get involved only in solutions contributed by bio-printing to problems not resolved by classical medicine [FLE 11].

In a universe controlled by the liberal system, it is necessary to legitimize (like in a speculative bubble) the reality of the "bio-printing" project that, as in the case of nanotechnologies, has a dialectic of promise implying the production of expectations/promises and the construction of the field in the same movement. The future of the technology must thus continuously be negotiated by all the producers of promises (from the researcher to the financer and then to the manufacturer) who have convergent interests in the operation [SEL 07, VAN 00].

Alliances, partnerships, even buyouts of bio-printing start-ups should be formalized at an advanced stage of the innovation process, particularly in clinical phase II or III (but this stage has not yet been achieved today). Risky-research is not very attractive to private investment notably concerning the uncertainties and mortality rates that characterize this kind of technoscientific project. The same logic presides today in the construction of the business model of nano-medicine firms strongly dependent on external investment (capital-risk).

It is at this phase of the research and R&D process that public aids and supports become crucial. They opt for a model convergent with investors' expectations [SAD 14] and markets that tend to require lower development cost budgets, fewer risks (or with mastered risks) and generating profitability that is best in a short processing horizon. This repositioning imposed by the lack of external financing is not moving towards all the efforts accepted by France in the construction of the medicine of the future and its international competitiveness in the field of healthcare. Evolutions are thus slower than philosophers had imagined, but with the risk of financing that is strongly oriented towards lucrative applications.

4.3.3.4. *Is time on our side?*

4.3.3.4.1. "Thou shalt not kill"

Breaking away from the needs expressed, conception processes gradually define the artifact that must respond to the chosen objectives (needs and constraints) through successive choices with more and more detailed approaches (and thus ones that are more and more irreversible in the limited term of the action). The possible (health) risks are studied (in principle) to respect the regulation in place in terms of risk management [BOU 13].

This reality, rather classical in innovation, leads to greater concern over the possible emergence of the technology, breaking down the necessary doors to the desired development. But it does not seem like it will still be easily possible to propose robust medical applications for "true" bio-printing. Thus, some may think that they have time. Several other arguments, more external in nature, may lie at the origin of the limitations mentioned:

– Interdisciplinarity: the fact of being at the crossroads of several disciplines poses the question of scientific identity, of its recognition and evaluation. Engagement at an interface can actually penalize a researcher, his/her ability to publish being notably reduced, at least at first. From this arises a difficulty with the corresponding recruitment and its follow-up. His/her practice in laboratories should be encouraged, but really is neither sufficiently valued nor well financed.

– Integration: the fact of developing new bio-printing processes for medicine is one thing, and the fact of integrating the development of matters and materials over the lifecycle into the development of processes is another. It would be best for interdisciplinary works to be developed so that, beyond the medical stage, there are no risks linked to the technology: during conception, use by medical professionals and, finally, in waste (environmental problems). This situation is more or less translated by the concept of "safe by design" [LEG 15] that still requires great exploration for bio-printing.

– Blindness: when early signs appear, no one really has experience with the hazard associated with the future and this leads, in the complex and opaque world in which we live, to neglecting the credibility of rare events (hence the interest of anticipation) [HER 09]. An opacity on these phenomena can, in some cases, be sought by organizations that have great interest in masking facts [PAL 12].

4.3.3.4.2. Medium-term ethics

Development in bio-printing research should be pursued in a supported way (permanent growth of the number of publications), but the expected visionary stakes have not been reached, beyond some quickly mentioned and rather uncreative biomedical applications (primarily prostheses). These latter serve as pioneers to convince others about the economic reality of bio-printing, and to do this, they based themselves on matters where care is a priority (and thus largely accepted by the population). The question of the cursor's position is normally not posed between aspects of "good" or "bad", beyond the authorization of a possible market launch. If the first hopes (or first fantasies) have led to a number of media fairy tales, philosophers and ethicists are still not in control of the subject (which efficiently allowed the horizon to be limited); it must be noted that the lack of ethics (beyond risk aspects that need to be better explored) is not abundant. The reasons for this state of affairs are numerous, as was quickly mentioned in this volume, but particularly because there would be a need to invest in a very great complexity associating technoscience, biology and medicine in a meta-interdisciplinary framework, which would demand an unheard of human and financial effort, without any knowledge of whether the chances of success were great.

For Deloitte [DEL 15a], the accelerated rhythm of technological change (i.e. shortening the cycle of technologies) represents a challenge of synchronization. By regulating too early, there is a risk of preventing; by regulating too late when a service is overly generalized, it is possible for it to become impossible. New business models imitating the economy of sharing were developed by the "makers" (and tomorrow by bio-hackers?), which disturb traditional business models and pose unanticipated questions of regulation. The rapid drop in bio-printing costs could change the landscape both on the side of companies capable of realizing it as well as that of "clients" who may access it.

Yet, based on experience, the regulator ends to ignore a phenomenon until it becomes massive. "Innovation is in direct competition with regulation, which is a problem both for the public actor, who often reacts too late, and for innovators, who develop models that can be undone by regulation, when even their revenue and client base have become massive [...]. On the other hand, faced with the demultiplication of innovation, the regulator tends to expect innovations to be imposed to react. The problem is that this phenomenon can go very fast without delivering solutions on its own" [GUI 15].

This means that we must not wait until it is too late (even more so as bio-printing machines, though still quite "rustic", are developed in Fab-Labs, but ones that are evolving in number and sophistication, thus playing sorcerers' apprentices by practicing self-medication, following in the footsteps of piercing and tattoo shops), but reflect before acting, if only by examining the most urgent situations at present as those linked to information and communication technologies (that are not the object of major challenges yet), nanotechnologies, or tomorrow morning, the revolution linked to communicating objects. But, it is perhaps only because bio-printing technologies will use their contributions that they might appear suspicious and therefore be the subject of an ethical debate!

4.3.4. *Conclusion*

If a system's operation can be explained scientifically by relations of causality, its construction is not understood without its end, nor without humans. According to Georges Canguilhem [CAN 65], "A machine is made by humans and for humans, with ends to be obtained in sight, in the form of effects to be produced". What should then be thought of the possible importance of bio-printing in disturbing human evolution theories thanks to the exploration of design ends that must be "mastered"? How can we separate individual choices from societal choices? How can we know if individual choices engage Society? How are we able to trace a borderline between enhancement and repair, with specific ethical statues [NAA 05]?

In this décor, researchers' mission can no longer be limited to the simple production of new knowledge and the transmission of the acquired knowledge. It is also not, in scientific expertise, quieting the worries of opinion, rejecting all "profane" judgment, expertise necessarily remaining the "privilege" of the scientists themselves [STE 13]. It must be part of the different humanist charters of research operation (existing even at the Europe-wide level; see, e.g. EU, [EU 05]). The social and economic impact of scientific innovations directly affecting humans grant researchers collective responsibility to participate in the necessary citizen debate on the stakes and priorities of scientific policy (ethics, risks for humans and the environment, etc.).

It is a matter of allowing scientists to face their social responsibility by expanding their initial training to the historical, philosophical, sociological and economic dimensions of their activity (as well as the destination of their creativity and the transfer of their know-how) or by encouraging their interactions with components of the Humanities and Social Sciences. By leaving the disciplinary realm, researchers need to "protect" this new form of responsible research to avoid oppositions and non-receiving ends when they must honestly reject certain facts and knowledge in the names of revisited ethics to have a will and effective support from the responsible authorities to see this cultural change take place, possibly opposable to simple disciplinary competition [STE 06]. What would become of bio-printing applied to medicine in these domains, the reducing factor of researcher evaluation, in these conditions [BIA 12]? But yet, this could be useful for a doubting society that (still) trusts research and must face powerful consumerist interests capable of disturbing the social imagination in a lasting way, even if the repair aspects seem strongly significant in medicine, as illustrated by the two previous chapters, with some challenges for tomorrow (see Figure 4.5).

In an attempt to pre-conclude through reflection, that of Andrieu [AND 13], let us consider what he writes, "By defending the thesis of hybridization, we are describing here a trans-corporal experience that integrates technological and biological aspects into a new human unity through dynamic interaction. This does not reduce the human subject to the mechanization of his/her handicaps". But at the same time, bio-printing technologies allow us to foresee "killing death" through hedonistic fantasies of a beautiful and competitive body (see Figure 4.5 according to Ayache, [AYA 16]) with the possibility of approaching these words from Munier [MUN 13]: "If the body is the place of death or illness, no longer the condition of Man's existence, but that of his limits, then, once the body is erased, death, illness [...] would no longer have a *raison d'être*" (see also Alexandre, [ALE 10]). It would not only be a matter of improving living conditions, to improve life expectancy, but of guaranteeing each individual a longer life, up to 120 or 150 years, and why not simply eradicating death, as expressed in this figure. For Benasayag [BEN 94], "the bearers of ethics demand, for example, the nomination, practically by decree, of real areas that are 'sanctified', as if the sacred could be created through arbitrary nomination [...]. Then, it would become a manipulatable consumer good like the others".

Killing death?

Figure 4.5. *Bio-printing, a path towards immortality?*

In this general framework, the EGE poses the question of creating technological improvement that could lead to forms of social segregation. The possibility of the industrial use of implants as a means of "creating [...] more performative bodies with economic ends poses the question of the limits to be imposed on such a use".

Well, is not it necessary to go back to the foundations and leave the ambiguities and non-choices? In this framework, Chavel [CHA 11] writes, "There are, first of all, cases where the situation is exceptional and affects the limits of the human experience, the one that makes the effort to imagine must recognize his/her ignorance, inability to conclude. The imagination is powerless, and it is thus good for it to be in a position where it is concerned with one form of respect of an exceptional situation that pushes human capacities to their limits, and because the moral attitude is then precisely to pronounce in a 'Rousseauist' gesture, a prudent and respectful 'I don't know' in the face of certain exceptional situations, which is a clear form of suspending judgement". To finish this part, let us leave it to Great Britain's Academy of Sciences in its reflection on medical innovation, one that integrates a number of questions posed on the prudence towards acting in the medical field!

"What is radically new is that it is henceforth possible to rewrite a face entirely", states Anne de Marnhac [DEM 11]. To be in line with the canons of beauty [VIG 14], surgery thus allows a long nose to be shortened, cheekbones to be raised, the lips to be repumped and even the eyes to be reworked. "We have full power over the subject, capable of redesigning one's genetic heritage to correspond to that famous ready-to-wear Western beauty". Bio-printing can return in this matter to the "perfection" that records it in a digital society. By making thousands of faces slide by on social networks, notably through the practice of the "selfie", there is a risk of conforming to one norm. "When Emma Bovary [an important figure in the novel written by French novelist Gustave Flaubert in the 19th Century] was only going to see some women in her life, we endlessly take in new images, offering, against the dominant model, a plurality of other faces". But following the example of Boris Vian [VIA 99], in "To Hell with the Ugly", it is still possible for some to demolish the tyranny of appearances. "We feel the rise of a desire for authenticity. In a society in crisis, the truth of the face-to-face and of the expressiveness of a face, in the way that Levinas speaks of it, appear as a need". Wait and see.

At the same time, with the dropping prices of 3D machines (even for bio-printing) and the possibility for around € 700 to have bio-engineering devices [L'EX 16], as well as the existence of open-source data (from Wever *et al.* [DEW 15]), it will be possible to keep living cells alive and to "practice DIY in your own garage" using fermented foods and drinks, but also to attempt to print tissues. This is the project of Canadian Julie Legault [LEG 16] with her laboratory of microorganisms that she makes open to anyone. The organic matter manufacturing kit Amino, created by MIT, allows work to be done with living cells and their life conditions to be managed. Its goal is to bring about better understanding of scientific advances via simple experiments, but with no control.

Ultimately, how can we handle the ethical questions of bio-printing? Can we make scientific knowledge and citizens' opinions converge? Can we put contradictions between the desired (or its opposite) and the possible to the test in a world that overexploits the field of artifactual prostheses? How then can we think of the coexistence between solidarities and materialist individualism? How can we intelligently write about the changes in the relations between humans and Nature? In the possible inversion of human finality? In our identity characterized by a geometry that has become variable in time and space? In our alienation, our subordination to technology and to those who possess it? The list is long like our path to define our freedom, so long as that goal remains attainable!

Each individual can appreciate the split in opinions between Europe and the rest of the world. The old continent hides behind principles and the *status quo* of prohibition when Asia is seen rushing zealously into the bio-economy that we must be aware of. Some, like Nick Bostrom, the founder of the Future of Humanity

Institute and director of the Strategic Artificial Intelligence Research Centre, show to what degree Americans show themselves progressively in agreement with technical propositions that they might have refused twenty years earlier. "And explaining what weighs on this change: fear of a risk, transition cost, evolutionary adaptation, affective impacts. It is implicitly understood that Man calculates the most efficient way to quickly move the masses. This is an interest that he shares with William Sims-Bainbridge, his colleague from the World Transhumanist Association (WTA)" [BRO 16].

In the United States, the Presidential Commission for the Study of Bioethical Issues charged the FBI in 2011 of developing a culture of responsibility in partnership with scientists of participatory biology. "The international collective DIYBio.org drafted a charter in 2011 to clearly confirm that 'biotechnology must only be used for peaceful ends, with respect for the rules of transparency, security, respect for the living, and responsibility, to promote citizen science'" [MOR 14]. Would not it be necessary to foresee this kind of operation, even if it will still be a while for bio-printing?

4.4. Governing bio-printing research: mastering convergence

First, additive manufacturing was created in the years 1984–1990 [AND 84]. If all 3D technologies are based on a principle of adding elementary volumes called voxels, this is linked to the spatially resolved photo-polymerization of a resin induced by light, advantageously (in 1984) from UV or IR lasers. It is called stereo-photolithography or more simply stereolithography (SL). The process involves polymerizing a layer of liquid monomer using light according to a digitized plan, then depositing a second polymerized layer according to the same principle, and so on. The part is thus gradually constructed. Today, stereolithography remains one of the seven basic additive manufacturing technologies.

Bio-printing (BP), drawing on additive manufacturing technologies, aims to manufacture living organs or tissues [GUÉ 17]. Relative to classical additive manufacturing technologies, printing biological elements adds a significant layer of complexity to processes because it is necessary to "intelligently" structure living materials, or not, mimicking the extracellular matrix and controlling the spatial distributions of different types of cells or biomolecules capable of playing a role in cellular differentiation, their growth and their death.

Secondly, nanotechnologies, governed by doing it, are considered by most developed states to be a factor of technical progress, a real grandiose form of technological utopism. When speaking of nanotechnologies, it is impossible to escape NBIC convergence, which emphasizes the growing interconnection between the infinitely small (N for nano), manufacturing the living (B for biology), computer sciences and technologies (I for information), and the study of the human brain (C for cognition), a true stake in studying and mastering complexity [ROC 01].

These two operations aiming at, in principle, rather disjointed goals, nevertheless have some common points: that of the convergence of disciplines and of technologies that require a specific eye to foresee that the associated promises can one day be fulfilled and for an effectively applicational vision on humanity. In the two situations, some think of the technical possibility of contributing to life. "The publicity of the neoliberal dogma has created ersatzes where the 'consumer', who can pay if he/she has the means, is temporarily given the illusion of being a superhuman who commands nature" [CIT 17]. From the social acceptability of the two modes of exploration through the science of fields providing diverse hopes to the implementation of governance strategies and pooling knowledge from different disciplines, this article, originally from the Nanoscale 3D Printing Workshop, Singapore on December 1-2, 2016, attempts to define and propose paths of progress for the development of these two fields that, when stabilized, could (maybe in a faraway future) be associated with performing nanoscale 3D bio-printing [AND 16]!

4.4.1. *Return to 3D printing*

4.4.1.1. *Additive manufacturing*

As explained in Volume 1, 3D printing actually includes a whole series of processes that share the manufacturing of objects through the deposition of successive, extremely fine layers of matter (or more simply, additions like melted wires), which are gradually solidified by an energy source (e.g. laser). It is defined by AFNOR [AFN 11] as "all those processes allowing a physical object to be manufactured, layer by layer, through the addition of matter, based on a digital object".

"3D printing is thus considered – along with mobile Internet, the Internet of Things (IoT), cloud computing, big data, the automation of skilled jobs, cutting-edge robotics, or even advanced materials – one of the technologies linked to the digital that is likely to profoundly transform [...] today's methods of production and, consequently, economic models" [ING 15] (see André, [AND 17]). It is, in the sense of Star and Griesemer [STA 08], a new "boundary object", "plastic enough to be adapted to local needs and to the constraints of the various groups that use it, all while being robust enough to maintain a common identity from one site to another". Innovation and process improvement for rapid prototyping have been turned in different directions [AND 15] that range from processes to materials to software with desires to attain decametric space or to reach atomic scales. Moreover, after 3D printing, 4D printing is also emerging with the possibility of introducing functionalities that can be programmed in time and space. Figure 4.6 recalls the emerging domains of additive manufacturing [AND 17], domains that are part of the elements presented in this current volume.

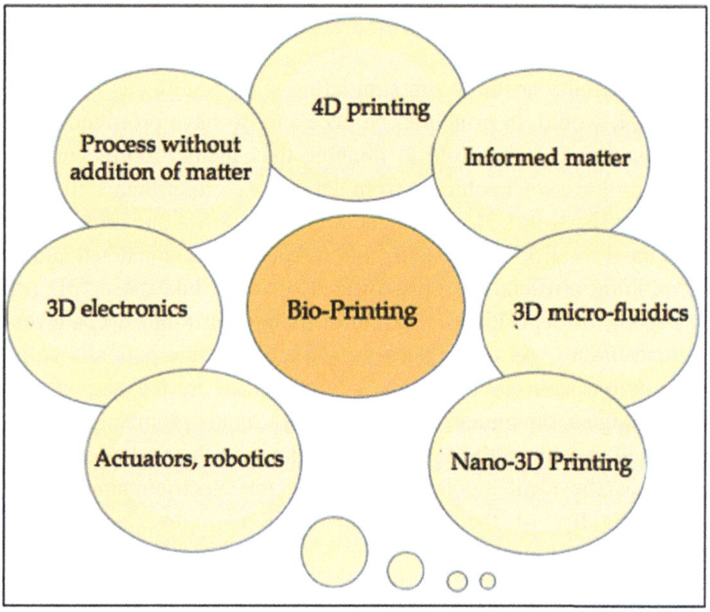

Figure 4.6. *Emerging domains in additive manufacturing*

4.4.1.2. *Bio-printing*

Bio-printing is an emerging subject stemming from additive manufacturing technologies [GUÉ 17]. It belongs to life engineering and integrates the physical and chemical sciences, mathematics and the principles of engineering to study biology, medicine, ethology and health: it aims to manufacture living tissues and organs. It must be stated that more and more sensational information is reaching the media (see, e.g. Loumé, [LOU 15]), pointing to the works of a team of researchers from the Carnegie Mellon University in the United States on 3D bio-printing [HIN 15] that could update a method allowing structures to be printed with soft materials serving to support the culture of cells. In these works, a sacrificial scaffold serves as the skeleton for the cells to provide the shape of a soft tissue. The authors have printed these soft materials, containing collagen and other biological fibers (alginate and fibrin), in a hydrogel support. The printed structure is then brought to body temperature to eliminate the hydrogel support without deteriorating the biological molecules. Thanks to this technology, the authors have managed to bio-print femur and coronary artery models. Furthermore, using magnetic resonance imaging (MRI) analysis, they have built a model of an embryo's heart. All that remains is for the cardiac cells to make the myocardium contract and an entire heart could be printed.

To create a partially artificial life simulating what Nature can make and even to go beyond this, it would, in principle, be necessary to have profound knowledge of the living. However, it is possible to imagine that, using "rudimentary" concepts, there will be progressive evolution from laboratory experiments to the repair of biological tissues [PAR 94]. However, the definition draws one remark: bio-printing cannot be adapted to all living systems, but to specific, personalized situations (e.g. a tissue) exploiting artificial systems. Instead of using ink, these "3D printers" of organs are capable of depositing a mixture of living cells ("bio-ink"), layer by layer, to form human tissues. As the organs are printed with the patient's cells (or with immunosuppressive stem cells), they are not rejected by the patient's body. The success of printing usable human organs therefore seems promising, so long as they can be developed to achieve the desired functionality in their microenvironment, that being defined as the local biochemical milieu, the electric/magnetic or chemical gradients, the rigidity of the substratum, its topography, as well as external mechanical demands. Behind bio-printing lies the idea of being able to model the living by thinking that it will be possible to go beyond the immense incompleteness of our knowledge of this world.

The original component of this concept resides in a principle of voluntarily and skillfully implementing matter to induce one or more of its spatial (and temporal) functions. On these bases, the engineer's initial vision was to make "simple", with the tools at his/her disposal. Using still "rustic", even simplistic, approaches, it is possible to manufacture parts using inert matter, in very diverse fields. However, it is necessary to consider the later goal of bio-printing: the cells positioned in the space must organize themselves, migrate and differentiate autonomously to form the desired functional tissues and to encourage the creation of a physiological microenvironment capable of maintaining the functionality of the cellular tissue.

The possibility of apparent spatial and functional determinism applied to the living milieu is introduced to the scientific action.

The milieu not only includes cellular heterogeneity, but also the establishment of the physical–chemical environment and paths of communication and exchange (e.g. vascular system, nervous system, etc.) necessary for the long-term survival of the tissue. Indeed, in a recent article, Munjai *et al.* [MUN 15] show the influence of the cellular environment on the behavior of the whole in terms of growth, differentiation, etc. It is thus a matter of making processes available to biologists that allow cellular suspensions, aqueous solutions or hydrogels mimicking the extracellular matrix to be deposited at the "right place" and the "right time" by limiting the different stresses that the cells can undergo through additive manufacturing processes. "The intelligence" of the process must aid cellular self-organization processes that naturally take place throughout embryogenesis, or even tissue remodeling. It is at this essential price, in order to go beyond proofs of concept, that bio-printing must emerge.

4.4.2. *Promises of NBIC convergence and bio-printing*

> *"We know well enough that the collective conscience is a miserable little plant that has a habit of fading when it is most needed" [EIN 53].*

4.4.2.1. *NBIC convergence*

A feeling of major scientific breakthrough can be correlated with the creation of the American NNI (National Nanotechnology Initiative) in 2001 and the reverberations it created in the field. In fact, nanotechnologies should be able to play a key role in practically every application sector: medicine, information and communication technologies, energy production and storage, new materials, manufacturing new structures, tooling, sustainable environment and development, security, etc. The acronym NBIC refers to four technologies, and nanotechnologies are technologies that allow progression towards performances that soon will take place at the molecular level, even if programming and modeling sciences are essential elements of the basic principles of this convergence [ROC 01].

The movement of human activities to the nanometric scale should allow industrial intervention at the heart of matter and the systematic research of the combinations between bits, atoms, neurons and genes, opening vertiginous perspectives: invisible and ubiquitous informatics (a computer in each object, communicating, moreover), a cyborg body increasingly hybridized with prostheses or supporting technical transplantations, modified creatures, brain–computer couplings, "molecular manufacturing" in the crosshairs, etc. The path borrowed since its origin by a humanity that is creating itself seems to be confirmed: a fascinating, even fun, path, but one that is narrow so long as the figures of crazy power or enslavement are present.

In this framework, discourse on convergence, its "great story", brings implications to be explained. It is tied to a vision of power (concentration, and secret), a relationship to nature (to be reconstructed), and above all else, a certain conception of that (its imperfection). It brings conceptions of knowledge that are specific (that is, in fact, brought back to engineering), and conceptions of the subject (cybernetized and reconnected) that are heavy with ethical and political consequences. Above all else, emergence foresees, without taking a step back, an upheaval of the conceptual categories between the human (based on neo-Darwinism), the animal and the artifact.

4.4.2.2. *BP convergence*

Bio-printing is based on the liberal promotion of a new conception of health and well-being focused on promises of commercializable innovations able to act in a profound way on fundamental biological processes (ranging from repair and maintenance to human enhancement) and, more realistically, on manufacturing supports representing human tissues to allow drugs to be tested (instead of working on animals and humans).

According to Lipson and Kurman [LIP 14], until now, Nature has remained much more skilled than humans and computers concerning choosing and using the different stem cells optimally! However, the vision is clear, supported by a public that is very interested in having access to specific care, maybe even to possibilities of "improving" human performances by awaiting the market launch of new products, services or systems that satisfy a reinforcement of the demand for "well-being", for which people will accept paying a certain price (notion of attractivity) and not the lowest price or the fastest production to access economically significant niches. Let us remember that, according to Nowotny, Scott, and Gibbons [NOW 03], "the domination of science and technology is […] a two-sided coin: through the production of "real" results and the creation of insatiable mirages. However, the mechanisms that bring out these desires, control access to their endless means of satisfaction, and regulate their diffusion come from the social universe".

Among the structuring factors important for supporting bio-printing, it is necessary to consider the demand of the "baby-boomers" in our fully evolving society. This group, which is great in number, has time and money, and wants to "remain/seem young". Yet, it has now been recognized that it is not generally the marketing exercises that awaken individuals' interest in a product, but rather people's intrinsic needs that create an effective demand. Then, the population's profile is crucially important for the choices made, bringing the other components of the social body into the dynamics of production or service created. This means that the opportune arrival of a new service is explained by the presence of a population willing to assume it, to use it to reach at least a part of its goals [DEB 07].

In this media (and scientific) war linked to the attractiveness of a process, Renouard [REN 15] reports the "hook" of the company BioBots, which conceives 3D printers capable of creating living tissues. Its products have been available on the market for a few months. In May 2015, BioBots, with sensationally promising software, presented a printer capable of creating living human tissues, reproducing Van Gogh's cut-off ear for that occasion and proposing the commercialization of 3D printers capable of creating life. In the case of success, an immense applicational field of this technology, which could have a potential worth of hundreds of millions or even billions of €/year, can be imagined!

4.4.2.3. *Differences and associations between promises*

Discourse concerning NBIC and BP fields stems from a strongly symbolic dimension and remains tributary to an inherited mythical thought. To seize all the profound stakes of these technologies, it is necessary to know their imaginary motivations, like trans-humanism [ALE 10, ALE 15]. Indeed, science and technology are, like every human practice, made up of images and ambivalent desires [CHI 09], with magical reminiscences. Owing to the primary subjects evoked in the symbolic NBIC and BP (the mastery and manipulation of inert and/or living matter, the transformation of Man, the heroization of the scientist, immortality), NBIC and BP stories can seem like the modern expression of a demiurgic challenge that Man is invited to accept, becoming himself the actor of a new religion: the religion of technology. These fields tend to build a new picture of matter and the body with interchangeable and programmable matters, the body able to be associated with information.

Table 4.4 gathers the characteristic elements of these two technologies, among which we can recognize "convergence" in terms of promises, oriented towards health, and direct or indirect forms of trans-humanism, even programmable immortality.

NBIC convergence	Bio-printing convergence
Quality of life	Health (repair, illnesses including cancer, etc.)
Enhancement of performance (and control)	Enhancement and recovery of performance
Man–machine relations	Regenerative medicine
Brain–brain interaction	
Genetic therapy	Research on model systems
Immortality	Immortality

Table 4.4. *Promises of NBIC convergence Bio-printing (BP)*

4.4.3. *Convergence*

The term "convergent technologies" aims to integrate separately developed components or concepts into a technical system. The cross-fertilization of convergent technologies could be a source of innovation [LAU 05]. For AFTU [AFT 06], it is indeed an interdisciplinary approach. The expression "convergent technologies" refers to the unfortunate encounter between innovations in the fields of microelectronics, bio-informatics, nanotechnologies and cognitive science. Technology advances to the boundaries of scientific knowledge, under the pressure of powerful commercial or political, and sometimes speculative, interests. The classical and almost "historical" example is that of NBIC convergence illustrated by Figure 4.7.

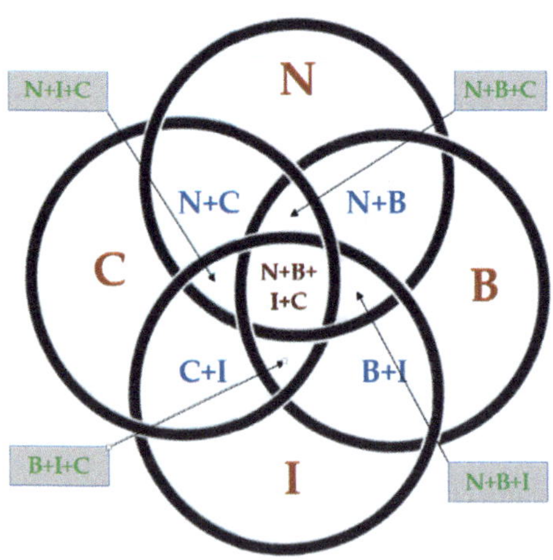

Figure 4.7. *Incremental/reductive process in NBIC "convergence". For a color version of the figure, see www.iste.co.uk/andre/printing3.zip*

This statement is validated by Cornu [COR 08], who writes, "Does technology, beyond its undeniable progress, help support the dream of making applications acceptable to the greater masses or does it rather give some fans of science-fiction legitimacy to influence research programs? Whatever the case may be, a 'convergence of interests' has allowed people to gather around this inspiring program". The basic idea of convergence is indeed to extract disjointed knowledge from the paths allowing industrial applications through the fusion of disciplinary knowledge, which induces coupling between scientific, economic and social aspects for successful innovation. Considering what was just explained, the two domains concerned in this part exploit the same disciplinary supports, as indicated in Figure 4.8.

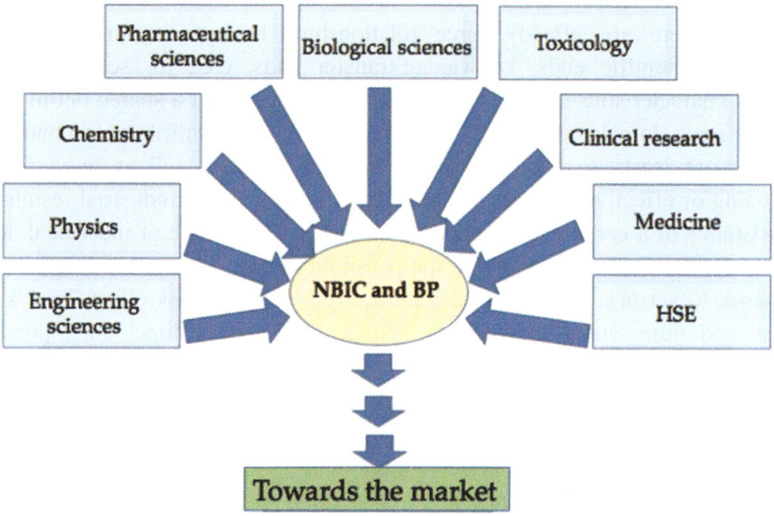

Figure 4.8. *NBIC and BP convergences and interdisciplinary approach. For a color version of the figure, see www.iste.co.uk/andre/printing3.zip*

4.4.4. *Comparisons*

4.4.4.1. *Quick return on interdisciplinarity*

Knorr-Cetina [KNO 99] defines epistemic cultures as machines to produce facts, which requires "skilled" scientists' ontologies of organisms and "machines" that result from knowledge stemming from reasoning through analogy, coming from different scientific and technical fields, masteries of the doors to be broken down, etc. Thus, the epistemology to be created between the virtual (realistic models) and the simplified real demands to be constructed through different disciplines with transaction zones; this iterative (non-stationary) epistemology results from an obvious interdisciplinary stake. Thus, presented as an eminently collaborative field, the NBIC and BP frameworks should offer a particularly interesting field of observation of interdisciplinarity. On this basis, disciplinary alliances must allow reflection on the best ways to explore these domains considered to be promising, without it being possible to know today if there will be an efficient relationship between NBIC/BP project and expected functionality one day. In fact, consulting different partners from different disciplines should be expressed through a diversity of schools of thought that is revealing of the dimension, complexity and tensions of the emerging domain that is the NBIC/BP framework. This normally must result from difficulties "cooperating" between strongly heterogeneous actors.

Indeed, there are already force relationships between technological ends, disciplinary scientific ends, knowledge-transfer ends, etc., for scientists. These separated characteristics are often based on an appearance of a shared definition. In fact, the new domain is probably defined, from a scientific or technological standpoint, at least as much by ends, "systems of meaning" as by a field of questioning or effective problems or a list of scientific and/or industrial results, etc. The existence of a certain fuzziness may be due to an absence of individual and/or disciplinary clarification, but also to the possibility of exploiting this non-stabilized framework to act in a way engaged with the risk of "red dress effects" or "sleeve effects" and other superficial effects. "Since there is no direct correspondence between disciplines and their objects of study and, on the other hand, disciplinization reduces comprehension space, the concrete study of objects requires the spread of a space of interdisciplinary mediation" [DUC 99]. In the case where this space is not found nor supported by the hierarchy in the broadest sense of the word, dysfunctions can be expected, even a breakdown of the dialogue between stakeholders. From this fact, proven in cases where the uncertain reigns [TEC 16], knowledge of the diversity of representations is a necessary prerequisite to positive, confident and honest exchange action (so long as this can exist).

4.4.4.2. *Facts concerning NBIC convergence*

A bibliographic analysis concerning the "hard science" aspects of the NBIC domain is presented in Table 4.5. It illustrates the weak interpenetration of disciplines. In the end, the expression NBIC hides a reality far removed from promises (and fears) related to N+B+I+C risks that are, for the moment, practically nonexistent on the component where every discipline interacts.

NBIC matters	N	B	I	C	Other fields
Neuron enhancement			+	+	
Nano-medicine					
– Diagnostics	+	+			Tooling
– Therapeutics	+	+			Medicine
– Tissue engineering		+			Engineering
– Nanometric motors, nano-robots	+				Robotics
Man–machine communication; artificial intelligence (AI)			+	+	Humanities and social sciences
Voice recognition			+	+	Humanities and social sciences
Nano-sensors	+		+		Medicine; automatons: IoT
Nanoelectronics	+				Electronics
Micro-fluidics	+				Process engineering; fluid mechanics
Protection of private life	+?		+		Big Data

Table 4.5. *Place of N, B, I, C components in scientific and technological activities today*

At the same time, the ethical and philosophical aspects surrounding a subject that brings fears and hopes is the object of a much more intense activity, as shown in Figure 4.9 (in the order of thousands of publications or works).

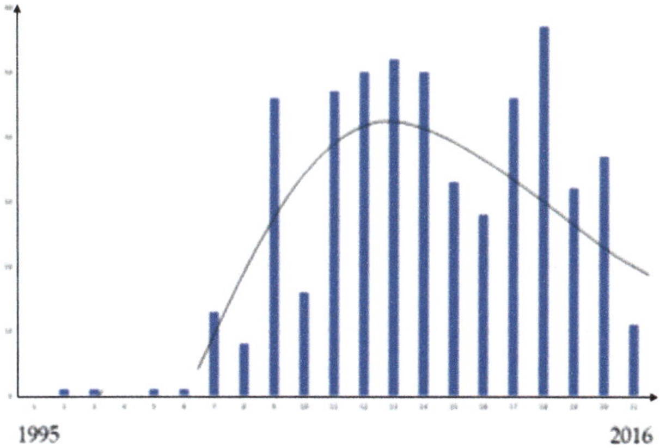

Figure 4.9. *Publications concerning the subject "NBIC convergence" for "ethical" aspects*

However, let us note that this situation specific to NBIC remains modest in the face of all the works on "nano" risks defined in terms of effects on health. Several statements can be made concerning these publications:

– the dynamic concerning NBIC aspects is practically simultaneous with the NNI relation in 2001, whereas activities in experimental toxicology come later, published with a certain delay, even if the increase in the number of published works remains the same;

– the number of works on "convergence" aspects is no longer increasing, and it seems to have a possible downwards trend;

– there is no longer a possible comparison today between the number of publications in toxicology (several thousand/year) and that associated with convergence (a few dozen).

On the other hand, without wanting to believe that the causal aspect is, to a certain degree, derisory and that it cannot be taken into consideration (which is in no way the author's desire), the fixation on this subject, largely reported on by the media, introduces a possible role of "enticement", a point of fixation possibly making citizens forget other, even more determinant changes that are taking place each day on their way of life or that could occur in their existence. So would a bit more or slightly fewer nanomaterials change undesired disturbances?

4.4.4.3. *Facts concerning BP convergence*

Two synthesis chapters (2 and 3) have been dedicated to this matter (see also Guédon, Malaquin, and André, [GUÉ 17]); they demonstrate real proofs of concept (creation of cell clusters for periods more than several weeks). But today, it does not seem foreseeable to master the bio-printing system robustly only with knowledge of its constituent elements, hence understandable trends for a comprehensive approach practically like the one practiced by engineers, considering phenomena judged to be preponderant in their model of understanding; it is thus necessary to escape "materialism of the promise". According to holistic theses, each cell influences the whole under construction, which, itself, is involved in a recursive way in the elementary element. But, how then to be sure that care is being taken with the principal variables, if the complex local interdependencies (cells/environment and vice-versa) exist? Yet, bio-printing objects are composites, systemic, unstable and under constraints (some of which remain unknown to us). Thus, for the author, whatever the scales considered, bio-printing is probably inaccessible to only mono-disciplinary approaches.

4.4.4.4. *Practice of interdisciplinarity*

Concerning bio-printing, the relatively "low" number of publications in this emerging field has allowed a study on the division of the publications in "focused" scientific journals. Figure 4.10 highlights the division that strongly supports a remark by giving an eminent place to engineering science and technology (50%), whereas the life science component remains below 20%. The borders between spectra of activity in scientific journals are more distant than for disciplines with the same application. The values found are therefore only indications, but the gaps between results are sufficient for the consequences to be significant, as explained below.

The question asked (and which will not be resolved here) is knowing if the bio-aspect constitutes a central element of publications or simply another on a new subject. However, if we are to believe Franco [FRA 15], a member of Hepatinov, "the majority of the actors in bio-construction belong, incidentally, to the domain of engineering, biomaterials, mathematics, and robotics, only leaving a small place for physicians and stem cell researchers". The problem of mediocre integration of knowledge is thus asked (and translated into the nature of scientific reviews).

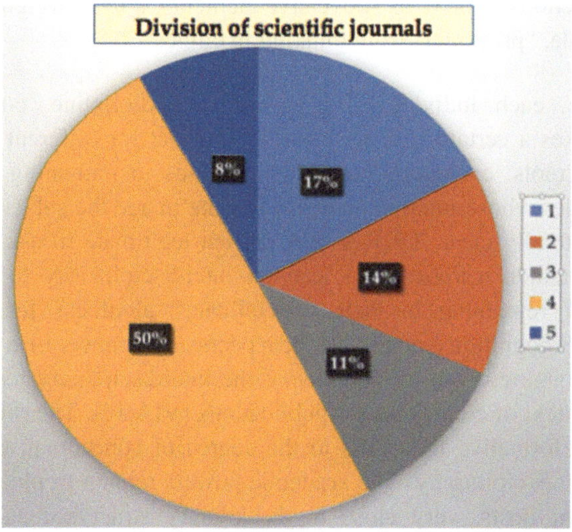

Figure 4.10. *Division of publications in the field of bio-printing by scientific journals. 1: Biology and medicine; 2: biomaterials and materials; 3: physics and chemistry; 4: engineering and technology (including bio-manufacturing); 5: miscellaneous. For a color version of the figure, see www.iste.co.uk/andre/printing3.zip*

4.4.5. *Epistemological questions*

In any case, the modeling aspects in bio-printing, like NBIC convergence, are hard to notice in the descriptions of the scientific papers based on their titles (and also in reading some hundreds of articles that could have seem pertinent). In this condition, it is clearly a matter of "finding its marks" by developing concepts, methods and processes, by seeking adapted materials. The absence of significant results concerning manufacturing real tissues still does not allow a vision borrowed from rationality "thanks" to the contributions of the "hard" sciences (even if this vision constitutes a promise on the part of start-ups and some scientists).

In the end, the publications attest to forms of induction or more prosaically "knowledgeable do-it-yourself (DIY)" in an emerging field, where the "hard" sciences play an important role (by expanding their numbers of financial sources)! For the author, it is (still) necessary to elaborate a distinction between power (providing a guide for action) and scientific authority (possibility of refutation); "knowledge does not make power" [LAS 07]. It is possible that the possibility to respond to news and desires may make one direction of research convincing, even if the approach does not clearly reveal its limitations, making debate difficult, as it allows contradictions to arise as subversive elements, a dead weight on thought or, for French people, "preventers from thinking in circles".

Furthermore, each individual, in his/her own discipline, conservatism and acquisitions, takes a certain distance from the subject for different reasons (know-how, available tools, evaluation, etc.), which leads to remaining aloof, avoiding adherence to roles whose internal coherence or the image the subject can contribute to them is not totally clear [DUB 94]. It is then legitimate to ask the question of knowing if certain researchers can lean on interdisciplinarity to attain financial support, but of remaining in their "disciplinary" aloofness for fear of losing credibility and recognition granted by their peers as members of a collective. This question of belonging, even the identity of the interdisciplinary researcher, is also asked in the context of competing for publications [VIA 16]. The rules of evaluation do not have performative indicators to the degree of conceptual, even theoretical, demand for interdisciplinarity. The academic powers remain in place, in "real life", despite recurrent posts, very close to disciplinary structurations, by essentially supporting almost-opening. This stability in the modes of action induces conservatism avoiding risk taking, particularly any time (!) budgetary restrictions are involved.

Independent of these aspects, the question is asked of means of tackling complex questions with areas of lacking knowledge (see Figure 4.11) (epistemology: on knowledge and logic, this is the branch of philosophy concerned with the theory of knowledge; studies of the nature of knowledge, the rationality of beliefs, and its justification) and of running a scientific and technological project. "What changes radically is that there is no longer combination, at first, of the knowledge from each discipline, but rather, a tendency to start assembling what is not known, 'the state of the non-art,' where the discipline could not give responses to the problems posed by the object that one seeks to know and construct. Each discipline seeks the area in which the object cannot be processed through its concepts and logic. This questions scientific practice to its core because this non-knowledge obliges us to see interdisciplinarity other than as a sum procedure. It obeys a logic of substraction (with no lack), which leads each discipline to be reinterpreted by others. Mastery is no longer the centerpiece. Not-knowledge is no longer marginalized; it lies at the center of the process" [SCH 10].

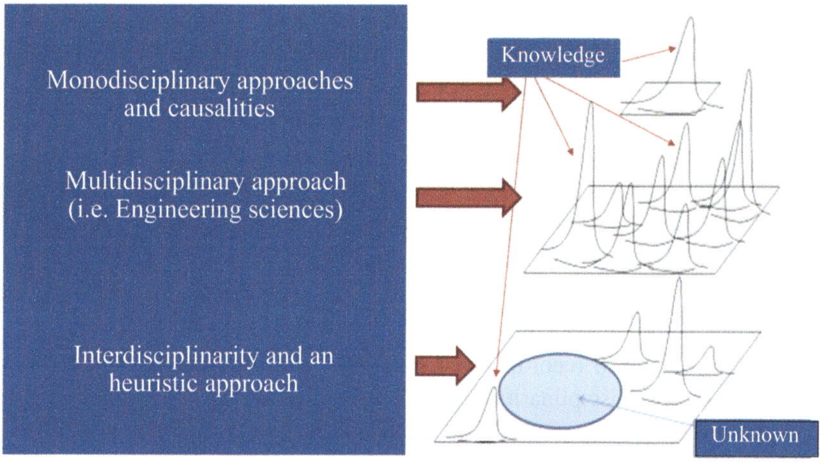

Figure 4.11. *Scientific approaches: cause–effect relations surrounding complexity. For a color version of the figure, see www.iste.co.uk/andre/printing3.zip*

4.4.5.1. *Epistemology*

Scientists' natural practice is to research whether mathematical rigor and algorithmic processes can ensure a conceptual body of doctrine at the interface between technologies, engineering science, information science and biology that could ensure new knowledge for the production of biological tissues and organs (BP) or NBIC systems, all while knowing that no specialist can, in principle, surpass a broader culture than that of his/her discipline. In a spirit of maximum empirical shoring (Occam's razor), the mobilizing idea is to establish a robust theoretical/empirical model whose surplus descriptive content is minimum in relation to what is indispensable to notice the phenomena.

In both cases, several disciplines are involved, which leads to "modeling" field mutualizations, probably with equations of different behaviors, multiple and singular interfaces, nonlinearities, scale changes (from sub-cellular components to organs) associated with potential fractal forms, etc. Complexity invades these domains by several paths: that of geometric arrangements, that of the complexity of the dynamic (growth, reduction of scaffolding, e.g. in BP) and those of functionality, robustness, cost, etc.

In agreement with Mathieu and Schmid [MAT 14], in-depth research can take very diverse forms based on some kernels of knowledge derived from experimental works or based on model objects supporting concepts (but which, in any case, will only find meaning through a truly integrational interdisciplinary approach). This implies an epistemological decision, which demands reflection for the best choice among different "method discourses" built upon various pillars: creativity, divergent thought, serendipity, exploration of complexity, proofs of concept, experimental validation, modeling, simulation, etc.

4.4.5.2. *Behavior of the interdisciplinary project*

But how to come to recording research in an interdisciplinary perspective? In several texts on interdisciplinarity, the idea according to which it is the complexity of certain objects of study [BOI 04] that calls practice interdisciplinary is recurrent. Indeed, the nature of the said objects is conceived as a dynamic system, "irreducible to a single dimension" [KLE 04]. The accumulation of knowledge and the crossing of views, the interactions between phenomena, undeniably characterize the objects of research: complexity has become an intrinsic value of knowledge and research in general.

In these knowledge-fusion activities, it is best to limit, as much as possible, the recourse to operations that increase the risk of losing mastery over the theorization process. Yet, the social communications at the heart of interactions necessarily mobilize common sense. It is difficult to "purify" the rational dimension of the project – management at the cognitive level – of its largely implicit affective and social components, and more fundamentally, of the symbolic dimension of every human process. Sociologies naturally play an eminent role in the exploration of cross-reflexive conditions; there is a need to know one another, to have common references, to rely on strong implicit shared foundations [LEM 12, BAT 12].

These difficult relational and cultural methods, with aforementioned limitations, can prevent (maybe definitively) the prediction of "the certain trajectory of an organ construction" over its development in as precise a way as the classical object created through additive manufacturing or the creation of an operational NBIC system. However, it could be advantageous to try to construct virtual objects [COL 13] to open new perspectives of understanding and management of the living and/or the infinitesimal.

The analysis of the interdisciplinary actors' games and the methods of project governance are essential in "convergence" operations. In research on bio-printing, it is important to reflect on the position of the actors in light of tensions between disciplines and risk of misunderstanding. Do these risks induce divergences of interest superimposed upon existing disciplinary splits or, to the contrary, do they encourage coming together, even consensuses? Are tensions and these risks the chance to renew existing policies and relationships between actors? Are we seeing, in this matter, the emergence of new practices for accessing knowledge, leading to technological and methodological innovations capable of conferring coherent content to scientific strategies? What, then, are the theoretical frameworks (paradigms) pertinent from a standpoint of articulating interdisciplinary dimensions? How can we rethink, in a mutualized process, the strategies of research development?

It is likely not a matter, in this case, of opening up to the proposal of a fixed binary model: "bottom-up" interdisciplinarity versus "top-down" interdisciplinarity. Each process of interdisciplinarity must mix, in a framework of trust, aspects of incitation, even exogenous injunctions and initiatives and/or internal responses to researcher collectives, motivated for research on boundary objects, but coming from different disciplinary horizons [ZUI 06].

4.4.5.3. *Consequences*

Bühlera, Cavaillé, and Gambino [BÜH 06] remind us, "for the most part, institutions, interdisciplinarity, is only conceivable if it does not question the foundations of the disciplines, better, if it comforts the disciplines where they are. Moreover, the methods of evaluation modify the ways of producing knowledge, increasingly oriented towards the satisfaction of methods imposed by the New Public Management than producing results better adapted to the service of the Nation". On this basis, Callon, Lascoumes, and Barthe [CAL 01] have introduced the notion of "laboratorization" of the world, revisited in Figure 4.12 below.

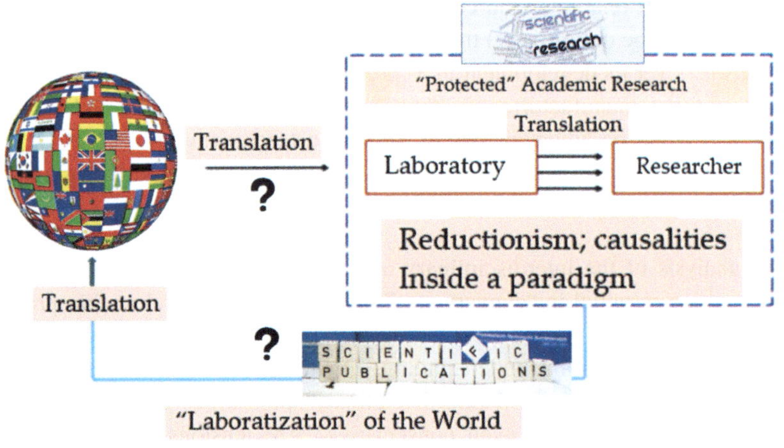

Figure 4.12. *Laboratization of the world. For a color version of the figure, see www.iste.co.uk/andre/printing3.zip*

Financing research has long been essentially based on a system of competitive resource allocation [TRI 07]; the stakes of interdisciplinary openness and convergence require more time to think and act, which is an important increase in both human and financial needs. In a context of stagnating financial support from National and European agencies, the clear result is a continuous reduction of success rates. The pressure induced by competition comes to a point that it can produce various negative effects: orientation of research activities towards subjects that are more "promising" and/or "productive" in the short term, more trendy, pressure to publish (with the risk of plagiarism), scientific fraud, etc. In this context, based on experience, it is also difficult to evaluate transdisciplinary projects, whether this is within a disciplinary commission or an *ad-hoc* transdisciplinary commission, itself generally made up largely of disciplinary experts who, in a context of financial limitation, only rarely take on the risk of integrating knowledge [ZAC 16, TEC 16].

Table 4.6 below summarizes the questioning associated with a "successful" epistemological approach associated with the two convergences, NBIC and BP.

Practices to challenge
Reflecting on the possible meaning of the hierarchy of disciplines
Problems associated with the cross-fertilization of research through interdisciplinarity (social engineering, mutualization, tooling, concepts, etc.)
Reflecting on the (over-specialized?) training of researchers
Reflecting on the evaluation of research (and researchers) on risky subjects; rebuilding trust
Supporting financing for risky creative operations
Supporting the exploration of complexity
Moving towards productive "slow science"

Table 4.6. *To challenge practices to explore NBIC and BP convergences*

4.5. Conclusion

It is generally accepted that scientific research produces predictive scientific theories and applications. These theories must contribute to the definition of strategies for intervening in the real. They can thus generate unheard of objects and new representations, ideas and theoretical tools that can overflow from their field of origin (e.g. trans-humanism resulting from the estimated potential of NBIC convergence and/or bio-printing). So long as they were produced, these objects could, in turn, become objects of knowledge and analysis of the possible risks that are associated with them. But in the case of the NBIC and BP convergences, the representations, fantasies, constructions or images created in the absence of knowledge escape the scientific domain. They have led to the premature creation of standpoints capable of providing references (that are certainly criticizable) to give an opinion on a complex domain of convergence that is still under construction for various reasons explained in this chapter. What representations of NBIC concepts result from this situation? In a probably caricature-like way, they are those of a world polluted with carbon nanotubes, asbestos, trans-humanism, truncated public debates, senseless promises, etc. The (recent) history of the "nano" domain and NBIC has an influence on the representations that can be made of a practically untouched domain from a scientific standpoint and one that is already greatly polluted negatively (NBIC) and in terms of desirability (BP) by diverse ideologies.

"To improve the conditions of their market launch, an obvious strategy to increase the desirability of objects involves magnifying their psychosocial halo by assocating them with various attractive and stimulating images [...]. The effect obtained is the intensification of their visibility, 'brilliance' analogous to a diffusion effect that would increase the apparent size of the object" [BON 09]. Even if everything that is NBIC or BP can be considered a harbinger of modernity and solutions for the planet, the initial rejection or attraction cloud the identity of each realization, which can be, through enlargement and amplification, profoundly counter-productive. It would thus be a matter of bringing out the two convergences as something other than a homogenous set appreciated or rejected by one group.

However, complexity is a challenge for project managers. Repeated failures of projects, however well planned, are proof of this. Why do we fail when everything seems to be made according to the rules of the art? Usual interdisciplinary project management methods are adapted to foreseeable situations, where the uncertain is considered only at the level of risk management or changes. On the basis of Descartes' teachings, the difficulties are often divided into as many parcels as are necessary for their comprehension. This is well illustrated by the divided structure of the project, which, if it is significant, must not, by the fact of these divisions, simplify the problems in a reductive way, completely ignoring if it is crucial to compose with the complexity of the real (see responses to an ANR (French National Agency for Research) or H2020 (EU Framework Program) call for tender). The model seems to be an efficient tool of translation into an acceptable language, but the question of its neutrality remains. By seeking to find, at any price, a common basis of work between disjointed knowledge, is there not a risk of smoothing out the interdisciplinary landscape, of standardizing science towards a mathematical logic, and of bringing about domination of the information sciences [HER 15]?

These remarks attempt to show the interest (and difficulty) of thinking of the elucidation of logics of constituting scientific knowledge in complex domains such as NBIC and BP convergences, because these have an influence on the representations associated with them. "Interrogation on the method could thus be an indispensable phase for the spread of [the researcher's] ethical responsibility" [DEI 10]. In an attempt to conclude, Figure 4.13 below tries to serve as a (2D!)

synthetic reminder so that the convergences (including NBIC and BP) can be explored by science (so long as it has time, people and means). However, this is independent of the author's scant power, even if he believes that the chances of succeeding, for reasons of epistemology and daily science practice, seem to him very slim today.

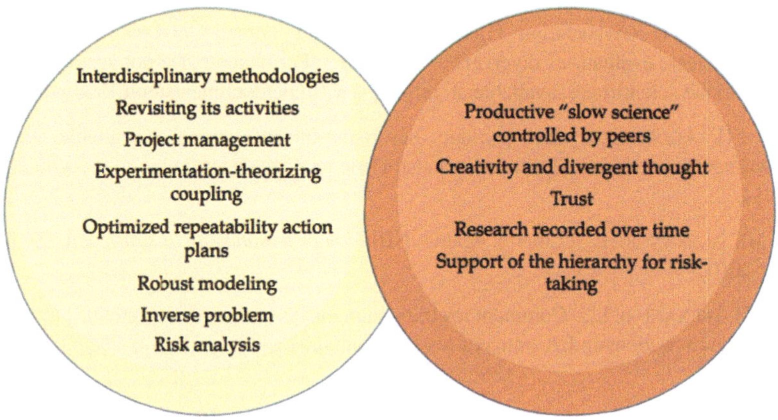

Figure 4.13. *Recommendations necessary for exploration of NBIC and BP convergences*

Just to conclude in an open manner, let us not forget these words from Benasayag [BEN 94]: "As for researchers, naivety is a bit more complex. They maintain that their models and automatons 'still' do not have all the qualities and possibilities of natural life, but there is a reason why they give it the name 'artificial intelligence' and 'artificial life.' There is indubitably a bit of Frankenstein in these honest researchers, because why not call these machines 'device', 'Schmilblick', why not speak of 'construction' rather than 'creation,' a word with strong connotations that, if it does not refer to Frankenstein, reminds us of a famous biblical figure". Difficult to undertake while remaining sensible!

> *"From now on, biologists will not only have to live with complexity, but also love it". (Klein quoted by Atlan [ATL 11])*

4.6. Bibliography

[ABE 94] ABEL O., "La responsabilité incertaine", in *Esprit Les équivoques de la responsabilité*, vol. 11, pp. 20–27, 1994.

[AFN 11] AFNOR, *NF E 67-001 October 2011: Fabrication Additive*, Vocabulaire, Paris, 2011.

[AFT 06] AFTU (Association Pour Une Fondation Travail-Université), "Technologies convergentes, évaluations divergentes", Education Permanente, De nouveaux enjeux pour la politique scientifique, available at: http://www.ftu.be/documents/ep/EP-24.pdf, 2006.

[AGO 14] AGORAVOX, "Le mariage des nano-bio-info-cogno", available at: http://www.agoravox.fr/culture-loisirs/etonnant/article/le-mariage-des-nano-bio-info-cogno-26349, 2014.

[ALE 10] ALEXANDRE L., "Liberté, égalité, NBIC", *Les tribunes de la santé*, vol. 29, pp. 75–83, 2010.

[ALE 15] ALEXANDRE L., "Comment les trans-humanistes ont gagné nos cœurs", available at: http://www.wedemain.fr/Comment-les-transhumanistes-ont-gagne-nos-coeurs_a903.html, 2015.

[AND 84] ANDRÉ J.C., LE MÉHAUTÉ A., DE WITTE O., "Dispositif pour réaliser un modèle de pièce industrielle", *Brevet français*, vol. 84, no. 11, p. 241, 16 July 1984.

[AND 02] ANDRÉ J.C., MASSE R. "L'expertise, la science et l'incertitude: l'expertise scientifique ou la langue d'Esope", *Environnement, Risques & Santé*, vol. 1, nos 5–6, pp. 299–306, 2002.

[AND 13] ANDRÉ J.C., "Towards a Socially Responsible Research (SRR) charter in Engineering Sciences at CNRS level", *International Journal of Technoethics*, vol. 4, pp. 39–51, 2013.

[AND 15] ANDRÉ J.C., "N'est pas Prométhée ou Mary Shelley qui veut: une petite histoire de la stéréo-photolithographie…", *Actualité Chimique*, vols 397–398, pp. 99–101, 2015.

[AND 16a] ANDRÉ J.C., FROCHOT C., TOMEI F., Approche responsable et acceptabilité sociale des nano-médicaments, 2016.

[AND 16b] ANDRÉ J.C., "Santé connectée", *Environnement Risques et Santé*, vol. 15, pp. 380–385, 2016.

[AND 16c] ANDRÉ J.C., Bio-printing and NBIC Convergences: Promises, ethics, epistemological approaches", *Nanoscale 3D Printing Workshop, Conference*, Singapore, 1–2 December 2016, 2016.

[AND 16d] ANDRÉ J.C., LARGER L., Approche épistémologique du bio-printing pour essayer de moins mutiler la recherche sur un objet-frontière, CSI-INSIS-CNRS Report, 2016.

[AND 17b] ANDRÉ J.C., "Impression 3D: niches applicatives porteuses and Fabrication additive: Révolution ou simple évolution sociétale?", *Techniques de l'Ingénieur*, forthcoming, 2017.

[ARA 00] ARAGON Y., BERTRAND S., CABANEL M. *et al.*, "Enquêtes par Internet: Leçons de quelques expériences", *Décisions Marketing*, vol. 19, pp. 29–37, 2000.

[ARC 16] ARCAND P., "Le chantier sur l'acceptabilité sociale: une démarche innovante tournée vers l'avenir", *Éthique publique*, available at: http://ethiquepublique.revues.org/2428, vol. 18, 2016.

[ATL 11] ATLAN H., *Qu'est-ce qu'un modèle?*, Manucius, Paris, 2011.

[AUV 13] AUVERLOT D., HAMELIN J., PUJOL J.L., "Le principe de précaution: quelques réflexions sur sa mise en œuvre (DT)", available at: http://www.strategie.gouv.fr/blog/2013/10/principe-de-precaution-quelques-reflexions-mise-en-oeuvre-dt/, 2013.

[AYA 16a] AYACHE G., "Serons-nous un jour immortels?", available at: http://www.up-magazine.info/index.php/le-vivant/homme-augmente/6052-serons-nous-un-jour-immortels-2, 2016.

[AYA 16b] AYACHE G., "Mark Hunyadi: Comprendre le monde, c'est déjà le transformer", available at: http://www.up-magazine.info/index.php?option=com_content&view=article&id=5685:mark-hunyadi-comprendre-le-monde-c-est-deja-le-transformer&catid=103:analyses&Itemid=542&utm_source=newsletter_251&utm_medium=email&utm_campaign=up-nl-vnew-quot#sthash.q1K8kptw.dpuf, 2016.

[BAD 09] BADIOU A., *Second manifeste pour la philosophie*, Champs-Essais/A. Fayard, Paris, 2009.

[BAK 16] BAKER M., "Is there a reproducibility crisis?", *Nature*, vol. 533, pp. 452–454, 2016.

[BAT 12] BATTARD N., "Convergence and multi-disciplinarity in nanotechnology: laboratories as technological hubs", *Technovation*, vol. 32, pp. 234–244, 2012.

[BAY 15] BAYON S., ANDRÉ J.C., Socially Responsible Research, but whom to trust? The example of renewable energy" *Organizational Creativity international Conference*, Nancy, p. 12, 26–27 March 2015.

[BEC 09] BECK U., *World at Risk*, Polity Press, Cambridge, 2009.

[BEN 94] BENASAYAG M., "Frege aujourd'hui: peut-on penser la recherche scientifique?", in TINLAND F. (ed.), *Ordre biologique, ordre technologique*, Champ-Vallon, Seyssel, 1994.

[BEN 13a] BENAMOUZIG D., BORRAZ O., JOUZEL J.N., "Contribution opérationnelle des sciences humaines et sociales à l'expertise en santé-alimentation-santé-travail", Contrat R&D 2010-CRD-18 ANSES, Paris, 2013.

[BEN 13b] BENSAUDE-VINCENT B., *L'opinion publique et la science; à chacun son ignorance*, La Découverte, Paris, 2013.

[BEN 15] BENVIN R., "3D blood vessel bio-printer useful for 3D printing human organs – the secret of immortality", available at: http://bioprintingworld.com/3d-blood-vessel-bioprinter-useful-for-3d-bioprinting-human-organs-the-secret-to-immortality/, 2015.

[BER 53] BERNANOS G., *La liberté, pour quoi faire?*, Gallimard, Paris, 1953.

[BER 55/56] BERNARD C., "Leçons de physiologie expérimentale appliquée à la médecine au Collège de France", Baillière et fils, Paris, 1855–1856.

[BER 90] BERNARD J., *De la biologie à l'éthique*, Buchet/Chastel, Paris, 1990.

[BER 10] BERNS T., BLÉSIN L., JEANMART G., *Du courage*, Ed. des belles lettres, Paris, 2010.

[BES 07] BEST J., "Why the economy is often the exception to politics as usual", *Theory, Culture and Society*, vol. 24, pp. 87–109, 2007.

[BIA 12] BIANCHINI L., "Les indicateurs de l'évaluation de la recherche: de l'impact factor à l'h-index", available at: http://www.mysciencework.com/news/6290/les-indicateurs-de-l-evaluation-de-la-recherche-de-l-impact-factor-a-l-h-index, 2012.

[BOI 04] BOIFFIN J., "Territoire: agronomie, géographie, écologie, où en est-on? Le point de vue d'un chercheur agronome", *Natures Sciences Sociétés*, vol. 12, pp. 307–309, 2004.

[BON 09] BONTEMS V., "Actualité d'une philosophie des machines; Gilbert Simondon, les hadrons et les nanotechnologies", in LE ROUX R., BONTEMS V., Braffort P. (eds), *"Les machines, objets de connaissance", Revue de Synthèse*, Springer, Paris, 2009.

[BOU 13] BOUTARIC F., "La méthode de l'évaluation des risques sanitaires en France: représentation, évolution et lectures plurielles", *Vertigo*, vol. 13, pp. 1–20, 2013.

[BRE 90] BRETON P., RIEU A.M., TINLAND M., *La technoscience en question*, Champ Vallon, Paris, 1990.

[BRO 16] BROWAEYS D., "Intelligence Artificielle et Trans-humanisme", available at: https://iatranshumanisme.com/tag/dorothee-benoit-browaeys/, 2016.

[BÜH 06] BÜHLERA E.A., CAVAILLÉ F., GAMBINO M., "Le jeune chercheur et l'interdisciplinarité en sciences sociales - Des pratiques remises en question", *Natures Sciences Sociétés*, vol. 14, pp. 392–398, 2006.

[BUT 10] BUTORI R., PARGUEL B., "Les biais de réponse – impact du mode de collecte des données et de l'attractivité de l'enquêteur", available at: http://hal.archives-ouvertes.fr/docs/00/63/62/28/PDF/AFM_2010_Version_DA_finitive_Raph_28_FA_vrier.pdf, 2010.

[BUT 16] BUTTIN N., SAFFRÉ B., *Activer l'économie circulaire*, Eyrolles, Paris, 2016.

[CAL 01] CALLON M., LASCOUMES P., BARTHE Y., "Agir dans un monde incertain – Essai sur la démocratie technique", Le Seuil, Paris, 2001.

[CAN 65] CANGUILHEM G., *La connaissance de la vie*, Vrin, Paris, 1965.

[CAU 06] CAUNE J., "Les discours sur les nanos: construction d'une nouvelle science et/ou controverses sur le progrès?", *Quaderni*, vol. 61, pp. 25–37, 2006.

[CBH 15] CBHS, "3D bioprinting: printable bones and organs changing the future of medicine", available at: https://www.cbhs.com.au/news/2015/03/12/3d-bioprinting-printable-bones-and-organs-changing-the-future-of-medicine, 2015.

[CCC 12] CCCERA (Comité Consultatif Commun d'Ethique pour la Recherche Agronomique), *Avis*, INRA-CIRAD, Paris, 2012.

[CCN 13] CCNE, "Recours aux techniques biomédicales en vue de "neuro-amélioration" chez la personne non malade: enjeux éthiques", available at: http://www.ccne-ethique.fr/sites/default/files/publications/ccne.avis_ndeg122.pdf, 2013.

[CHA 13] CHATEAURAYNAUD F., DEBAZ J., CHARRIOU J.P. *et al.*, *Une pragmatique des alertes et des controverses en appui de l'évaluation publique des risques*, ANSES-EHESS, Paris, 2013.

[CHA 11] CHAVEL S., *Se mettre à la place d'autrui; l'imagination morale*, Presses University of Rennes, Rennes, 2011.

[CHE 07] CHEVASSUS-AU-LOUIS B., *L'analyse des risques: l'expert, le décideur et le citoyen*, Quae, Paris, 2007.

[CHI 09] CHIFFLET S., L'imaginaire technoscientifique: cas de la convergence NBIC, available at: http://www-llb.cea.fr/Phocea/Vie_des_labos/Seminaires/index.php?id_type=19&type=19&y=2009&id=1663 and Le récit de la convergence NBIC: vers une nouvelle cosmogonie?, Thesis, Stendhal University, Grenoble, 2009.

[CHN 12] CHNEIWEISS H., *L'Homme réparé; espoirs, limites et enjeux de la médecine régénérative*, Plon, Paris, 2012.

[CHO 06] CHOPPLET M., Les nanotechnologies entre utopie et contre-utopie, *Quaderni*, vol. 61, pp. 7–80, 2006.

[CHR 91] CHRÉTIEN C., *La science à l'œuvre*, Hatier, Paris, 1991.

[CIT 17] CITOUR C.E., Qu'est-ce que l'Humain? La science brouille les frontières, available at: http://www.agoravox.fr/tribune-libre/article/qu-est-ce-que-l-humain-la-science-188249, 2017.

[CIT 13] CITTON Y., *Pour une interprétation littéraire des controverses scientifiques*, Quaë, Paris, 2013.

[CLA 13] CLAVERIE B., LE BLANC B., "Homme augmenté et augmentation de l'Homme" in KLEINPETER E. (ed.), *L'Humain augmenté*, CNRS, Paris, 2013.

[COB 01] COBANOGLU C., WARDE B., MOREO P., "A comparison of mail, fax, and Web-based survey methods", *International Journal of Market Research*, vol. 43, no. 4, pp. 405–410, 2001.

[COL 13] COLLINS D.H., HUZURBAZAR A.V., "Multi-state stochastic processes: a statistical flow-graph perspective", *International Statistical Review*, vol. 81, pp. 78–106, 2013.

[COL 14] COLLINS S.F., "Bioprinting in changing regenerative medicine forever", *Stem Cells and Development*, vol. 23, Supp. 1, pp. 79–82, 2014.

[COM 03] COMBLIN J., *Le néolibéralisme; pensée unique*, L'Harmattan, Paris, 2003.

[COM 06] COMETS, Enjeux éthiques des nanosciences et nanotechnologies, available at: http://www.cnrs.fr/comets/IMG/pdf/10-ethique-nanos.pdf, 2006.

[COR 08] CORNU J.M., "Nouvelles technologies, nouvelles pensées? La convergence des NBIC, Fyp, Limoges and Des technologies convergentes… Vers quoi?", available at: http://alltrends.over-blog.net/article-25668522.html, 2008.

[COU 00] COUPER M.P., Web surveys: a review of issues and approaches, The Public Opinion, available at: http://www.goethe-university-frankfurt.de/47929851/Couper_2000.pdf, 2000.

[CPP 10] CPP, La décision publique face à l'incertitude – clarifier les règles, améliorer les outils, MEEDDM-CGDD, Paris, 2010.

[CRO 60] CROWNE D., MARLOWE D., "A new scale of social desirability independent of Psycho-pathology", *Journal of Consulting Psychology*, vol. 24, pp. 349–354, 1960.

[DEB 07] DEBONNEUIL M., *L'espoir économique: vers la révolution du quaternaire*, Bourin, Paris, 2007.

[DEC 01] DECROP G., DOURLENS C., VIDAL-NAQUET P.A., "L'opacité des scènes locales" in GUIBERT B., BERLAN-DARQUÉ M. (eds), *L'environnement, questions sociales*, Odile Jacob, Paris, 2001.

[DEI 10] DEI-CAS E., "Éthique du vivant et épistémologie de la biologie", *Éthique & Santé*, vol. 7, pp. 24–30, 2010.

[DEL 15a] DELOITTE, "The regulator of tomorrow", available at: http://dupress.com/articles/US-regulatory-agencies-and-technology/?id=us:2sm:3tw:dup1178:eng:dup:nickolson:061115:author, 2015.

[DEM 11] DE MARNHAC A., *Beauté – Histoire, florilège et astuces*, La Martinière, Paris, 2011.

[DEW 15] DE WEVER B., GOLDBERG A., ESKES C. *et al.*, "Open source – Besed human tissue model: a new gold standard for non animal testing through openness, transparency and collaboration promoted by rhe Alexandra Association", *Applied in Vitro Toxicology*, vol. 1, no. 1, pp. 5–9, 2015.

[DID 08] DIDIER C., *Penser l'éthique des ingénieurs*, PUF, Paris, 2008.

[DOD 15] DODDS S., "3D printing raises ethical issues in medicine", available at: http://www.abc.net.au/science/articles/2015/02/11/4161675.htm, 2015.

[DRO 95] DROUART A., "Aux sources de l'eugénisme français", *La Recherche*, vol. 277, pp. 648–654, 1995.

[DUB 94] DUBET F., *Sociologie de l'expérience*, Le Seuil, Paris, 1994.

[DUC 99] DUCHASTEL J., LABERGE D., "La recherche comme espace de médiation interdisciplinaire", *Sociologie et Sociétés*, vol. 31, pp. 63–76, 1999.

[DUP 02] DUPUY J.P., *Pour un catastrophisme éclairé; quand l'impossible est certain*, Le Seuil, Paris, 2002.

[DUS 09] DUSSAIX A.M., "La qualité dans les enquêtes", *Revue MODULAD*, available at: https://www.rocq.inria.fr/axis/modulad/archives/numero-39/Tutoriel-Dussaix/Dussaix-39.pdf, vol. 39, pp. 137–171, 2009.

[EGE 05] EGE, "Opinion on the ethical aspects of ICT implants in the human body", available at: http://europa.eu/rapid/press-release_MEMO-05-97_en.htm?locale=en, 2005.

[EGL 14] EGLISE CATHOLIQUE, "Nanotechnologies et bioéthique: éléments de compréhension", available at: http://www.eglise.catholique.fr/page.php?Id=8563, 2014.

[EIN 53] EINSTEIN A., "Lettre à Max Born" cited by KORN H., BERVCHE P., BINDER P. (2008), *Les menaces biologiques; bio-sécurité et responsabilité des scientifiques*, PUF, Paris, 1953.

[EU 05] EU, "Charte européenne du chercheur", available at: http://ec.europa.eu/euraxess/pdf/brochure_rights/eur_21620_en-fr.pdf, 2005.

[FLE 11] FLEITOUR G., Nano-marché pour la nano-médecine, available at: http://www.usinenouvelle.com/article/nano-marche-pour-la-nanomedecine.N147037, 2011.

[FRA 15] FRANCO D., "Après le séminaire Bio-Printing et cell patterning", available at: http://www.centre-hepato-biliaire.org/sites/chb.prod.saegir.cyim.com/files/files/PDF/Hepatinov/HepatinovNewsletter25_Mai2015.pdf, 2015.

[FRE 02] FREITAG M., "De la terreur nazie au meilleur des mondes cybernétique", available at: http://www.dogma.lu/txt/MF-Totalitarisme.htm, 2002.

[FUK 04] FUKUYAMA F., "La Fin de l'homme: Les Conséquences de la révolution biotechnique", Folio, Paris, 2004.

[FEL 14] FELTZ B., *La science et le vivant; philosophie des sciences et modernité critique*, De Boeck, Brussels, 2014.

[FRA 15] FRANCO D., "Après le séminaire bio-printing et cell patterning", available at: http://www.centre-hepato-biliaire.org/sites/chb.prod.saegir.cyim.com/files/files/PDF/Hepatinov/HepatinovNewsletter25_Mai2015.pdf, 2015.

[FRE 99] FREITAS R., *Nanomedicine*, vol. 1, Landes Biosciences, Georgetown, 1999.

[FRE 03] FREITAS R., *Nanomedicine*, vol. 2, Landes Biosciences, Georgetown, 2003.

[FRE 13] FRESSOZ J.B., PESTRE D., Risque et société du risque depuis deux siècles, in BOURG D., JOLY P.B., KAUFMANN A. (eds), *Du risque à la menace; penser la catastrophe*, PUF, Paris, 2013.

[FUS 75] FUSTER S. (CASAMAYOR), *La tolérance*, Gallimard, Paris, 1975.

[GAL 00] GALAN J.P., VERNETTE E., "Vers une quatrième génération: les études de marché "on-line"", *Décisions Marketing*, vol. 19, pp. 39–52, 2000.

[GAL 03] GALAM S., "Modelling rumors: the no plane Pentagon French hoax case", *Physica A*, vol. 320, pp. 571–580, 2003.

[GAL 08] GALAM S., *Les scientifiques ont perdu le Nord*, Plon, Paris, 2008.

[GAN 04] GANASSALI S., MOSCAROLA J., "Protocoles d'enquête et efficacité des sondages par Internet", *Decisions Marketing*, vol. 33, pp. 63–75, 2004.

[GIE 12] GIEZENDANNER F.D., "Enquêtes: Principaux biais dans la formulation des questions", available at: http://icp.ge.ch/sem/cms-spip/spip.php?article1765, 2012.

[GRA 03] GRAS A., *Fragilité de la puissance: se libérer de l'emprise technologique*, Fayard, Paris, 2003.

[GRI 15] GRINGRAS Y., "Dérives et effets pervers de l'évaluation quantitative de la recherche: sur le mauvais usage de la bibliométrie", *RIPME*, vol. 28, pp. 7–14, 2015.

[GUÉ 17] GUÉDON E., MALAQUIN L., ANDRÉ J.C., *Bio-Printing*, Techniques de l'ingénieur, no. RE268 V1, p. 23, Paris, forthcoming, 2017.

[GUI 11] GUILLAUD H., "La programmation du monde: où nous mène la convergence technologique?", available at: http://www.internetactu.net/2011/07/19/la-programmation-du-monde-ou-nous-mene-la-convergence-technologique/, 2011.

[GUI 12a] GUILLAUD H., "Les dispositifs créatifs en question: les limites de la créativité collective", available at: http://www.internetactu.net/2012/11/16/les-dispositifs-creatifs-en-questions-22-les-limites-a-la-creativite-collective/, 2012.

[GUI 12b] GUILLEBAUD J.C., *Une autre vie est possible; comment retrouver l'espérance*, Pocket, Paris, 2012.

[GUI 15] GUILLAUD H., "Comment réguler l'innovation", available at: http://internetactu.blog.lemonde.fr/2015/09/11/comment-reguler-linnovation/, 2015.

[HAB 02] HABERMAS J., *L'avenir de l'aventure humaine: vers un eugénisme libéral?*, Gallimard, Paris, 2002.

[HER 07] HERBERT M., Que se passe-t-il lorsque les répondants à un questionnaire tentent de deviner l'objectif de recherche? Le biais du répondant: conceptualisation, mesure et étude d'impact, *Proceedings from the AFM Congress*, Aix-les-Bains, France, 2007.

[HER 09] HERTWIG R., EREV I., "The description-experience gap in risky choice", *Trends in Cognitive Science*, vol. 13/12, pp. 517–523, 2009.

[HER 15] HERVÉ D., RIVIÈRE M., "L'interdisciplinarité s'invite dans les systèmes complexes: les journées de Rochebrune", *Nature Sciences Sociétés*, vol. 23, pp. 54–60, 2015.

[HIN 15] HINTON T.J., JALLERAT Q., PALCHESKO R.N. *et al.*, "3D printing of complex biological structures by freeform reversible embedding of suspended hydrogels", *Science Advances*, vol. 1, e1500758, 2015.

[HUN 15] HUNYADI M., *La tyrannie des modes de vie; sur le paradoxe moral de notre temps*, Le Bord de l'eau, Lormont, 2015.

[HUN 16] HUNAYADI M., "Mark Hunyadi: Comprendre le monde, c'est déjà le transformer", available at: http://www.up-magazine.info/index.php?option=com_content&view=article&id=5685:mark-hunyadi-comprendre-le-monde-c-est-deja-le-transformer&catid=103:analyses&Itemid=542&utm_source=newsletter_251&utm_medium=email&utm_campaign=up-nl-vnew-quot#sthash.q1K8kptw.dpuf, 2016.

[ING 15] INGELAERE R., "Innovation technologique et performance industrielle: l'exemple de l'impression 3D", available at: http://www.lecese.fr/sites/default/files/pdf/Avis/2015/2015_07_innovations_techno_impression_3d_.pdf, 2015.

[ISA 23] ISAAC J., "Paradoxe sur la science homicide" (cited by A. Kaspi (2002)), *Jules Isaac ou la passion de la vérité*, Plon, Paris, 1923.

[JON 90] JONAS H., *Le principe responsabilité*, Champs Flammarion, Paris, 1990.

[JON 97] JONAS H., "La technique moderne comme sujet de réflexion éthique", in NEUBERG M. (ed.), PUF, Paris, 1997.

[KAM 15] KAMEN M., "Bioprinting human tissue may yield cruelty free drugs and immortality", available at: http://www.wired.co.uk/news/archive/2015-10/16/gabor-forgacs-wired-2015, 2015.

[KAS 02] KASPI A., *Jules Isaac ou la passion de la vérité*, Plon, Paris, 2002.

[KAU 93] KAUFFMAN S.A., *The Origin of Order; Self-organization and Selaction in Evolution*, Oxford University Press, Oxford, 1993.

[KLE 04] KLEIN E., *Les Tactiques de Chronos*, Champs Flammarion, Paris.

[KLE 08] KLEIN E., *Allons-nous liquider la science? Galilée et les indiens*, Champs-Sciences, Paris, 2008.

[KNO 99] KNORR-CETINA K., *Epistemic Cultures: How the Sciences Make Knowledge*, Harvard University Press, Cambridge, 1999.

[KUN 90] KUNDERA M., *L'insoutenable légéreté de l'être*, Folio, Paris, 1990.

[KUN 05] KUNDERA M., *Le Rideau*, Gallimard, Paris, 2005.

[KUR 03] KURZWEIL R., "Human Body Version 2.0", available at: http://www.kurzweilai.net/human-body-version-20, 2003.

[KUR 06] KURZWEIL R., *Serons-nous immortels?: Oméga 3, nanotechnologies, clonage...*, Dunod, Paris, 2006.

[LEX 16] L'EXPRESS, "Les dix innovations technologiques les plus prometteuses du monde en 2016", available at: http://lentreprise.lexpress.fr/high-tech-innovation/les-dix-innovations-technologiques-les-plus-prometteuses-du-monde-en-2016_1757510.html, 2016.

[LAG 12] LAGADEC P., *Du risque majeur aux méga-choses*, Préventique, Bordeaux, 2012.

[LAR 10] LAROUCHE J.M., PIRON M. (eds), "Responsabilité sociale et éthique de la recherche", *Revue internationale Ethique Sociétale et Gouvernementale*, vol. 12, 2010.

[LAR 17] LARRÈRE C., LARRÈRE R., *Bulles technologiques*, Wildproject, Paris, 2017.

[LAS 07] LASCH C., *La révolte des élites et la trahison de la démocratie*, Champs-Essais, Paris, 2007.

[LAU 05] LAURENT L., "Nanotechnologies et technologies convergentes: quel avenir à vingt ans?", Le Banquet, available at: http://www.revue-lebanquet.com/nanotechnologies-et-technologies-convergentes-quel-avenir-a-vingt-ans/, 2005.

[LEB 90] LE BRETON D., *Anthropologie du corps et modernité*, Coll. Sociologic d'aujourd'hui, Paris, 1990.

[LEB 07] LEBAS F., FREDSABLE A., "Stelarc performance", available at: http://lesnodulesetranges.blogspot.fr/2007/04/texte-stelarc.html, 2007.

[LEB 13] LE BRETON D., *L'adieu au corps*, Métaillé, Paris, 2013.

[LEG 15] LEGRAND P., ANDRÉ J.C., "Pour une approche "safer by design" en nano-médecine", *Micro & Nano-médecine "Avancées, tendances et perspectives"*, OMNT, Paris, 2015.

[LEG 16] LEGAULT J., "Designing for pleasure demands a different approach from designing for utility", available at: http://julielegault.com/amino/, 2016.

[LEL 11] LELU B., EASTES R.E., *Les scientifiques jouent-ils aux dés?*, Le Cavalier Bleu, Paris, 2011.

[LEM 02] LE MAREC J., *Ce que le terrain fait aux concepts: vers une théorie des composites*, HDR, Paris Diderot University, Paris, 2002.

[LEM 11] LE MÉNESTREL M., RODE J., "25 – why did business not react with precaution to early warning?" *Risks and Regulatory Policy*, pp. 639–651, OCDE, Paris, 2011.

[LEM 12] LE MAREC J., FAURIE M., "Communication et réflexivité dans l'enquête par des chercheurs sur des chercheurs", *Symposium Proceedings*, available at: https://halshs.archives-ouvertes.fr/hal-00671219/document, 2012.

[LEM 14] LEMOINE L., GAZIAUX E., MÜLLER D., "Dictionnaire encyclopédique d'éthique chrétienne", *Revue des sciences religieuses*, vol. 88, pp. 133–134, 2014.

[LÉO 10] LÉONETTI J., *Quand la science transformera l'humain*, Plon, Paris, 2010.

[LER 11] LEROUX T., *Le laboratoire des pollutions industrielles: Paris 1770-1830*, Albin Michel, Paris, 2011.

[LEV 03] LEVAL G., *Eléments d'éthique moderne*, Groupe Maurice-Joyeux, Paris, 2003.

[LIO 10] LIOGIER R., "La vie rêvée de l'Homme", in LIOGIER R. (ed.), *De l'Humain; pensée et artifice*, Actes Sud, Arles, 2010.

[LIP 06] LIPOVETSKY G., *Le bonheur paradoxal; essai sur la société d'hyper-consommation*, Gallimard, Paris, 2006.

[LIP 14] LIPSON H., KURMAN M., *Impression 3D: la prochaine révolution*, Wiley & First Interactive, Paris, 2014.

[LOU 15a] LOUMÉ L., "Bientôt l'impression d'un cœur en 3D?", available at: http://www.sciencesetavenir.fr/sante/coeur-et-cardio/20151028.OBS8505/video-bientot-l-impression-d-un-c-ur-en-3d.html, 2015.

[LOU 15b] LOURDANI H., "La recherche sur l'allongement de la vie progresse", available at: http://www.diplomatie.gouv.fr/fr/politique-etrangere-de-la-france/diplomatie-scientifique/veille-scientifique-et-technologique/etats-unis/article/la-recherche-sur-l-allongement-de-la-vie-progresse, 2015.

[MAG 13] MAGNIEN T., "Les nano-biotechnologies en médecine: questions éthiques", available at: http://www.huffingtonpost.fr/thierry-magnin/ethique-biotechnologies_b_3421072.html, 2013.

[MAL 16] MALGOYRE A., "L'éthique, place respective du médecin et du scientifique face à des impératifs militaires de commandement et d'opérationnalité", *Etudes de l'IRSEM*, "L'Homme augmenté; réflexion sociologique pour le militaire" available at: http://www.defense.gouv.fr/content/download/451219/7109695/file/Etude_IRSEM_n42.pdf, vol. 42, pp. 26–35, 2016.

[MAT 14] MATHIEU N., SCHMID A.F., "Reconsidérer le lien entre modélisation et interdisciplinarité", in MATHIEU N., SCHMID A.F. (eds) *Modélisation et interdisciplinarité; six disciplines en quête d'épistémologie*, Quae, Paris, 2014.

[MCG 09] MC GOEY L., "Pharmaceutical controversies and the performance value of uncertainty", *Science as Culture*, vol. 18, pp. 151–164, 2009.

[MCG 12] MC GOEY L., "Strategic unknowns: towards sociology of ignorance", *Economy and Society*, vol. 41, pp. 1–16, 2012.

[MIN 07] MINSKY M., "New prospects of immortality", available at: http://edge.org/q2007/q07_16.html#minsky, 2007.

[MIR 13] MIRATON D., Culture de l'innovation et éducation: retour d'expérience industriel", in CHEVALLIER-LE GUYADER M.F., GIREL M. (eds), *Partager la science; l'illetrisme scientifique en question*, Actes Sud, Paris, 2013.

[MIS 09] MISSA J.N., "Recherches technoscientifiques sur l'être humain et éthique du future", available at: http://www.belspo.be/belspo/organisation/publ/pub_ostc/Eth_code/Missa_fr.pdf, 2009.

[MOR 90] MORIN E., *Science avec conscience*, Fayard, Paris, 1990.

[MOR 92] MORAVEC H., *Une vie après la vie*, Odile Jacob, Paris, 1992.

[MOR 14a] MORIN L., "Biohackers, l'internationale des savants fous", available at: http://www.liberation.fr/futurs/2014/11/09/l-internationale-des-savants-fous_1139780, 2014.

[MOR 14b] MOROZOV E., *Pour tout résoudre, cliquez ici; l'aberration du solutionnisme technologique*, FYP, Paris, 2014.

[MUN 13] MUNIER B., Le Golem ou les vertiges d'un Homme fabriqué, in MUNIER B. (ed.), *Technocorps; la sociologie à l'épreuve des nouvelles technologies*, F. Bourrin, Paris, 2013.

[MUN 15] MUNJAI A., PHILIPPE J.M., MUNRO E. *et al.*, "A self-organized biomechanical network drives shape changes during tissue morphogenesis", *Nature*, vol. 524, pp. 351–355, 2015.

[MUS 13] MUSSO P., "Le Techno-corps, symbole de la société technicienne", in MUNIER B. (ed.) *Technocorps; la sociologie à l'épreuve des nouvelles technologies*, F. Bourrin, Paris, 2013.

[NAA 05] NAAM R., *More than Human*, Broadway Book, New York, 2005.

[NCB 16] NCB, "Conducting research and innovation in the context of global health emergencies: what are the ethical challenges?", available at: http://nuffieldbioethics.org/wp-content/uploads/Research-in-GHEs-note-for-web.pdf, 2016.

[NEE 16] NEELY E.L., "The risks of revolution: ethical dilemmas in 3D printing", *Science and Engineering Ethics*, vol. 22, pp. 1285–1297, 2016.

[NIE 89] NIETZSCHE F., *La naissance de la tragédie*, Gallimard, Paris, 1989.

[NOR 90] NORDON D., *L'intellectuel et la croyance*, L'Harmattan, Paris, 1990.

[NOW 03] NOWOTNY H., SCOTT P., GIBBONS M., *Repenser la Science*, Belin, Paris, 2003.

[NWO 14] NWO, "Science organization present new evaluation protocol for research", available at: http://www.nwo.nl/en/news-and-events/news/2014/science-organisations-present-new-evaluation-protocol-for-research.html, 2014.

[OGI 09] OGIEN R., TAPPOLET C., *Les concepts de l'éthique; faut-il être conséquentialiste?*, Hermann, Paris, 2009.

[PAL 12] PALAZZO G., KRINGS F., HOFFRAGE U., "Ethical blindness", *Journal of Business Ethics*, vol. 109, pp. 323–338, 2012.

[PAP 13] PAPINOT C., "Erreurs, biais, perturbations de l'observateur et autres "mauvais génies" des sciences sociales", *SociologieS*. available at: http://sociologies.revues.org.bases-doc.univ-lorraine.fr/4534, 2013.

[PAR 94] PARROCHIA D., "Le statut épistémologique de la "vie artificielle"", in TINLAND F. (ed.) *Ordre biologique; ordre technologique*, Champ-Vallon, Paris, 1994.

[PES 10] PESTRE D., "Des sciences et des productions techniques depuis trente ans – Chronique d'une mutation", *Le débat*, vol. 160, pp. 115–131, 2010.

[PES 14] PESTRE D., "Du gouvernement du progrès technique et de ses effets", in PESTRE D. (ed.) *Le gouvernement des technosciences; gouverner le progrès et ses dégâts depuis 1945*, La Découverte, Paris, 2014.

[PMO 04] PMO, "La véritable histoire de NanoBio (pour ce qu'on en sait)", available at: http://www.piecesetmaindoeuvre.com/IMG/pdf/Veritable_histoire_NanoBio.pdf, 2004.

[POU 06] POUTEAU S., "Biotechnologies: la vie en morceaux et l'éthique en péril", *Journal International de Bioéthique*, vol. 17, pp. 15–27, 2006.

[R&D 15] R&D TAXSAVERS, "The R&D tax credits aspects of 3D bio-printing", available at: http://3dprintingindustry.com/2015/12/07/rd-tax-credit-aspects-3d-bioprinting/?utm_source=3D+Printing+Industry+Update&utm_medium=email&utm_campaign=f8d067328b-RSS_EMAIL_CAMPAIGN&utm_term=0_695d5c73dc-f8d067328b-64583257, 2015.

[REN 15] RENOUARD G., "L'impression 3D peut désormais créer la vie", available at: http://www.atelier.net/trends/articles/impression-3d-desormais-creer-vie_438092?banner=1, 2015.

[RES 73] RESTAK R.M., *Pre-meditated man: Bioethics and the control of future human life*, The Viking Press, New York, 1973.

[RIC 90] RICŒUR P., *Soi-même comme un autre*, Le Seuil, Paris, 1990.

[RIC 94] RICŒUR P., "L'éthique, entre le mal et le pire", in Fonds Ricœur (2014) "L'incapacité dans la philosophie et les sciences sociales", available at: http://fondsricoeur.fr/uploads/medias/actualite/jfr-ricoeur-et-incapacite.pdf, 1994.

[RIC 13] RICŒUR P., "Post-face", in LENOIR F. (ed.) *Le principe de responsabilité; entretiens sur l'éthique*, Fayard, Paris, 2013.

[ROC 01] ROCO M.C., BAINBRIDGE W.S. (eds), *Societal implications of nanoscience and nanotechnology*, Springer, Dordrecht, 2001.

[ROC 02] ROCO M.C., BAINBRIDGE W.S., "Converging technologies for improving human performance: integrating from the nanoscale", *Journal of Nanoparticle Research.*, vol. 4, pp. 281–295, 2002.

[ROC 13] ROCO M.C., BAINBRIDGE W.S., "The new world of discovery, invention, and innovation: convergence of knowledge, technology, and society", *Journal of Nanoparticle Research*, vol. 15, p. 1946, 2013.

[SAD 14] SADI N.E., "La nano-médecine à la recherche d'un modèle viable dans un environnement contraignant", available at: http://lecercle.lesechos.fr/economie-societe/social/sante/221190500/nanomedecine, 2014.

[SAR 04] SARTRE-DOUBLET G., "Morale, éthique, vertu et devoir : liberté et servitude", available at: http://www.vox-populi.net/article.php3?id_article=293, 2004.

[SCH 10] SCHMID A.F., "Epistémologie et éthique technologique", available at: https://hal-univ-lyon3.archives-ouvertes.fr/file/index/docid/707832/filename/epistemo_ethique_schmid.pdf, 2010.

[SCH 44] SCHRÖDINGER E., *Qu'est-ce que la vie? De la physique à la biologie*, Le Seuil, Paris, 1944.

[SCI 14] SCIENCE ET AVENIR, "Quand l'imprimante façonne le vivant", available at: http://www.sciencesetavenir.fr/sante/quand-l-imprimante-faconne-le-vivant_13022, 2014.

[SEL 07] SELIN C., "Expectation and the emergence of nanotechnology", *Science, Technology & Human Values*, vol. 32, pp. 196–220, 2007.

[SFE 95] SFEZ L., *La santé parfaite*, Le Seuil, Paris, 1995.

[SIC 06] SICARD D., *L'alibi éthique*, Plon, Paris, 2006.

[SIL 97] SILVER L., *Remaking Eden: Cloning and beyond in the Brave New World*, Avon Books, New York, 1997.

[SON 00] SONIGO P., "Cellules en liberté", in KUPIEC J.J., SONIGO P. (eds.) *Ni Dieu, ni gène; pour une autre théorie de l'hérédité*, Points Sciences, Paris, 2000.

[STA 08] STAR S.L., GRIESEMER J.R., "Ecologie institutionnelle, traductions et objets-frontières: des amateurs et des professionnels au musée de zoologie vertébrée de Berkeley 1907–1939", in LAHIRE B., ROSENTAL C. (eds.), *La cognition au prisme des sciences sociales*, Ed. des Archives Contemporaines, Paris, 2008.

[STE 06] STENGERS I., *La vierge et le neutrino; les scientifiques dans la tourmente*, Les Empêcheurs de penser en Rond, Paris, 2006.

[STE 10] STEINER P., "La transplantation d'organes: un nouveau commerce entre êtres humains?", Revue du MAUSS 1/2010, available at: http://www.cairn.info/revue-du-mauss-2010-1-page-455.htm, pp. 455–462, 2010.

[STE 13] STENGERS I., *Une autre science est possible! Manifeste pour un ralentissement des sciences*, Les empêcheurs de penser en rond, Paris, 2013.

[SUR 16] SURVEYMAGAZINE, "Comment réaliser une enquête par questionnaire?", available at: http://www.surveystore.info/NSarticle/enquete-par-questionnaire.asp, 2016.

[SUS 09] SUSSAN R., "NBIC, Les nouvelles frontières de la pensée", available at: http://www.lesinfluences.fr/NBIC-Les-nouvelles-frontieres-de.html, 2009.

[SUS 12] SUSSAN R., "Bienvenue chez les post-humains", available at: http://www.scienceshumaines.com/bienvenue-chez-les-posthumains_fr_28222.html, 2012.

[TAL 07] TALEB N.N., *The Black Swan – The Impact of the Highly Improbable*, Random House, New York, 2007.

[TEC 16] TECHNOPOLIS GROUP, *Case Study Review of Interdisciplinary Research in Higher Education Institutions in England*, available at: http://www.hefce.ac.uk/media/HEFCE,2014/Content/Pubs/Independentresearch/2016/Two,reports,on,interdisciplinary,research/Technopolis-Case%20study%20of%20interdisciplinary%20research%20in%20HEIs%20in%20England.pdf, 2016.

[THE 13] THE ECONOMIST, "Prometheus unbound; researchers have yet to realize the old dream of regenerating organs; but they are getting closer", available at: http://www.economist.com/news/science-and-technology/21580440-researchers-have-yet-realise-old-dream-regenerating-organs-they-are, 2013.

[TOP 13] TOPÇU S., *La France nucléaire; l'art de gouverner une technologie contestée*, Le Seuil, Paris, 2013.

[TRI 07] TRICORE A., "Le financement compétitif de la recherche scientifique en Europe et aux États-Unis", *Lieux communs*, vol. 10, pp. 69–91, 2007.

[VAN 00] VAN LENTE H., "Forceful futures: from promise to requierement" in BROWN N. (ed.) *Contest Futures: A Sociology of Prospective Technoscience*, Burlington, Ashgate, 2000.

[VAN 15] VAN SONG N.P., "Michel Benasayag: hybridation homme-machine sous conditions", available at: http://www.up-magazine.info/index.php?option=com_content&view=article&id=5009:hybridations-homme-machine-sous-conditions&catid=103:analyses&Itemid=542&utm_source=newsletter_131&utm_medium=email&utm_campaign=up-nl-vnew, 2015.

[VIA 99] VIAN B., *Et l'on tuera tous les affreux*, Le livre de poche, Paris, 1999.

[VIA 16] VIANIN A., "Quand l'interdisciplinarité naît de la croisée des regards", L'Interdisciplinaire (2016) page 1, available at: http://www.ihqeds.ulaval.ca/fileadmin/fichiers/fichiersIHQEDS/Publications/LintErDiSciplinaire/IHQEDS_Journal_No10_lres.pdf, 2016.

[VIG 14] VIGARELLO G., *Histoire de la beauté. Le corps et l'art d'embellir de la Renaissance à nos jours*, Le Seuil, Paris, 2014.

[WAL 08] WALTER F., *Catastrophes; une histoire culturelle XVI-XXIème siècle*, Le Seuil, Paris, 2008.

[WEB 59] WEBER M., *Le savant et le politique*, Plon, Paris, 1959.

[WIE 71] WIENER N., Cybernétique et société, Collection 10-18, Paris, 1971.

[WIK 15] *WIKIPEDIA*, "Humanity+", available at: http://fr.wikipedia.org/wiki/Humanity%2B, 2015.

[WUR 77] WURMSER L., *The mask of shame*, J. Aronson Inc., Baltimore, 1977.

[ZAC 16] ZACCAI E., TIMMERMANS B., HUDON M. *et al.*, "L'évaluation de la recherche en question(s)" Académie royale des Sciences, des Lettres et des Beaux-Arts de Belgique", *Bruxelles Quarterly*, vol. 64, pp. 4, 464–494, available at: http://penserlascience.ulb.ac.be/IMG/pdf/penserlascience_2015_actes_complets.pdf, 2016.

[ZHU 17] ZHU W., QUA X, ZHU J. *et al.*, "Direct 3D bioprinting of prevascularized tissue constructs with complex microarchitecture", *Biomaterials*, vol. 124, pp. 106–115, 2017.

[ZUI 06] ZUINDEAU B., "Le réseau comme forme d'organisation de l'interdisciplinarité "par le bas "l'exemple de Développement durable et Territoires fragiles", *Natures Sciences Sociétés*, vol. 14, pp. 286–292, 2006.

5

Questions of Epistemology and Modeling

"Technoscience is the study of that intricacy of the technical and scientific, it is interrogating what brings them together (which is known as their convergence), is it also looking closely at their practical project: 'domesticating matter,' 'manufacturing life'? What conceptions do the technical actions of such ambitions call on?" [LAR 17]

"The fact remains that the construction of a tissue or the differentiation of an organ, macroscopic phenomena, must be considered the integrated result of multiple microscopic interactions due to proteins and depending on their properties of stereospecific recognition". [MON 14]

"Engineers hate complexity. I hate emerging properties. I don't want the airplane I want to take tomorrow to have some emerging property in flight". [DRE 10]

"We don't have the right to demand that a promise be kept; we can only hope for this as a virtue of primal ethics, because promising is already betting on time, on the permanence of the subject that will not have been versed in folly or succumbed to death. But promising is believing that language involves you and that a truth is at risk in everything that is said, rather in the manner and moment when this will have been said, in the intention of the body and voice, in contained presence and the surrounding silence". [DUF 11]

"Modern natural history, witness of obscure battles, is not netural; its models and observations form the plot of contradictory certainties that sometimes allow science to be used to forge opinions, to comfort options and social convictions". [CAR 86]

"By focusing it on the 'useful,' there is a specific risk of compromising the very fertility of the social contribution of science". [GUI 15]

"Several disciplines of [Engineering Sciences] are thus, today, broken down, not on the academic level where some excel, but in relation to what has fed their foundation and give them pertinence. The industrial models that inspired them are sometimes outdated by several decades. For some disciplines, it is urgent to renew their foundations, to requestion basic hypotheses, explicit and implicit alike. In the industrial world, there has been a change in the productive paradigm [...]. It is best to return to the site to renew the models. Interdisciplinarity, in this context, becomes a modality of research that makes sense". [VIN 00]

"The pretention of science to dissolve into the anonymity of the mechanical, physical, and chemical environment, those centers of organization, adaptation, and invention that are living beings... Hence the insufficiency of all biology that [...] would like to eliminate any consideration of meaning". [CAN 65]

"This refers to saying that a correctly posed issue does not necessarily have a solution". (Turing, according to Levin [LEV 10])

5.1. Introduction

"The body ceases to be the key concept that surrounds the idea of life. This appears more and more to be an original, complex organism of chemical and thermodynamic phenomena in self-organized systems (with its own generative device localized in the genes), and at the same time, eco-organized, drawing on the energy and, even better, as all animal life does, from organized complexity, in an environment that, itself, is living or an ecosystem. Cybernetics, systems theory, and automaton theory are already being applied and cover various aspects of the operations of living beings". [MOR 74]

> "It is the intrinsic activity of the system that determines how we must describe its relationship to the environment, which thus engenders the type of intelligibility that will be pertinent to understand its possible histories [...] The notion of sensibility is found to be associated with that of instability because, in this case, it is a matter of the system's sensitivity to itself, to the fluctuations of its own activity". [PRI 09]

> "What changes radically is that there is no longer combination, at first, of the knowledge from each discipline, but rather, a tendency to start assembling what is not known, 'the state of the non-art,' where the discipline could not give responses to the problems posed by the object that one seeks to know and construct. Each discipline seeks the area in which the object cannot be processed through its concepts and logic. This questions scientific practice to its core because this non-knowledge obliges us to see interdisciplinarity other than as a sum procedure. It obeys a logic of substraction (with no lack), which leads each discipline to be reinterpreted by others. Mastery is no longer the centerpiece. Not-knowledge is no longer marginalized; it lies at the center of the process". [SCH 10]

> "Paradigm shifts make scientists, in their research field, see everything from a different perspective. Insofar as they only have access to the world through what they see, we can be led to say that, after a revolution, scientists react to a different world". [KUH 62]

> "One of the most spectacular aspects of this new behavior is the formation of unbalanced structures, when irreversible phenomena play the fundamental role. One of the most spectacular aspects of this new behavior is the formation of unbalanced structures that only exist as long as the system dissipates from energy and continues to interact with the outside world". [PRI 93]

The subject of mastering bio-matter embedded in these five quotes resonates with a foundational article from 1929, quoted by Eric Cances [CAN 16], in which P.A.M. Dirac wrote, "the underlying laws of physics necessary for a mathematical theory covering a large part of physics and the totality of chemistry are thus completely known, and the difficulty is only that the exact application of these laws leads to much more complicated equations to be solved. It is thus desirable for methods to be developed for the approximation of quantum mechanics models, which would allow the properties of complex atomic systems to be explained without many calculations" (numerous challenges must still be risen to be able to perform simulations through these methods of complex systems due to the scourge of the dimension and the multiplicity of scales) [MOL 12].

The experimental results on bio-printing, presented in Chapters 2 and 3, come from experiments that are not, despite saying it, either thought to be or experienced as integral; the emphasis is placed on singularities rather than processes. The expression convergence indicates a decisive sorting activity for (proper) diagnosis. In fact, it is marked by the sign of indecision, the unkept promise, in short, virtually the undecidable. In this notion, an implicit unit is presupposed to produce a uniting notion about which it is possible to think that it may be associated with solid epistemology. Can we agree with Revault d'Allonnes [REV 12] when she writes, "everything seems to change frenetically, whereas, in reality, nothing moves in a petrified and immobile world"? In the book "The Leopard" [TOM 06], the idea was very much the same: "everything must change so that nothing changes…". Just the appearance of change?

Tissue engineers have long attempted to produce correctly vascularized tissue in laboratories that is robust enough to replace damaged tissue, wiping away numerous failures to achieve results qualifying as a proof of concept (but this is the price to be paid in risky research). Other teams have played the bio-printing card more prudently, but they have had to limit themselves to an insufficient portion by producing extremely fine tissue. Attempts to print thick layers were negative, as sandwiched cells did not have access to sufficient oxygen or nutrients, nor were they able to properly dispose of carbon dioxide and other wastes products. They thus ended up dying. To bypass this problem, nature has a network of minuscule blood vessels with thin walls, which nourish everything while removing waste. This is a schema selected by attempting to couple bio-printing and microfluidics to allow the irrigation of bio-constructs.

Is something truly new, a breakthrough, being explored, or is this, in a trend associated with real social demands, 3D printing with a "false nose"? "The novelty of modern times thus has a very particular emphasis. It is not only about an awareness of the time that is delineated from the past and institutes the border between 'here' and 'now,' but a breakthrough that, by refusing the authority of models, takes away all of their value as examples: all antecedence is declared defunct and obsolete" [BLU 99]. Waves have been made in the sea of scientific incompleteness.

In the different works illustrating the possible satisfaction of extraordinary promises, published activities are essentially descriptive, without any attempt at understanding in the sense of mathematical models, based on the finest knowledge possible of biological systems and their interactions with the environment being scientists' focus. In this chapter, the mission is to escape a "bottom-up" approach through trial and error in an attempt to propose plans of action allowing synergies between modeling and experimentation (top-down approach). Figure 5.1 illustrates the situation as it is perceived by the author today.

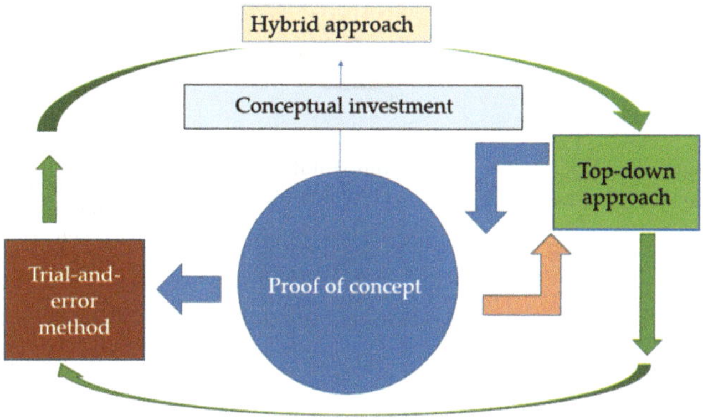

Figure 5.1. *Proof of concept output in bio-printing. For a color version of the figure, see www.iste.co.uk/andre/printing3.zip*

Several types of interrogation emerge in the scientific and technological evolutions related to bio-printing:

– using the bio-printed system as it is built for scientific studies on cellular migration, proliferation, etc. (this stems from the author's modest expertise), as well as using it for industrial toxicology assays or in cancer research and personalized medicine;

– seeking to experimentally produce the most simple biological systems possible for easier creation, but studied to be adapted to real biological environments (e.g. skin grafts; see Chapter 3);

– trying to model the behavior of bio-printed cells. Considering the results of the literature and the complex interdependencies between parameters (see attempt at representation in Figure 5.2), the task seems difficult today. This is indeed what motivated the present reflection, which aims to start seeking better familiarity with the literature on different kinds of tissues to reduce this complexity as much as possible. This "simplification" task still has a long way to go;

– "playing" on both approaches at once (hybrid approach).

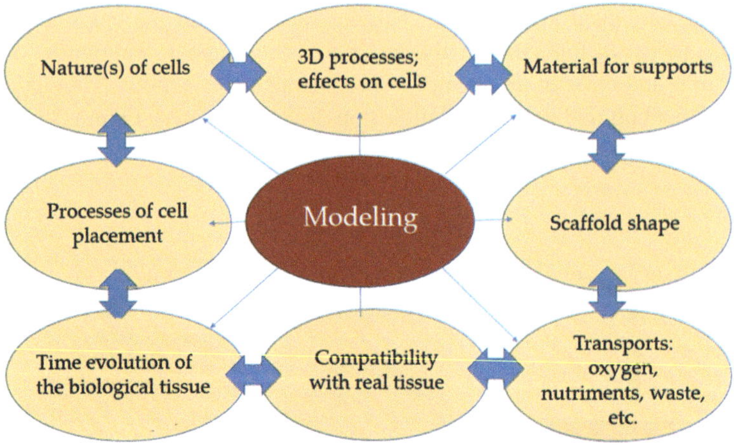

Figure 5.2. *Interdependent difficulties of complete bio-printing modeling*

The question posed is indeed that of having robust information on cellular behavior, allowing progress to be made in modeling aspects. Following discussions that took place with biologists, to revisit an image (provided in part by Watzlawick [WAT 91]), are we not in a situation where an observer (knowing neither the game nor the language of the players) has, for a given time, an observation of failing players: if he/she is given time, it will be possible to isolate regularities, to figure out that the goal is defeat, etc. In short, he/she must be able to recognize, after a certain time, the rule of the game and himself/herself become a player. If his/her time is counted, it may be possible to find some rudiments of the operation, or nothing at all. If it is now a matter of crazy players who apply a rule with anarchic evolutions, the observer will not be able to discern a legitimate trend model. This is also the case if he/she cannot see "the things behind things" (consider Plato's allegory of the cave). This is indeed one of the dilemmas of the emergence of bio-printing as a scientific domain that is discovering if there is one or more still-hidden mechanisms (and thus mechanisms to be discovered) or if, to the contrary, no matter the time dedicated to it, the observer will never be able to understand the intimate functioning of the game of differentiation and cell growth in a bio-construct. As an example, Chapouthier, in 2012, writes, "Observation of the living world clearly shows that complexity results from the repeated application of two large principles: the principle of 'juxtapositing' entities of the same order of complexity and the principle of 'integrating' these intities into more 'complex' wholes, of which they then become part [...]. Thus, for example, identical cells can first be juxtaposed to provide tissues of identical cells, then integration leads to organisms, where cells have different functions: liver cells, muscle cells, nerve cells, etc." (see also [DOR 10]).

From a bio-printing modeling standpoint, we can consider a system evolving over time and originating from a given shape at the moment 3D printing took place to achieve another shape (dynamic approach), normally researched, or impose different kinds of constraints (chemical, mechanical, energy, etc.) that will bring about deformation upon an already-stable shape (thermodynamic approach). In the former case, it is best to introduce time elements allowing the evolution to be defined at a given time; in the latter, thermodynamic elements and those with chemical potential are introduced. These are classical problems that, even with non-living matter, are always difficult to resolve, if only in the case of badly conditioned linear systems (see difficulties of deconvoluting signals due to noise and measurement errors; see [MUG 08]); this will be even more difficult when using nonlinear approaches with noised signals. It is essential for this to be at stake because mastering the inverse problem that must interest the healthcare professional may constitute an integral means of creating an object from an easily formable matrix whose deformation function of time or energy provided could allow a functional tissue or organ with a complex shape to be created. "The notion of determination depends on the inference of the cause based on the effect. Epistemologically speaking, a valid assertion can only be made concerning the cause, based on the effect, if a single sort of cause may have produced the effect in question. Yet, in the case of determination, this is precisely what happens" [BER 84].

Here, it is indeed a matter of examining whether there are one or more responsible processes allowing this goal to be achieved, of attempting to temporarily remove proofs of concept and not to get involved in these words from Larrère and Larrère: "With this naturalization of the dynamics of science and technology, there is an escape from the conception of technology in terms of means and ends – which implies a project and thus a subject – to tackle a conception of technological development in terms of processes with no subject, with no possible mastery. Taken over by a number of scientists, this discourse, whose goal is to declare all contestation as old-fashioned and useless, has the consequences of freeing the technoscientific networks involved in the conception and diffusion of such innovations of all responsibility" [LAR 17].

It is thus to be expected, in this final chapter of Part 2, concerning a possible "rational" approach to bio-printing, to "trudge through" an exploration of a fascinating subject, but one that is difficult to take into one's own hands (at least for the author!) and to involve the reader in understanding doors to be knocked down before bio-printing can become a routine process in reparative medicine (or even more?). As there is a risk of discussing ontology, Figure 5.3 synthetically recalls the general goal of this chapter, which is to be able (and with precision) to model the production of biological tissues (and tomorrow or later, even organs).

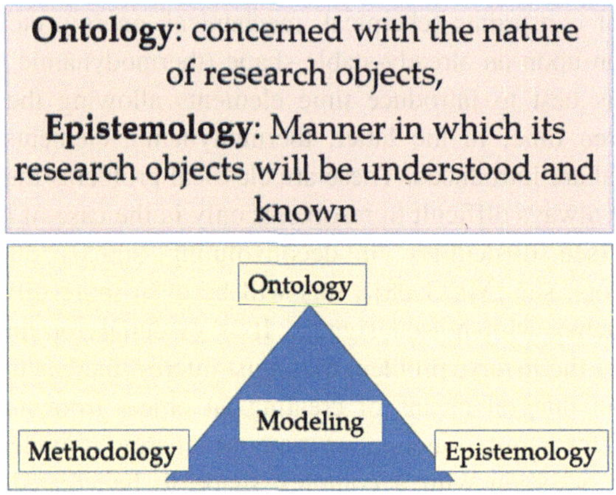

Figure 5.3. *Basis of reflection for modeling*

"We must not forget that epistemology is the part of philosophy that aims to critically study the postulates, conclusions, and methods of a particular science, considered from the standpoint of its evolution, in order to determine its logical origin, value, and scientific and philosophical significance, as well as the fundamental concepts, theories, and results of diverse sciences to determine their origin, value, and objective significance" ([NAD 99], see also [DHO 06, NAD 16]). The recognized epistemology is unlikely to be able to correctly evaluate the place of science in the imaginary, because this risks overvaluing the discriminating power of reason. Similarly, an in-depth analysis of the literature on bio-printing may lead to the inverse effect with tendencies to despair because the task seems immense, as so many strangers appear overnight. Precious on-site because it allows the research to take some distance from his/her "objects", epistemology is both a healthy practice but also one that can reveal itself to be a rather vain theory, as it risks not helping to grasp the reality and complexity of scientifically unstructured domains like bio-printing. In short, is the case as hopeless as these few phrases lead us to believe? We must look deeper into this question, as usual. But this time, the literature on bio-printing, considered for its "modeling" and/or "simulation" aspects, is particularly scant. Only a few teams have looked at the question (apparently, fewer than 10–15) and these since roughly 2010. The experts in the field must either be sought with a magnifying glass or exist (maybe without knowing it) in other disciplinary domains.

With all due respect to the majority of intellectuals, reason is not qualified to criticize itself, because it is self-indulgent, particularly in the framework of collective actions where the justification mechanism is fully at play (particularly when seeking financial support). But what must be demanded is a debate, even a vigorous one, and not just on information, otherwise researchers' interest will not reach the substance of decision-making, delegated to technocrats of a "higher level", associated with societal pressure. It is true that there is a debate between proponents of the bottom-up approach (pioneering experimenters, PE) and those of integrative reflection (top-down risks appearing, on open subjects, in a non-stabilized heuristic approach to be a discussion without robust foundations through the impossibility of deciding and moving on). The latter hopes to find meaning in a heuristic approach defined in Figure 5.4 (taken in part from the previous chapter); we will call these "heuristician" experts or HEs; through their modalities, they foresee becoming "boundary-crossers". Debating is, however, attempting to go beyond certain paralyzing stereotypes, applicational desirabilities, competitions, emotions, and the search for recognition, even "rat races".

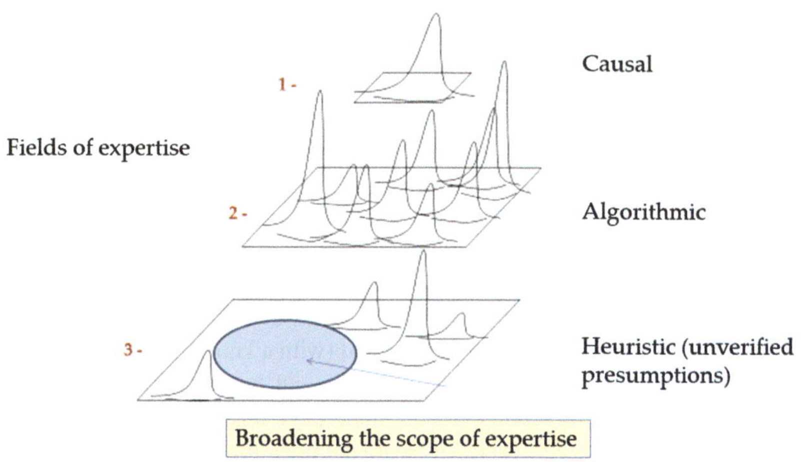

Figure 5.4. *Fields of expertise (the Gaussian approximations represent scientific acquisitions resulting from PE work)*

According to Durand [DUR 13] quoting de Saussure, a system is "an organized whole, made of solidary elements that can only be defined in relation to one another as a function of their place in that whole". The question asked in this reflection concerns the decomposition of a system into specialized sub-systems (e.g. at different spatial scales) in an attempt to advance (see Figure 5.5).

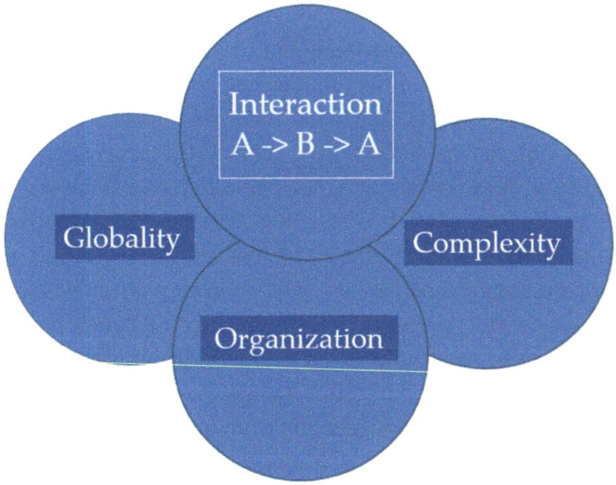

Figure 5.5. *Complex system*

5.2. The PE approach (seen by a possible divergent, somewhat of an HE) [AND 16]

Living matter can be ordered, meaning that its shape can, at least in an initial approximation, be described geometrically. It is essentially on these foundations that the publications concerning bio-printing are situated. By preparing shape elements through additive manufacturing (including inert matter and living matter), it is possible to think that, if one is perseverant, it will be possible to (soon) achieve organ repairs by introducing, where and when necessary, living cells that will know how to develop "as they should" for a given end (with a known environment). In this "paradigmatic" context, bio-printing could be found between "classical" additive manufacturing, a field of (process and materials) engineering excellence, and biology, even "natural" life sciences (if we do not draw too close to a tautology).

Relative to classical additive manufacturing techniques, printing biological elements adds a very significant level of complexity to processes because it is necessary to "intelligently" structure living or non-living materials, imitating the extracellular matrix, and to control the spatial distributions of different types of cells or biomolecules that either communicate among themselves or participate in the transmission of information and that can play a role in cellular differentiation, cell growth, apoptosis, etc. (e.g. in a recent article, Munjai *et al.* [MUN 15] show the influence of the cellular environment on the behavior of this entity in terms of growth, differentiation, etc.; they are followed by Caiozzo *et al.* [CAI 16], who show that modifications of the physical characteristics of an artificial 3D extracellular matrix in which differentiated mature cells are cultivated (modifications in the biochemical composition and the rigidity of their microenvironment) influence the generation of iPSC or induced Pluripotent Stem Cells). It is thus a matter of providing biologists with processes allowing cellular suspensions, aqueous solutions or hydrogels to be deposited by limiting the different stresses that cells can undergo through additive manufacturing processes. However, at the same time, mechanical vibrations can have positive effects on the expression of certain biomolecules. The engineer then seems helpless.

The approach on varied levels involves substituting a given system, particularly a living one, a different environment presenting "precisely what is needed" in terms of analogies with the first, so as to infer credible and repeatable conclusions because they are more reduced and thus, in principle, easier to study. This multi-level approach to the living should be analyzed using concrete examples to identify what can be explored (knowledge of the real and possible): robust, current knowledge and present abilities of bio-printing. Through scale elements, it is already possible to predict the behavior of DNA and the other nanometric constituents of living matter, intracellular components, individual cells, cellular interdependencies, the microscopic architecture of organs, the organs and some of their sub-sets, their shapes and interactions with their environment, etc. However, when an object attains a certain complexity, how can we know if the smallest known and representative description is not the object itself?

With comprehensive approaches (on small voxels, these being, let us recall, the "elementary bricks" for constructing an object layer by layer), the comprehensive system analysis methods could give rise to a "black box" type modeling that will neglect at least some of the elements whose voxels are constituted, keeping only the main parameters, which are considered to be reasonably related to – and representative of – the studied system. "Systematic modeling bases its originality on its ability to respect this dialectic constituent of all complexity: becoming by functioning and functioning by becoming, by maintaining its identity" [LEM 06]. Every level very likely has its own specificities in terms of functioning, interaction with its environment and structure. From this comment emerges the following question (that must be validated): is fractal perfection (same behavior and properties, whatever the scales) applicable to the real living world?

Moreover, it is necessary to recall that we are always working on a corpus of extremely incomplete knowledge whose degree of generalization at larger or smaller scales can be questioned. However, the idea is to study the control parameters (temperature, various flows, various heterogeneities, spatial divisions, etc.) to try to examine the existence of attractors and their deformation(s) induced by the environment. Indeed, "most of the processes responsible for the functioning of the living, at every level – from the cell to the ecosystem – result from tens, hundreds, even thousands of factors: presence or absence of an ion, of a molecule, of a nutrient, of a gene, structure of a protein, etc. Deepening their understanding therefore implies mastering tools likely to represent and analyze these complex interactions. This is what today is known as "network biology" [VAN 12].

The laws of biological growth are associated with numerous dynamic phenomena, nonlinear due to the richness of the already observed scenarios. The appropriation of the domain into an interdisciplinary approach (involving all the concerned disciplines; see previous chapters dealing with the notion of convergence) is obviously a key element to attempt to move the subject forward (somewhat following in the footsteps, maintaining all proportions, of the conceptual work by Carnot in thermodynamics). The culture of nonlinear dynamic systems (NLDS), certainly complex, is still not very diffused, even (and particularly?) among the HEs.

Let us recall that, in linear systems, modifications of the parameters lead to quantitative changes, but they do not modify the behavior (stationary condition) of the system. In NLDS, a small variation in certain parameters, the so-called control parameters, can, in well-defined conditions, approaching a critical value, provoke a total behavior change in the system's balance. That is what is called bifurcation. Figure 5.6 represents such a "bifurcation in a fork".

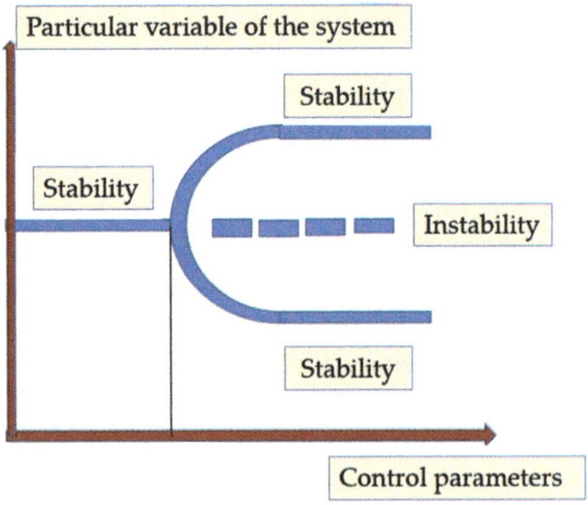

Figure 5.6. *Bifurcation in a fork*

"Thus, what seems *a priori* to be 'simple' is in fact the result of the combined evolution of various characters, moving towards real optimization of means. Therefore, this is something very complex, from a biological standpoint, and the first vision of the simple ancestral disappears through this reversal due to evolving reasoning" [LEG 08]. Figure 5.7 from Delignières [DEL 15] reminds us, in a two-dimensional representation, of the situations associated with possible bifurcations (achieving different states through the same law of behavior with multi-stable states).

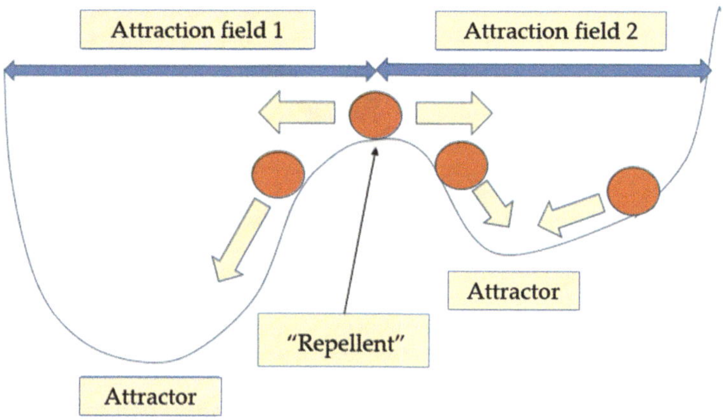

Figure 5.7. *Schematic representation of a landscape of attractors. For a color version of the figure, see www.iste.co.uk/andre/printing3.zip*

In this figure, the stability levels are in the catchment areas while the "repellents" correspond to the areas of maximum instability; in the figure, the ball can go either to the right or to the left according to the very weak impulse that it receives. The essential characteristic of the attractor is the stability of the order parameter with a strong tendency towards reproducibility. According to Delignières [DEL 15], bifurcations are announced by critical fluctuations, that is, a significant evolution of the order parameter approaching transition (as in a modification of the orientation of the molecules in an organized system once the transition temperature is exceeded). Later, it could be possible to find an applicable approach if, during these evolutions, the number of attraction fields and their qualities are not greatly disturbed. In other words, is it possible to show the possibility of having, as Thomas *et al.* [THO 04] propose, "chaotic labyrinths" that constrain the complexity? For bio-printing, starting at a given position of the cells in a given environment, does the system's evolution always converge towards the same solution? In any case, the information resulting from the experimental process to be undertaken should provide data (if possible, robust data) on the living systems studied and on their evolution dynamics, with the potential existence of catchment areas. If a single one exists, there is a possibility of thinking that the evolving system could be approached with the help of a deterministic model. But will this be a general case? If we refer to the recent paper by biologists [OLL 16], who do not illustrate this potentiality of having a single path of bio-manufacturing in a biological system, there is room for doubt (see Figure 5.8).

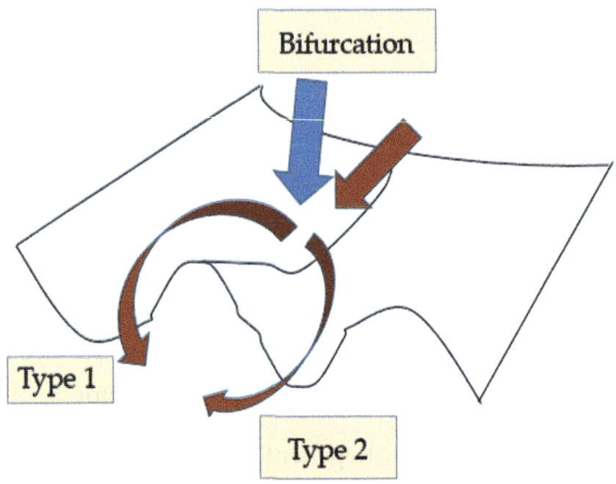

Figure 5.8. *Bifurcations in a biological system. For a color version of the figure, see www.iste.co.uk/andre/printing3.zip*

The creation of "custom-made" heterogeneous controlled 3D systems may effectively be a means of (slightly) approaching reality in the sense that it allows a certain elevated complexity (so long as at least part of a vision of reality is maintained and subsequently a rather robust modeling basis) to perform experiments that effectively reinforce the concept (or that will contradict it) [LAU 13, SÈV 05, ABB 15, CHA 11]. Such a scientific approach must allow knowledge of whether a certain degree of predictability can be achieved in manufacturing a biological organization. It is then a matter of having experimental study tools to validate a conceptual framework allowing the planning of a satisfying solution to the biological question, particularly if it must one day be applied to humanity. The PEs have already had to define experiments whose respective statuses, the questions they are meant to answer, the methodologies, ambitions, etc., are normally defined.

The stake would thus be to perform large-scale scientific research in order to have a knowledge base allowing a causalist credibility of the bio-printing project to be qualified vis-à-vis printing organs (and which organs). Thus, to take an example, if a system made up of a voxel (or several) can undergo an uncontrolled bifurcation by controlling external parameters, how could we, with credible likelihoods, foresee the pertinence of an operation possibly aiming one day at a graft?

5.3. The HE approach

Many fields escape "rational verification" in the natural sciences, is this also the case for the hundreds or thousands of publications on bio-printing, with different conditions, different cells, etc.? However, this conception of rationality seems ruinous to some: it could only be the twin of the arbitrary and the obstacle to seeking, continuing to publish beautiful proofs of concept. Consequently, to avoid this kind of pitfall, the HE project can/must take shape: if this image of reality corresponds well to the attempt discussed below, then it will be a matter of starting with this conception and attempting to develop it.

The goal of all modeling is, for every "professional" who is somewhat a technician, to provide a parametrable system (when possible) with a view to optimizing it relative to an end, thanks to a wise choice of input parameters, a "stable" and robust enslavement in its temporal (even spatial–temporal) dynamics. The environment of bio-printing systems requires the knowledge of independent and interdependent variables that condition its state at a given time. The more complex this system is, the more it must be classically described with the help of a number of coupled equations, if possible with the most limited number possible considering what is measurable (and measured). For the engineer, the model that will hopefully be elaborated is generally situated between efficiency and truth, where there is a hope for an "acceptable" compromise between faithfulness to reality and simplicity of implementation. But what are the approximations that make the model operational according to instructions imposed by users? To achieve such a goal, it is necessary to pool the knowledge from the different disciplines concerned, with all the associated difficulties (see section 4.4 of the previous chapter).

By proposing an interdisciplinary approach, the HE project does not aspire to make the approach play any other role than that of being at the service of each discipline, by setting itself the goal of underlining, for example, the complex process that leads to formulating an explanatory or interpretative hypothesis, allowing a relation of dialogue and mutuality to be installed. Let us add that the HE project outlined in this way presupposes the adherence to the standpoint of metaphysical ignorance. With this in mind, the project should have the rule of motivating critical reflection on directive ideas, founding intentions, directing concepts, and the methodological, paradigmatic, anthropological, epistemological and ethical presuppositions of each discipline. But, at the same time, it must accept humbly restricting itself to a task of clarification, of wisely clearing up, which could not aspire to later clarity or a fundamental truth.

Technologies are only visible elements of a broader whole, the physical and conceptual constituents of which merge only because of a common history [TIN 94]. What was just recalled is that we wish to make a comparison between bio-printing and "classical" additive manufacturing processes and that additional barriers are added, some of which stem from the conceptual [GOD 04, GOD 11, GOD 12, GOD 15]:

– a complex choice of supports (scaffolding) for the development of systems using living cells;

– there are many biological phenomena involved in the development process and their laws are far less well understood than the elementary laws of physics, for example;

– cells with variable origins, function of the chosen end;

– the interactions with the environment are complex and continuously regulate cell growth;

– evolutions of cellular systems, complex functions of factors from the spatial, chemical and mechanical environment, interactions with the support, various transfers, etc.; there are many tightly interwoven levels of organization, making it extremely difficult, even arbitrary, to decompose such a system into simpler and more independent elementary sub-systems;

– does the change in scale respect principles of autosimilarity;

– is the intermediate growth of cells before deposition detrimental to the goal (see [GUI 10]);

– the initial conditions are not measurable with an arbitrary precision;

– interdependencies between these large primary domains; etc.

Living matter being positioned in space, it is possible to think that structures, more or less self-organized, must be associated with certain types of physical or physicochemical conditions that surround them [MYO 11, MOR 04]. The ultimate goal is to control the interactions with the environment for the desired cellular growth, that is, being able to maintain matter, energy and information flows necessary to the metabolic processes they house. "It is thus that we expect, in this framework, to see the appearance of morphologies like the materialization of areas of greater likelihoods. That is to say that we are dealing with a very general theory of self-organization, particularly at the level of morphogenesis and morphological evolution" [CHA 12].

It is possible to call on heuristic resources and a hierarchical vision of the questions associated with bio-printing marked by the mechanist paradigm inherited from the Aristotelian tradition that has prevailed in scientific thought to today. For some years now, another community has regarded this question of managing/ modeling from another paradigm, that of the self-organization of living systems of close systems: multi-agent, "holonic", "heterarchical", and bionic systems [RAM 88].

However, if we place ourselves in a broader evolutionary perspective, as was already mentioned, it does not seem possible to predict the functional, spatial, and temporal elements of shapes by deducing them from the application of the simple laws that we could have identified, and this, on different scales (from the components of a cell to the tissue; see Figure 5.9, inspired by Martins, Ferreira, and Vilela [MAR 10]). More precisely, upon experimentation, if these laws are necessary, they are not enough. They form the inevitable background of an evolution that is developing in an unapparently deterministic way through the interaction of natural systems among themselves and with their environment. The sciences of "morphogenesis" thus would only be valid in the limitations of ordinary or macroscopic physics. Yet, this is only an approximation, in all likelihood requiring the elimination of linear determinism.

Today, it does not seem (to the author) foreseeable to master the bio-printing system only through the knowledge of its constituent elements, hence understandable demands for a comprehensive approach as practiced by engineers, taking into consideration the phenomena judged to be preponderant in the model. According to holistic theses, each cell influences the whole under construction, which, itself, in a recursive way, gets involved in turn in the elementary entity. "There is a series of actions and reactions, that is, we have a system of balance between different forces rather than a phenomenon that may be reduced to a cause and the effects of this cause" [PAR 65]. But then how to be sure that we are dealing with the principle variables if complex local interdependencies (cells/environment and vice-versa) exist? Yet, the objects of bio-printing, as they have been presented, are composite, systematic, unstable, and under constraints (some of which are still unknown). Thus, as shown in Figure 5.9, whatever the scales considered, bio-printing is probably inaccessible to only monodisciplinary approaches ("contingent structuring produced by a particular historic process of parcellizing science", according to Legrand [LEG 02]).

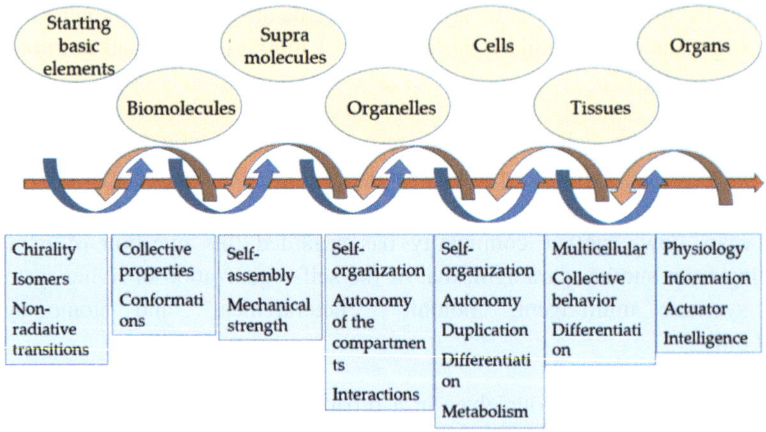

Figure 5.9. *Internal and external interactions between spatial domains (starting basic elements: nucleotides, amino acids, sugars, fatty acids; biomolecules: DNA, RNA, proteins, polysaccharides, lipids; supramolecular structures: membranes, ribosomes; organelles: nucleus, mitochondria). For a color version of the figure, see www.iste.co.uk/andre/printing3.zip*

In light of the incompleteness of data, during the cellular differentiation of embryonic stem cells or while reprogramming cells into stem cells, Richard *et al.* [RIC 16] observed a heterogeneity from one cell to another in the expression of certain genes, which suggest that this variability plays a role in differentiation. This result, which recalls the principles introduced by Turing in 1952 and seems to show nonlinear phenomena, poses an epistemological problem. Furthermore, according to Lentini *et al.* [LEN 17], cells constantly exchange messages and "vectors" are molecules synthesized by their genes, which spread to the outside by crossing cell membranes. When these cells reach another cell, they penetrate their membrane, reach the DNA and activate genes. These then produce "response" molecules, which in turn lead to the activation of genes in the first cell and so on. Fourel *et al.* [FOU 16] confirm, through another example, that cells perceive their environment via physicochemical stimuli that orient the differentiation processes. The extracellular matrix, made up of proteins, polysaccharides and proteoglycans, controls the bio-activity of cells. These are capable of perceiving the physicochemical properties of their environment and adapting to them by varying their shape, migratory behavior or ability to proliferate. One goal for the cell is to differentiate, that is, to specialize so as to acquire properties of a particular biological tissue. Understanding how cells integrate the physicochemical stimuli that orient differentiation processes is a stake, notably for the repair of damaged tissue via bio-printing.

"The set of genes designs the landscape, and the probability of following a given path is generally very high, the other trajectories or chreodes being rather unlikely, which does not eliminate their existence. A single genome thus allows several of these chreodes, which means that there are reserve phenotypes, distinct from the majority genotype, and new environmental conditions can favor one of these reserve phenotypes…" [PRO 12].

As a reminder, according to Sheldrake [SHE 03], Blanchoin *et al.* [BLA 14], and Heard [HEA 13], crucial words (found in the literature) are defined as follows:

– Entelechy: phenomenon containing within itself its end and its goal.

– Chreode: morphogenetic field (cré: it is necessary, odos: path).

– Epigenesis: theory of development through the progressive elaboration of shapes.

– Motility: reference to the capacity to move spontaneously or through reaction to stimuli and actively by consuming energy during the process.

5.4. Complexity and bio-printing

The natural vision is to research whether mathematical rigor, algorithmic procedures, can ensure a conceptual body of doctrine at the interface between technology, engineering science and biology capable of ensuring new knowledge for the production of biological tissues and organs, all while knowing that no specialist can, in principle, exceed a broader culture than that of his/her discipline. In a desire to do the simplest thing possible, the mobilizing idea is to develop a robust theoretical/empirical model whose excess of descriptive content is minimal in relation to what is necessary to notice phenomena.

In the case of bio-printing, several disciplines are involved (see convergence in the previous chapter) which lead to "modeling" domain mixtures/mutualizations likely having different equations of behavior, multiple and singular interfaces, nonlinearities, scale changes (from subcellular components to organs) associated with potential fractal shapes, etc. [CHA 12, NOT 02]. Complexity invades the field by several means: that of geometric arrangements, the complexity of dynamics (e.g. growth or reduction of scaffolding) and that of functionality, all likely with great incompleteness.

5.4.1. *Complexity?*

Under the name "systematic movement", Abdelmalek [ABE 04] gathers a set of activities concerning the dynamics of natural and cultural systems. They depend on a certain number of presuppositions, the most important of which are:

– existence of common general, transdisciplinary laws governing complex and strongly interactive systems;

– relational (or cybernetic: internal or external interactions) nature of the laws;

– "holistic" character of certain properties, in the sense that they concern the whole system. Some emerging properties only have existence and meaning at the level of the system as an indivisible whole; the degree of autonomy depends on the structure in space and time and on the logical organization of the whole system involved (see Figure 5.9).

Finally, the existence of general laws and those of invariants would not imply that the systems be deterministic and predictable. Quite the contrary, the systems would be very sensitive to the play between local contingence and relational need. The systematic approach is, thus, a grid that prepares for the paradigm shift. But, in the end, what is understood by complexity? Ricard [RIC 03] considers a set made up of two sub-systems, X and Y, for which he introduces functions H(X,Y), H(X) and H(Y) expressing the number of degrees of freedom.

If H(X,Y) = H(X) + H(Y), the properties of the whole can be deduced from the "single" components X and Y (it behaves as a resultant from the union of the two sub-systems). If H(H,Y) is less than the sum H(X) + H(Y), the potential wealth of the whole results from a reduction in terms of the degree of freedom following an integration phenomenon; if H(X,Y) is greater than this sum, new properties emerge from the association. These two situations are considered complex (and not complicated in the sense of Ferrari [FER 02]).

In the case of bio-printing, this complexity appears in situations like:

– coupling (space, "inert" materials, living materials, flow of matter and energy) is translated by various information playing a role in the desired dynamic; the "living" part is capable of regulating this information as a function of the symbols emanating from outside its environment (and maybe also within it); the complexification increases if the scaffolding is resorbable;

– the principle of superposition is not valid due to interdependences;

– is an experiment with a given population of N cells identical to two experiments performed with N/2 cells?

– the system's dynamic behavior can be described by its local dynamic properties, which play on the other sub-sets (e.g. cell growth) and by its collective properties characteristic of the global system;

– the system has an organization that is neither fuzzy nor totally organized and this at different spatial scales;

– the system is open, exchanging matter and energy (and information) with its environment; it is thus not balanced;

– the system has its history and its state depends on this (chaos);

– as a consequence, such a system presents nonlinear effects.

This problem becomes slightly more complicated, according to Ricard [RIC 03], because in these systems, where there can be successions of bifurcations, threshold events, etc., the number of interdependence parameters to be considered increases exponentially (combinatory explosion). To try to go beyond this pessimistic statement, Nicolescu [NIC 02] reintroduces the concepts of level of reality, previously developed by Heisenberg in 1942, of which there are three:

– state of the objectivable elements, independent of the knowledge processes (*a priori*);

– state of the elements inseparable from the knowledge process;

– state of the elements created in connection with the knowledge process (applications, effects, ethics, etc.).

However, this work poses the hypothesis that the variables influencing the global system are under control; in any case, this is an exercise that must be performed, even if it means inserting it into reflection with retroaction. However, even if the system has a nonlinear dynamic behavior, can conditions be found so that the trajectories of tissue printing does not diverge, or be strictly out of control? Various situations should be foreseen:

– there is an "attraction area" such that all the trajectories remain within a domain that is "acceptable" for the final use;

– there are initial conditions leading to diversified trajectories but ones that end in the aforementioned acceptable domain;

– it is possible to introduce a corrective command structure affecting the dynamics of the system, as agents of growth (see remark below);

– the system admits a stable trajectory (apparent causal model).

NOTE 1.– Figure 5.10 presents a system with a counter-reaction. However, as the experimental measures are imbued with random errors, even biases, this imprecision adds a degree of uncertainty to the model. It is possible to envisage the contribution of a model of uncertainty Δ; then, the model and its structure being defined, the goal of the command will be to ensure the system's stability to achieve the goal of the bio-printing operation. Would we know, in these conditions, how to reduce the sensitivity of the command structure in the presence of unknowns, parametric variations and possible disturbances affecting the system in its spatial–temporal and functional evolutions [CIN 02, CIN 05]?

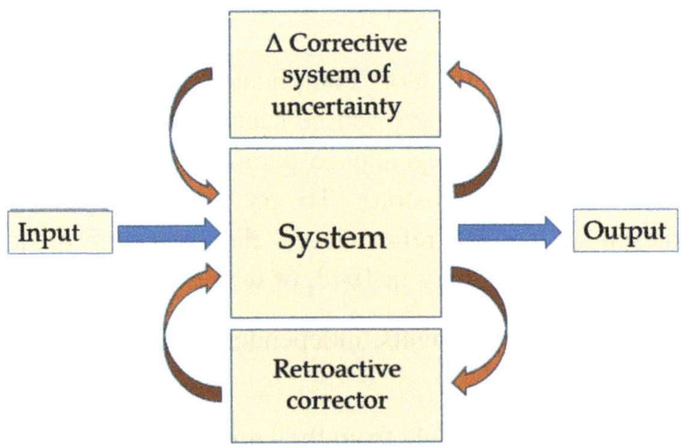

Figure 5.10. *Control model used in automation. For a color version of the figure, see www.iste.co.uk/andre/printing3.zip*

For Olle-Villa *et al.* [OLL 16], the biological system they propose approaches this description of control presented in Figure 5.8 with data collections and specific actuators: measurers (measurement of the gaps between instructions and reality) like metabolic paths, ionic channels, glycolysis; actuators like hormones; the physical system being played by the blood; sensors like enzymatic kinases; etc. So long as there is control of knowledge of the biological system, it is thus possible to find analogies between systems resulting from automation with its enslavements and biology (if it is not only that it is more complicated that making two electric trains cross paths; see also Ramasubramanian and Taber [RAM 08]).

NOTE 2.– a nonlinear system introduces functions (i being between 1 and N, N representing the maximum number of functions) that are not linear and stability conditions can be attained for values of the parameters x_j (j being between 1 and M, M being the maximum number of system influence parameters). To study the stability of the said system around these positions of stability, it would be best to "lightly" modify the x_j around the stability values; then if the goal is not disturbed, the system can be considered stable (which is not the case of the ball in Figure 5.7, which can go left as well as right). This principle makes the hypothesis that all x_j are known and that it is possible to follow the evolution of bio-printing in its spatial–temporal and spatial–functional dynamics. It also implies that it is possible to incrementally modify parameters; if it is easy to foresee for physical–chemical and even hydrodynamic aspects, nothing says it is foreseeable for the biological aspect.

For Kirschner and Anseth [KIR 13], at least two domains must be distinguished, that of partly controllable "physical–chemical" space (chemical composition, physical and mechanical conditions, external constraints (temperature, exposure to various electromagnetic fields, etc.)) and that of the living. This association between blocks (inert vs. living) "obviously" brings interdependences between these two domains into play. They believe that the physical–chemical factors play on the enzyme secretion, pH, protein interactions with the environment, etc. The equivalent of the chemical potential can thus be expressed as a potential energy specific to each of the species present in the environment and which translates the rather irreversible global effect on the chemical and/or biological space of the molecular and microscopic interactions with its environment (other macromolecules of the species, biomolecules of other species present in the system of cellular aggregates).

In agreement with Mathieu and Schmid [MAT 14], the necessary in-depth study can take very diverse shapes based on some nuclei of knowledge resulting from PE works (experimental approach) or on model objects, concept supports (but that, in every form, will only find meaning through an interdisciplinary approach, if only to bring out the principal influence factors/parameters). This implies an epistemological decision which forces reflection for the best choice between different "discourses on method":

– How to perform robust modeling without a comprehensive approach, since each isolated discipline does not have the power to shed light on the goal? But which common language to use to advance without being superficial or reductive? The question is also asked of defining a reasonable set of variables of influence that must concern a model system, representing a simplifying complexity system, to be able to explore it by making the hypothesis (a bet) that the competence acquired on this reduced system can serve as a foundation for extrapolation for more complex systems, closer to reality. This is partly what, from an experimental standpoint, is attempted by working, for example, on printing cartilage, which is a "simple" tissue organization with modest vascularization (see Chapter 3).

– How to assemble knowledge stemming from disjointed disciplines? Do we start with one discipline to find an optimum, then will we consider a variation of the parameters on a second axis to access a "more optimal" optimum, etc.? This implies the existence of a single optimum and, in this case, whatever the path, long or short, it is possible to achieve it, as has been shown for several decades by methods of signal processing. But this forms abstraction of local optimums, of good precision, and of the existence of numerous nonlinearities and recursivities, possibly induced by the living. This vision has, for some, the advantage of playing with interdisciplinarity through juxtaposition, without needing to seek major knowledge integration. Figure 5.11 proposes an illustrative example of this statement in a two-dimensional space. The idea is thus to successively vary the variable to attain the local minimum, then change it, etc., hence this zig-zagging path presented on the right side of the figure. In a noiseless system, with all variables identified, if there is only one optimum, it is imagined that it is not impossible for this to be attained.

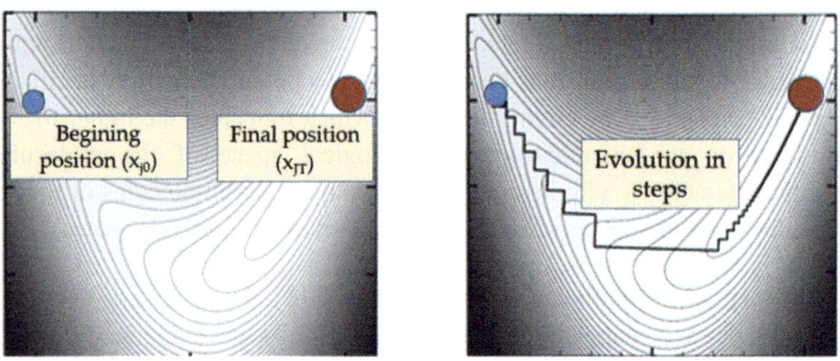

Figure 5.11. *Search for an optimum (only assumed) through successive changes in independent variables: alternating path with unidirectional search for the local optimum. For a color version of the figure, see www.iste.co.uk/andre/printing3.zip*

– Finally, an attempt can be made to "play" on the "serendipitous" chance through a method of shots aiming at the influence variables and crossing them randomly to get the results and then control the robustness; after the largest number of assays possible, a cartography can be performed on the system following what was proposed in Figure 5.9 (so long as there is a measurable definition of the target function). Experiment plans can be exploited to try to reduce the number of shots (by posing the hypothesis that there is a small idea of the approximate position of the sought-after optimum and so long as it exists, hence the interest in an activity performed in close collaboration with bio-printing biologists).

– The concept of "phronesis" was created by Aristotle. He refers to a mode of knowledge and action required when the mechanical application of scientific knowledge, routines or even standardized protocols could be disproved by the complexity and singularity of the situation or the problem to be resolved (this is just a bit more in-depth than the empiricism enlightened by common sense). This complexity and this singularity have as consequences the irreducible uncertainty of the results of human scientific action and consequently the particular difficulties encountered in formalizing these actions. It thus cannot hope for universality. Recognizing that a situation stems from practical wisdom leads to accepting a degree of the subjectivity. However, this does not allow just anything to be done. Demonstrating "phronesis", on the contrary, has several implications:

- analyzing a case in its complexity before making decisions for action, trying to keep a global vision of the topic;

- allowing collective deliberations on this case;

- remaining attentive to signals that could be indicators of problems or hidden difficulties;

- seeking a justifiable balance between demands that the complexity of these cases can bring into conflict ("efficiency" of a bio-printing experiment).

For all that, if there can be an agreement on the understanding of an approximate evolution, will we have achieved the goal of the project that is to create "custom-made" operational biological tissue (which is still far from the resolution of the inverse problem)?

– The experimental verification of the hypotheses posed to construct the model must be founded in a cycle of comings and goings for experiment-calculation-based verification, validation that questions the validity of experiments, their specific precisions, the validity of the calculations with their programming errors, the existence of bifurcations, and discrete approximations of generally continuous mathematical expressions. The restitution of a set of singular experiments is necessary to access the average, which aims to define the model. Thus, the indetermination and application of a large number of laws can contribute to excluding from the field of models the singular events or those involving non-locality and dependence vis-à-vis verification processes. The epistemology of the approach by model can thus show itself as singularly limiting to the explanatory ambitions of bio-printing science(s); but it allows scientific knowledge to be considered in a more balanced and rigorous framework [PER 03].

– Several types of complexity have been revealed through the practice of modeling: the complexity of the arrangements in space and the complexity of the dynamics of growth, transport of matter and energy, etc. The number of cellular and inert edifices possible from simple elements, their random or deterministic arrangement with structural scales obviously introduce diversity that is difficult to control. Could modeling use fibrous, laminated aggregates alternately with structures of resorbable nonliving matter? The result would be the obligation to model mixtures of spatial domains having different equations of behavior, multiple and singular interfaces, etc. The mathematical tool of choice to rise from small scales to larger ones could be the homogenization and identification of "fractal" shapes, that is, repeatability or differentiation of multi-scale shapes [PER 03, PAV 06, HEA 13].

5.4.2. *Initial reflection for action*

These reasons mentioned above can prevent (possibly definitively) predicting "the safe trajectory of organ construction" in the course of its development as precisely as a classical object created through additive manufacturing (which also poses some small difficulties). However, it could be advantageous to try to construct virtual objects [COL 13] to open new perspectives of understanding and management of the living. It would become possible, for example, to:

– study the interaction between biophysical/biochemical shapes and processes (analysis of the flows internal to the elementary voxel, even a few voxels, for example, or exchanges with the environment) on the basis of a detailed realistic representation;

– analyze and better understand the coupling of different biological phenomena involved in the operation or growth (water flow, nutrient flow, flow of waste, exchange of information via hormonal flows between different parts of the elementary bio-construct, even, why not, by combining scales, the human over the course of cell growth);

– compare the structure of several typical tissues/organs on structural and quantitative bases;

– better consider the non-deterministic side of growth mechanisms by identifying the different sources of noise or, lacking knowledge of data, with the help of appropriate stochastic models;

– exploit adequate data in terms of spatial and temporal resolution to model the dynamics of the networks of molecular and genetic interactions in the context of displacements and cellular divisions in the body (for now, rather absent);

– exploit, for precise cases, the "suggestive" strength of models that, on a basis of finer and finer formalization, allow experimental verifications (detection of absurdities and feasibilities);

– find support "[no longer in a scientific theory, but] in a wealth of knowledge that the model, and even more, the software, are capable of integrating as a function of knowledge and intervention goals" [ARM 05];

– translate the modeling approach by a multiplicity of models and modeling methods with the necessary confrontation with different standpoints to find arrangements to seek successive coherences.

The techniques for developing virtual cell growth models are still young (even if there are many methods, as indicated by Figure 5.12, inspired by [PAV 12]). To fix ideas, would we know how to: measure and reconstruct simple tissues/organs in 3D in a detailed way with computer aid, through a method of "3D tracking"; construct precise biophysical models (interception and mechanics models); construct first models quantitatively simulating growth, nutrient circulation, and waste in the growing organ, on the basis of imperfectly known rules of operation; realistically simulate the qualitative or quantitative development of an organ and visualize it through infography? There is the stake! Mathematical tooling can, for example, be based on networks: of "nodes", representing the different physical and living elements in a state of interaction, with links whose magnitude corresponds to the influence of one node on another. "These models, which are approximations of fragments of the living, allow biological mechanisms to be simulated to formulate predictions, which leads to experimentation, to confront them with the reality of the living" [PAV 12].

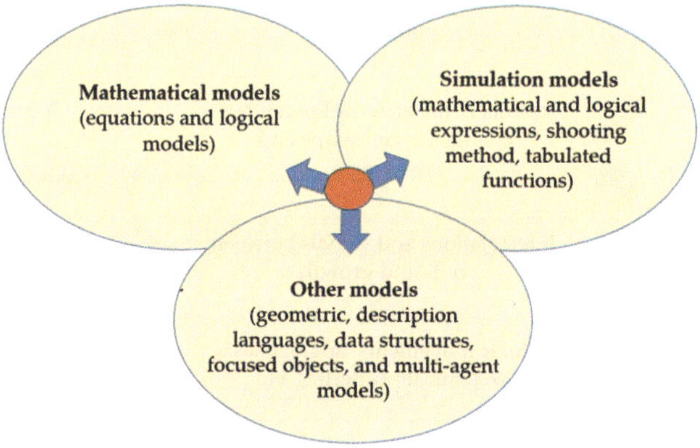

Figure 5.12. *Formal operational models (only representing part of a complex system). For a color version of the figure, see www.iste.co.uk/andre/printing3.zip*

Independent of this general context, it is essential to raise the state of knowledge on the subject of bio-printing (scales by scales, as proposed by Martins, Ferreira, and Vilela [MAR 10] and Figure 5.13; see also [GRE 17, OKU 16] and, to a lesser degree, [LEJ 17]) and sciences that compete with its development, through the most exhaustive review possible of the scientific literature (the knowledge represented by Gaussian approximates in Figure 5.4). Figure 5.15 from Green and Batterman [GRE 17] illustrates the interactions to consider for modeling with scale changes in mind.

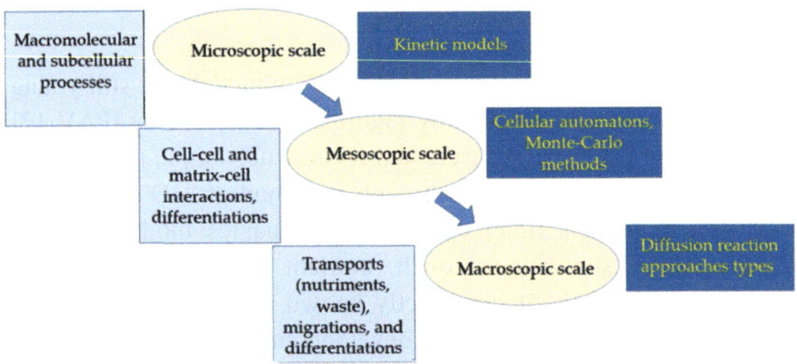

Figure 5.13. *Spatial scale models. For a color version of the figure, see www.iste.co.uk/andre/printing3.zip*

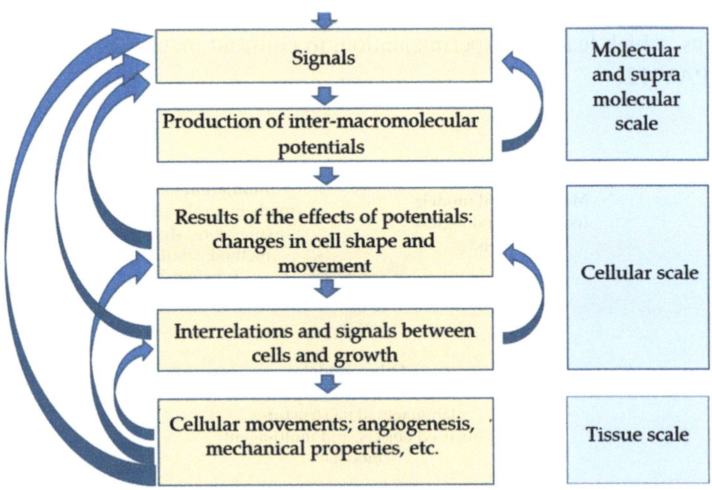

Figure 5.14. *Interrelations between spatial scales. For a color version of the figure, see www.iste.co.uk/andre/printing3.zip*

On the basis of robust knowledge, it is then possible to construct networks representing the multiple factors affecting differentiation and cell growth, the potential disappearance of the supports on which the printed cells develop, etc., as well as their (nonlinear?) interactions. By studying these networks, particularly by simulating their behavior, it could be foreseen to determine, for example, which nodes play a preponderant role in the upset between two stable states. It may then be possible to make predictions on certain aspects of the mechanisms that experimentation has not yet shed light on. "A central element to construct a causal biological network – in which the relations between nodes are of a causal nature – is recourse to conditional dependencies resulting from the theory of probabilities, to infer causal relationships, in the context of a systematic source of disturbances" (Schadt quoted by Vanderginste [VAN 12]). Will networks constructed in this way be content with drawing interrelations between the factors involved in the processes implied in bio-printing, by robustly reproducing the causal logic of the biological reality studied, or will they not work as a "simple" correlative approximation? This is what will allow us to know if it is possible to bring about exploitable predictions or to remain in a state of undecidability in bio-printing.

For Gascuel [GAS 10], the robustness of biological networks raises awareness of the limitations of the space of biological systems on which we can act for ends concerning intervention, control, or regulation. "Only a global understanding of the interaction networks would allow new trails to be blazed for [possible] developments in human therapy. To these ends, elucidation of the links between genotype and phenotype through network analysis constitutes a significant stake of the current systems' biology". For the engineer engaged in this undertaking, it is indeed a matter of investing in complexity by trying not to drown in it; that is to say that it will be necessary to research singular elements that may represent the dynamic system, trying to find local systems considered to be autonomous [HEU 98], to optimize these numbers, to use methods of compression [DEM 14], and trying to find sets like "numbers without dimension" to (potentially) allow the multi-scale approach presented in Figure 5.15. Furthermore, as indicated by Dallon [DAL 10], it should be possible to work on certain specific spatial and partially autonomous domains by then combining them to achieve greater scales. This reductionist effort is indispensable if we want to escape the incantations and maybe one day manage to resolve the inverse problem.

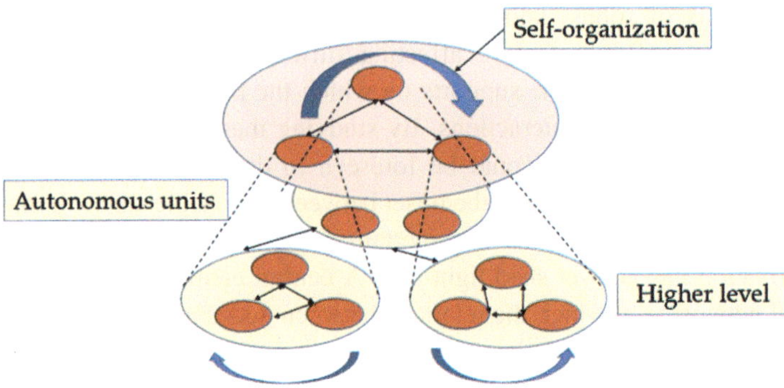

Figure 5.15. *Autonomous dynamic systems. For a color version of the figure, see www.iste.co.uk/andre/printing3.zip*

According to Rashevsky [RAS 54], it is possible to assume that there are uniform topological transformations: the relational graph of the functions of all living beings can then be deduced from the same primordial graph using the same topological transformation if the living is sufficiently uniform: consequently, only the parameters of this uniform transformation change. According to Rashevsky, the passage to bio-topology would thus provide the language that allows a sufficiently abstract point of view to be expressed, from which the unity of the living can be conceived exactly [VAR 13].

One of Rosen's results [ROS 58] is that the system cannot always be maintained in its initial topology: it will move, sooner or later, to a topologically less complex sub-system (sub-graphs) or will even be totally destroyed. Rosen also shows the existence of a central component for every system of this type, that is, a component whose destruction directly leads to the failure of the whole system [VAR 13]. Rosen will follow the positions of René Thom on connected matters (refusing the hegemony of discrete models). On this basis, he maintains in 2000, with good reasons but without certainty, that any mechanism (in the mathematical sense) can be perfectly emulated by a universal Turing machine (a classical computer with an infinite memory); it could possibly be concluded, by sophism, that any phenomenon with a physical manifestation is or will be simulable by a computer!

These theses, these promises, involve assuming that, under the pretext of a physical being, a phenomenon can be reduced to a mechanism in the mathematical sense. Could any computational formalization be overly reduced to characterize the living in bio-printing processes and allow veritable alternative solutions of formalizations to be fed, ones likely to face the on-site data validated and resulting from the PE? How could they trust the views of the HE?

5.5. Return to complexity

These first paragraphs bring about the creation of living tissues via bio-printing through an obligation to control complex systems. The fundamental question asked is knowing if it is possible to achieve effective mastery of all the physical, chemical, biological, and temporal processes in question (before possibly one day evoking the solution of the inverse problem) for a given end. When we speak of complexity, we invariably stumble upon conceptual difficulties, which generally leads to the lateralization of paradigmatic questions to limit ourselves in experimental demonstrations, the accumulation of whose results are thought to allow a validated trace of a tissue's history from its creation through bio-printing to its function. Then, for some, "biology is, at best, a province of physical science, a dependence that can only advance by applying methods of physical science" [ROS 85].

The properties of living systems could potentially be inscribed in the structure of molecules and in the structure of the genes that code proteins. There is then no need for biological complexity or the emergence of new properties that could reveal themselves at the macroscopic level, as these properties would already be implicitly present in the molecules. This old idea is not satisfactory because it is based on the existence of successive thermodynamic balances during stereospecific recognitions of macromolecules. Yet, organisms are systems situated out of thermodynamic balance and the gaps in a system in relation to its state of balance are likely to produce unexpected behaviors. Thus, for around twenty years now, researchers have progressively come to the idea that the complex has its own reality and that it cannot be reduced to the simple [MOR 13].

The approach through complexity is a breakthrough that recognizes the exceptional and problematic character of the living, the part of the indetermination of its limits, the chaotic character of phase or regime changes, the plurality of balances or levels of reality, the heterogeneity of elements, their interdependencies and interactions, the indirect or delayed effects, the mass effects, or the emergence phenomena of collective properties. It may be useful here, in order to illustrate complexity and its study, to invoke the necessary acceptance of uncertainty in understanding the so-called "complex" problems, thus giving greater value to the qualitative than the quantitative (what is important is the global, qualitative function and not its "metric" framing in very, even overly, precise instructions, because precision would be inaccessible in the sense of past hyperdeterministic science, also hyper-reduced to complicated but non-complex problems). However, if one day we wanted to transplant a bio-printed organ, will the qualitative be enough?

The epistemological problem (how robust knowledge of reality is acquired) has been a much greater concern for philosophers since Plato than for most scientists [WAT 88]. He writes, "we find ourselves faced with a lock that must be opened to reach the loot that he [the savant] hopes to take". This question poses the issue of the existence of several keys, even a master key, or to the contrary, the absence of possibilities to open an anti-Pandora's box. If scientists are not able to rationally explain facts, knowledge becomes open for discussion, useless and devalued to the rank of ideology, "dancing in the rain" (which can be standardized, with precision, without knowing whether it will have an effect on it).

Spinoza, like Nietzsche, aims at joy, the pleasure to find, the power to act as ends to gain in perfection and thus to go beyond oneself. Spinoza believes in the possibility of going from the imagination to intuitive and/or rational knowledge of humanity (even of the researcher!). On the other hand, certainly reductively, Nietzsche believes that there are deficiencies in the abilities to access knowledge. For Spinoza, the miracle does not exist, "things" are necessary and created by a cause. Nietzsche announces a battle for new men, for knowledge and the love of ideas. These new men will be those "who seek in everything what must be overcome" [NIE 07]. For this philosopher, these men have abandoned all fear and live dangerously. The "Übermensch" [superman] is, from this fact, a warrior of knowledge, of joyful science, that which doubts to the point of destroying old illusions and trying to build new values. It is indeed a matter of not being trapped in old walls (paradigms and conservatisms, which must then, so as not to consider them, be well known), to be capable of freeing oneself from them (with the constraints that are known) to participate in the construction of others, even more modestly, to construct new levels. In this partial dichotomy is indeed a question already mentioned above, that of a(n) (im)possible description of the approaches linked to bio-printing.

In the process, it is a matter of showing that some hypotheses are wrong (see [MCC 65]) and avoiding the development of a model (which would make the system's blackbox whiter), as indicated in Figures 5.15 and 5.16 (from [MON 08]), which would only concern the ordering and organization of data and the published results into frameworks of comfortable convenience and adaptability. However, as pointed out by Maturana [MAT 70], "The *a priori* affirmation according to which objective knowledge constitutes a description of what is known [...] calls on the questions 'What is knowledge?' and 'How do we know it?'" In this regard, Kant [KAN 85] wrote, "Nature [...] is the sum of all subjects of experience". How, with an activity of appropriating the real far from evidence, to deal with this inevitable and ultimately unsolvable dilemma of organized complexity? It then becomes reasonable to think that, at best, we will not only approach the "real" reality by proposing one or more possible models of knowledge elaborated through the experience of scientists from different horizons. "As daring, strong, and beautiful as it may be, and as closed as it may seem, a system has no less of a fatal imperfection: it cannot itself prove its own logic and coherence" [WAT 88].

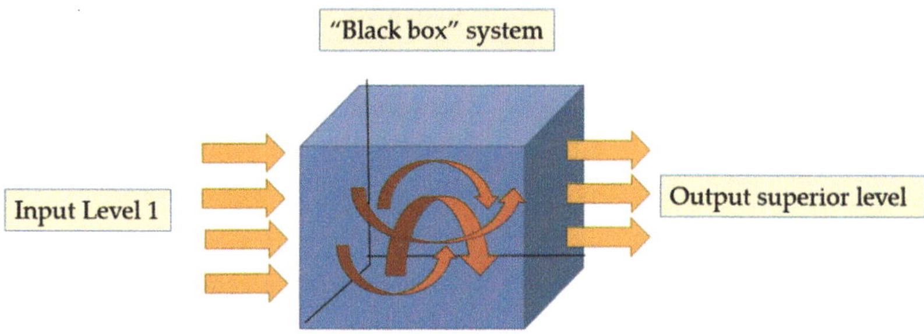

Figure 5.16. *The question of modeling. For a color version of the figure, see www.iste.co.uk/andre/printing3.zip*

Classical physics, in its Laplacian ideal, had a goal [LAP 86], that of determinism; when scientists replaced the deterministic laws with statistical laws, the general standpoint was not disturbed; von Bertalanfy [VON 12] even believes that it was reinforced. Riedl [RIE 88] reminds us that, by continuity, epistemology has long considered the evolution of systems, including those of the living, to result from a process of knowledge accumulation within a paradigm, ultimately disturbed by manifest impossibilities. "Chemical and biological shapes are repeated not because they are determined by immutable laws or eternal shapes, but due to the causal influence of previous, similar shapes" [SHE 03]. In short, a shape, in an "inertial determinism", would be reproduced because matter has taken on a habit of this.

Complex systems can be broken down into different levels of interaction that allow simple elements to be combined into more evolved components that themselves give rise to organized and hierarchized structures that interact strongly among themselves and with their environment (see Figure 5.15). The structures that emerge then cannot be made up only using entities put into play. The interactions are often nonlinear and generally contain retroactive loops. The more organized the structures are, the more time they need to appear (see Bennett's concept of depth [BEN 88]). At the global level, these systems are characterized by the emergence of non-observable phenomena at the scale of the system's constituent elements: an outside observer will grasp and understand the system differently than an observer within the system. It is thus characterized by the emergence at the global level of new properties and a dynamic of global functioning that is difficult to predict through the observation and analysis of the elementary interactions. Table 5.1 gathers systematic and analytical comparison elements between the two approaches [LAP 16], which illustrate the lacks to be made up for in modeling bio-printing.

Analytical approach	**Systematic approach**
Isolates: focuses on the elements	Connects: focuses on the interactions between the elements at all levels (information, matter, energy)
Considers the nature of the interactions	Considers the interactions between elements
Based on the precision of details	Based on the measurement of the global perception
Possibility of modifying one variable at a time	Modification of groups of variables interacting simultaneously
Independent of duration: phenomena in question are reversible	Integrates the duration and irreversibility of processes
Facts are validated through experimental proof in the framework of a theory (paradigm)	Facts are validated through comparison of the model's operation with reality (without being sure that the model represents the "real" reality)
Precise, detailed models, but difficult to use in action	Insufficiently rigorous models to serve as a knowledge base, but usable in decision and action
Efficient approach when interactions are linear and reliable	Efficient approach when interactions are linear, nonlinear, strong, and weak
Leads to teaching by discipline (juxta-disciplinary)	Leads to an obligation for interdisciplinary exchanges
Leads to action planned in detail	Leads to action according to goals
Knowledge of details, ill-defined goals	Knowledge of goals, fuzzy or unknown details

Table 5.1. *Differences between analytical and systematic approaches*

The stability of the levels of organization is, according to Zin [ZIN 03], objective. "It is, however, always relevant, including a certain precariousness, an indispensable instability to the measurement of its complexity. No complexity without energy dynamics, thermodynamic flow, cycles, or even simple 'noise' having a role of disturbance that sets into motion, lends energy, and frees, proves and weakens the established links, introducing at least fluctuations, vibrations, constant variations at the origin of its evolution and its adaptive capacities. It would be a mistake, though, to reduce everything to noise or energy, which represent a local effect, short term and with no memory, forgetting the inscription of information in the organization and weight of longer-term catastrophes. There are several duration scales that must be considered, cycles and disturbances of different magnitudes or periodicities, as well as a more or less large delay in 'response times' that can sometimes be very long. Without the intervention of an active end capable of remembering and adapting, a dynamic system is not durable and quickly breaks down under the effect of entropy. An organism is not reduced to a dissipative structure whereas, on the other hand, it opposes the dissipation of energy, even if the production of its internal organization inevitably has a reported energy and entropic cost on its environment (globally, the physical law of growing entropy is thus still respected)". Figure 5.17 from Heudin [HEU 98] illustrates the issue of organized complexity where, using an initial 3D creation, the system will undergo simultaneous processes of variation and stabilization allowing a goal to be achieved.

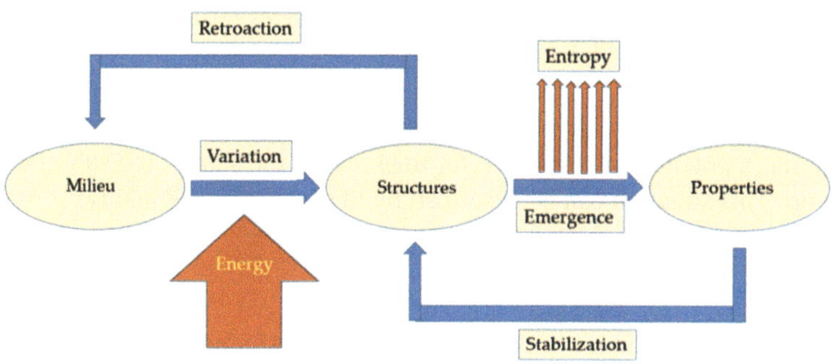

Figure 5.17. *Principle of variation – stabilization. For a color version of the figure, see www.iste.co.uk/andre/printing3.zip*

It is not so much the multiplicity of the components, nor even the diversity of their interrelations, that characterize the complexity of a system: so long as they are practically and exhaustively countable, we are in the presence of a complicated system, even a very complicated one, a combinatory count of which could allow, in principle, the description of all possible behaviors (and thus the prediction of its effective behavior at all times once the rule or program that governs them is known): in mathematical terms, we are in the presence of a "polynomial problem". It is the potential unforeseeability (*a priori* incalculable) of the behaviors of this system, particularly linked to the recursivity that affects the operation of its components ("by functioning, they transform", inducing emergence phenomena, which certainly are sometimes intelligible (as is sometimes shown, see Chapter 3), but not always foreseeable). The observed behaviors of living systems provide countless examples of this complexity and bio-printing is part of this.

5.5.1. *Complexity and system approach*

In the case of bio-printing, several disciplines are involved, which leads to "modeling" "intimized mixes" of domains, probably with equations of different behaviors, multiple and singular interfaces, nonlinearities and scale changes (from sub-cellular components to organs) associated with potential fractal forms. Complexity invades these domains by several paths: that of geometric arrangements, of the complexity of the dynamic (growth, reduction of scaffolding for example), and those of functionality, competition between cellular systems, reproduction, etc. [AND 16, GLE 08]. One associated issue concerns, for a determined action, the effect of a stimulus on it (e.g. growth agents), with a possible comparison between the magnitude of the input stimulus and its effect ([CUN 14] and Figure 5.13) with at least apparently contingent, spontaneous and intentional aspects (how does reading the genome allow us to predict the future of the biological system [PIC 15]?).

A biological system will be called complex if it presents emergence phenomena, that is, it has a potential richness greater than that of the sum of the sub-systems that make it up. This situation can be described by the following inequality:

$$H(x, y) > H(x) + H(y)$$

with x and y being variables of two sub-systems and H a mathematical function that expresses the potential richness of a system (in speaking of richness, we bring out the appreciative concept that would need to be defined more precisely, which is what will be attempted, without going too far however, in the following section). In statistical physics, this function is called the system's entropy [BEN 06]. Complex systems generally present a certain number of characteristics that make them more difficult to understand and manage than simple and complicated systems [GAL 01]:

– *Multiplicity of legitimate viewpoints*: for example, it is difficult to understand an adaptive system without also considering its context and taking into consideration "viewpoints" and interests of the different elements concerned, particularly seen from different disciplines (none holding the "truth"); but the search for data collection systems is, in this context, indispensable, the living environments of groups of cells are relatively mute.

– *Nonlinearity*: complex systems are not linear, in the sense that numerous relations between their elements are not, such that the magnitude of the effects is not proportional to that of the causes and that the repertoire of behaviors is very rich (chaotic behavior, multi-stability due to the existence of several states of equilibrium and regulation processes (see Figure 5.10), drifting processes, etc.). Nonlinearity plays a decisive role in the birth of counterintuitive behaviors typical of numerous complex systems.

– *Emergence*: this systematic property, expressed by the words "the whole is not reduced to the sum of the parts", means that the properties can only be understood in parts in the broader context of the whole and that the analysis (without residue) of this is not reducible to that of its parts. A truly new element can arise from the (nonlinear?) interaction of the system's elements.

– *Self-organization*: phenomenon described in Figure 5.18, through which interacting elements cooperate to produce a behavior and structures coordinated on a large scale (like the configurations created by dissipative structures). The appearance of a higher level structure is produced without the need for an organizing principle by proving a certain autonomy and robustness [DIM 06, OTT 04, FRO 04, BAR 03]; let us recall that, for Asby (quoted by [JAC 70]), self-organization is normally impossible: "a system cannot organize itself because this would mean changing its law of organization following another law that should itself be its organization…"; this is another debate.

– *Multiplicity of types of complexity*: combinatory between cells among themselves, on the one hand, and cells and scaffolding, on the other; morphological; functional.

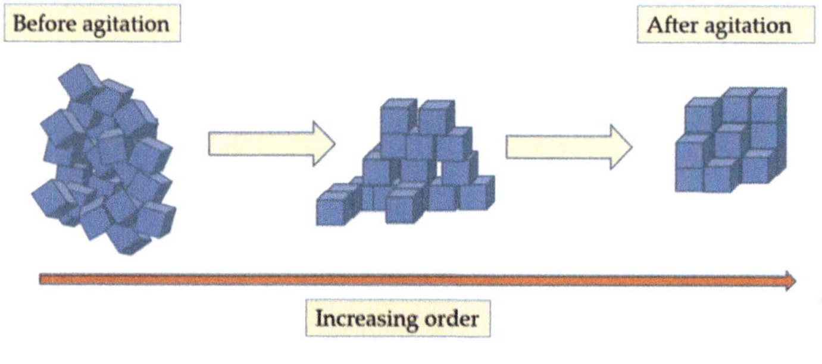

Figure 5.18. *Principle of a self-organizing system. For a color version of the figure, see www.iste.co.uk/andre/printing3.zip*

– *Multiplicity of scales*: many complex systems are hierarchical, in the sense that each element in the system is a sub-system of a lower order system, and the system itself is a sub-system of a higher order "supra-system". What is important is that in many complex systems, there is a powerful coupling between the different levels, such that the system must be analyzed or managed on different scales at the same time. However, systems at different levels of scale have different sorts of interactions and just as different characteristic rhythms of evolution. It is thus impossible to have a unique, fair and exhaustive point of view about one system, even at a single system level. Plurality and uncertainty are inherent to the behavior of complex systems.

– *Irreducible uncertainty*: there are, within complex systems, numerous sources of uncertainty. Some are reducible by averaging a complement of data and research, notably when the uncertainty is due to random processes (leading to a statistic or probabilistic analysis so long as this is adapted to reproducible experiments with controlled parameters) or to ignorance (absence of data or set of inadequate data, incomplete definition of the system and its limitations, incomplete or weak understanding of the system).

Although some of these characteristics (notably nonlinearity or uncertainty) can be found in some complicated or even simple systems, the fact is that a complex system can possess them all [PAV 12].

5.5.1.1. *Determinism and chaos*

Determinism reflects the following definition: "two experiments with exactly the same initial conditions and the same conditions at the limits must yield exactly the same results" and the mathematical model of a phenomenon will be considered deterministic if the conditions of existence and uniqueness of the solutions are satisfactory, which is generally the case for models using systems of differential equations [GLE 08, DAL 92, CON 11].

Chaos implies that disorder and determinism are tied to an order that obeys laws. This dialectic contradiction is desired, because it is a matter of describing phenomena in which there is a hidden order behind an apparent disorder, with disorder on one level and order on another (see Figure 5.15). Reductionism and the description of trajectories are no longer valuable in a phenomenon where the whole structure does not stem from the sum of the elements' evolutions. The overlapping of order and disorder is not the only characteristic of deterministic chaos. A crucial point is "sensitivity to the initial conditions" as can easily be imagined for a trajectory like that corresponding to the displacement structure in Figure 5.19 [SAA 96]. Another apparent contradiction exists: all while obeying laws, these phenomena are not predictable, because they are susceptible to brutal large-scale bifurcations [SPE 08]. These authors write that "in this framework, emphasis is not placed on the search for precise solutions to equations of the dynamic system (which, in any case, is often hopeless), but rather on the response to questions like 'Will the system converge towards a stationary state in the long term, and in this case, what are the possible stationary states?' or 'Does the system's long-term behavior depend on the initial conditions?' An important goal is the description of fixed points, or stationary states, of the system; these are values of the variable for which it no longer evolves over time. Some of these fixed points are attractive, which means that if the system comes near them, it will converge towards the fixed point".

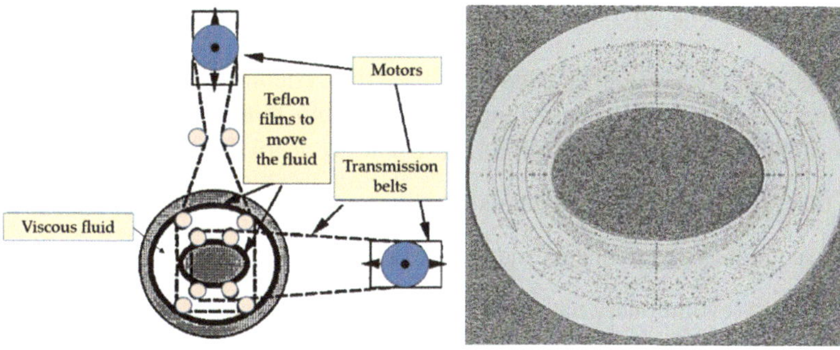

Figure 5.19. *Visualization of deterministic chaos: the chaotic trajectory of a fluid in a deterministic flux [SAA 96]. For a color version of the figure, see www.iste.co.uk/andre/printing3.zip*

Chaotic systems have complex behavior. They are attracted by a geometric figure of complex structure on which they seem to err towards chance, but without ever leaving it nor passing the same point twice. The attractors that characterize these systems seem to include both deterministic and random laws, which makes all long-term prevision impossible.

The state of a dynamic system can be described by the variables $x_i(t)$ with $i \in [1,n]$. By performing a study in the space of phases, it is easy to distinguish chaotic behavior: if the system is periodic, the points meet on a closed curve (attractor); if the evolution is random, the system's points randomly fill the space of the phases; no structure is impossible. As for bio-printing, the variables $x_i(t)$ should be known before being able to study the chaotic nature of the system. It does not seem, as shown in the previous two chapters, that this situation has been achieved.

5.5.1.2. *Reminders and definitions in systemics*

The object of the systematic approach is to elaborate a system of representation that allows complex situations to be grasped appropriately. Among the foreseen methods of exploration, a first stage aims to focus on defining the limitations of the system to be studied, to situate the system in its environment, to understand the nature and reason for exchanges the system makes with its environment, to have an idea of its inner architecture, principal components, and the nature of the relationships between its components, to know enough about the history of the system to better grasp its evolution (by writing "enough", the question is then asked of knowing the irreducible variables to be considered. But for this, it is necessary to accept the long duration of experiments with difficulties of "great publications" and the risk of not being referenced in the pioneering group).

In its principle, systemics classically uses the systemic triangulation method for this phase, which consists in asking questions from three poles allowing in-depth study of the system's representation. These poles concern the following aspects:

– functional (what is the system's role in its environment?);

– structural (its components and their arrangement);

– historic (evolutionary nature of the system).

This exploration must allow the different flows that run through the system, information flows as well as those of matter, to be identified (and so far as this is possible, measured). All the information assembled can then be translated by network graphics, charts, and diagrams allowing possible modeling to be foreseen. According to Yatchinovsky [YAT 12], the principles of the systemic approach (nearly those mentioned in Table 5.1) are the following:

– interdependencies: each sub-system must be considered in the context in which it interacts;

– totality: the grouping logic takes precedence over that of each element;

– retroaction or circular causality: effect of A on B that has an effect on A (see [TUR 52]);

– homeostasis: tendency of a destabilized system to return to its initial state;

– equi-finality: multiplicity of possible paths.

For Donnadieu *et al.* [DON 03], the characteristics of complex systems are:

– the "broad" definition by Jacques Lesourne [LES 82]: a system is a set of elements in a state of dynamic interaction;

– the "narrow" definition provided by Joël de Rosnay [DER 75]: a system is a set of elements in a state of dynamic interaction, organized as a function of a goal. This definition emphasizes the finality or goal sought by the system.

Numerous typologies have also been proposed:

– open/closed systems to their environment;

– natural/artificial/social systems;

– hierarchically organized/network systems.

An interesting typology, due to M. Bunge [BUN 79], is based on the supposed order of appearance of the different systems over time. From living systems, there is the emergence of creative self-organization. Such systems are labeled HCS, "Hyper-Complex Systems". These are open, relational, finalized systems requiring a variety of self-organizers.

The systems approach is the object of Figure 5.20, extracted from Donnadieu *et al.* [DON 03].

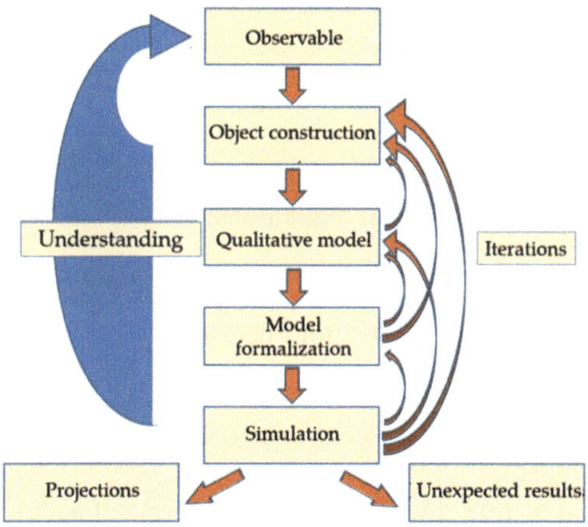

Figure 5.20. *Stages of the systematic process. For a color version of the figure, see www.iste.co.uk/andre/printing3.zip*

"Systemics, systems, system dynamics, systemic analysis, systematics, cybernetics, synergetic, complex, complicated, dynamic system, etc., so many terms that belong to the same semantic field, that refer more or less directly to systemics in general, but covering very different realities. A certain confusion reigns around these notions: quite often, one word is used for another, the same word, inversely, is used to refer to very different notions, systemics is confused with the methods and tools that are developed with it, and the systematic imprecision surrounding the use of the vocabulary in this domain contributes to perpetuating the ambient confusion around the notion of systemics" [CER 07].

To the author's knowledge, the first to get to a synthesis task is biologist Ludwig Von Bertalanffy who, in 1968, published a synthesis work entitled "General System Theory" (republished in 2012). He defines a certain number of concepts such as those of open systems, homeostasis, equi-finality, etc. He advocates a global understanding of the system, insisting on the importance of understanding the relationships between the different elements and not an analytical grasp of the system's elements (see Table 5.1).

The definitions of the concepts of arborescence and levels of organization [LEM 99, LEM 04] are part of a stage of constructing a system typology. From the first level, corresponding to the static and simple objects of physics and chemistry, to the ninth level, Le Moigne is involved in describing systems with growing complexity. Understanding the system represented by the last level implies that of all the lower levels. The articulation of a complex nine-level system (see also [DUR 13]) is thus defined as follows:

1) the phenomcnon is identifiable;

2) it is active;

3) it has regularities with internal regulations;

4) the system internally produces information that acts on it;

5) it has a sub-system with autonomous decision;

6) it includes memorization, a trace of the past (delay, threshold, etc.);

7) it coordinates decisions;

8) it can be "imaginative" by elaborating new forms of action;

9) the system is "able to decide on its decision".

Le Moigne [LEM 99] presents systemic modeling that fills the gaps of analytical modeling like the closedness of the model, thus reasoning on a single criterion. It starts with the hypothesis that to tackle a complex system, it is necessary to change registers, to go from the register of rescinding disciplinary knowledge to that of methods of enriching active knowledge. Intellectual inertia has the consequence of quasi-systematically using analytical models in complex organizations but probably finds its limits with bio-printing [NON 00].

5.5.1.2.1. Self-organization [ATL 11]

"Self-organization [FOX 09] is a mechanism or set of mechanisms through which structures are produced at the global level of a system from interactions between its constituents at a lower level of integration. The interactions between constituents are themselves locally produced without any reference to a preconceived global structure. To the contrary, it is an emerging property of the system and not a property imposed from outside the system". This definition leads to the introduction of the concept of the emergence of meta, even global, structures from interactions between "elementary" constituents on a local scale. These can be self-organizations of structures and/or functions.

5.5.1.2.2. Shannon's redundancy function R

R is defined by $R = 1 - H/H_{max}$, where H_{max} is the quantity of information that corresponds to a diversity system in the absence of redundancy. The organization plays on a sort of compromise between the quantity of information and the redundancy. Self-organization is associated with an increase in the quantity of information (internal programming? see, e.g. Kaern *et al.* [KAE 05]). For Rothstein [ROT 62], the system's order is defined by $R.H_{max}$, R being dependent on the system's constraints and H_{max} on the potential diversity of the said system (see also [ATL 06]).

Under the influence of disturbances, for high R values, the system can continue to function so long as a certain redundancy can remedy the losses of communication according to the principle of order through noise [VON 60] (see Figure 5.17). This principle in which the levels of integration are nested within one another poses the question of the quality of information (meaning and local transmission). Atlan takes the example of a cell affected by a "noise" that disorganizes it, which leads to diversifying relative to its neighbors, translated to a higher level through creation of possible variety and adaptation with possible feedback. In the presence of noise, the system decodes it and can elaborate new behaviors (see Figure 5.10).

Shannon's model constitutes a schema from which information allows the organization to adapt its behavior to each moment through regulation and transform and rebalance itself to be in thermodynamic coherence with the environment. Information gives rise to a process of permanent adjustment in the organization through canals (the system adapts through accommodation) and codes (the system adapts through assimilation) of communication in relation to a project.

Shannon dissociates the quantity and meaning of information: the quantity of information contained in a message can be measured independently of all reference to this message's meaning. Applied to biology, it implies that the quantity of information brought by a strand of DNA, an organelle, a cell, etc., does not depend on fine knowledge of the organism whose construction, development, and senescence are scheduled, at least partly, by this DNA. This idea, behind the apparent differences in behavior, the structures that connect them, poses the question of knowing if these differences ultimately hide the same phenomenon. The debate must be cut short [FRA 97].

5.5.1.2.3. Machine of states

This is a system made up of a certain number of states in which it can be found and of a law of spatial–temporal evolution directing its passage from one state to another under the effect of its environment. It can be formalized either from groups of differential equations or discretely, with, for example, a network of automatons [WOL 84]. In these operations of modeling the real a critical element appears, that of the models' degree of sub-determination for lack of understanding or ignorance of the functional constraints.

Numerous different models, with their reductions, can explain the states observed without it being possible to separate them in the absence of new experiments that could be likely to raise some ambiguities (insofar as the "right" variables of the system have been considered). These models bring out macroscopic structures from local constraints (see [THO 72, MAN 73, NIC 77, KAC 71]). However, will we know how to perform the determinant experiment allowing us to choose robustly between two different theoretical foundations in one example, using specific cells, chosen consensually between bio-printing specialists to move forward on this subject?

5.5.1.3. *"General system theory" [VON 12]*

5.5.1.3.1. Theory of compartments

This is a set of sub-units that has boundary conditions between which processes of transport (information, energy, etc.) take place. It can take different structures. The laws of so-called "classical" physics are essentially laws of disorder, the result of random events. Concepts like organization, directivity and differentiation are foreign to this physics of dissipation. Every living system is open to flows from different origins (matter, information, energy, etc.). It is naturally possible to disturb the system through commands that can act on the state of the system, in principle allowing it to be aided in achieving a desired final state. The underlying idea is to make the complex system created through bio-printing react thanks to stimuli to drive it continuously or discontinuously towards a quasi-stationary state under constraints. Le Moigne [LEM 99] writes that every person searches not for the general solution, but is content with a particular solution. "The expert is not the one who shows the convergence of a good algorithm, but who, through a well-reasoned ruse, has large collections of plausible heuristics". Around these aspects, there is a possibility of defining a possible roadmap for (and with) the PE.

5.5.1.3.2. The projective modeling of complex action

The basic concept of Systemic Modeling is action, and thus processes, starting with the question "what does that do?" whereas analytical modeling deals with the question "what is it made of?" (see Table 5.1). The process is temporal, at the interface of two kinds of transfers, one temporal (displacement in space) and one transformational (modification of the morphology, functionality). Thus, every complex system can be represented by a set of multiple reciprocal actions or by a global process that may be an interweaving of processes. It is by playing on the mechanisms and their magnitudes that behaviors can be defined (the variety of the system is the number of interrelations). If this number is too high for robust treatment, it is possible to work with sub-sets whose behaviors and knowledge of the interrelations between sub-sets should be made to emerge.

5.6. Bases of reflection on modeling

5.6.1. *Shooting or Monte-Carlo methods*

A complex system presents different types of equivalent behaviors from an energy standpoint, but it does not seem possible to classify them hierarchically [LIC 10]. The discussion on complexity developed in a transdisciplinary way by attempting to bring together the need to explain this concept through highly organized collective behaviors and the activity in physical, biological, cognitive, and social systems. Unfortunately, no clear definition has been achieved, so complexity appears as an antireductionist paradigm in search of a theory [LIC 08, GIU 11, THO 14, IZA 04, MOM 05].

According to Licata [LIC 16], "a complex system can belong to one of the three following categories:

– "information systems: example of 'perfect' Brownian movement (elastic shocks); no matter what sample is considered, it respects the same law;

– information compression systems: closed but non-isolated systems, submitted to a progressive weakening of correlations (entropy) and having a finite number of asymptomatic states (top-down operation);

– information amplification systems: open system that requires making new models get involved to characterize the emergent structures. The number of degrees of freedom must decrease (Haken's 'slaving principle'). Thus, 'the loss of information on the degrees of microscopic freedom is a necessary condition for organized structures to emerge.'"

It is still necessary to feel around in an attempt to advance, bio-printing likely belonging to the third configuration. It is thus that it may be proposed, to get engaged in a modeling approach, to work with a random shooting (Monte Carlo) method based on the laws of behavior closest to biological equality [SUN 13, FLE 12]. It is possible to make the hypothesis that the learning faculties of the artificial "cells" have a behavior associated with forms of self-organization with connections under the effect of the stimuli they exchange (would it then be necessary to talk about hetero-organization?). In biological systems, what is known? For example, during tissue formation, the cells can divide, differentiate, change shape or position, or die. This small number of mechanisms determines the size and shape of the tissues. If most of the genes that orchestrate these mechanisms have been discovered, do we know, do we have the means to learn about the physical forces that "sculpt" the tissues by making them functional? How do tissues acquire specific axes of polarity during morphogenesis [IBD 16]?

According to Kupiec [KUP 08, KUP 98], the cells differentiate using a probabilistic operation: according to how one of the genes expresses itself, the cell could acquire characteristics corresponding to a differentiated state. Cellular interactions could stabilize the genetic expression so long as the differentiated cell is viable. The genetic expression would then be fixed and, subsequently, the cells would no longer change state (mix of chance and selection; see [TUR 52]).

Licata [LIC 08] reminds us of the problem of mastering nonlinear dynamic systems by considering the existence or not of uni-modal relations in cell growth. After a large number of shots, it should be possible to examine the average behavior of a "cellular" system that develops in a given environment.

A connectionist model is characterized by three basic constituents: a network, an activation rule, and a learning rule [COL 95]. The network is made up of a set of cells initially placed by the operator via the bio-printer (see, in two dimensions, the example from Figure 5.21 whose principle stems from Neagu *et al.* [NEA 06]). "The activation rule of a connectionist model is a local procedure that each cell follows by updating its activation level as a function of the activation ontext of the neighboring cells. A genetic model depends on the evolution of a population in an environment according to preset survival criteria, function of the arrival of nutrients, oxygen, the presence of dead cells, the possibility of their elimination, etc. The general principle of the genetic algorithm to be developed depends on a continuous process that replaces the least adapted cells (going through apoptosis) by other cells coming from other, randomly chosen ones among the most adapted" [NEA 06].

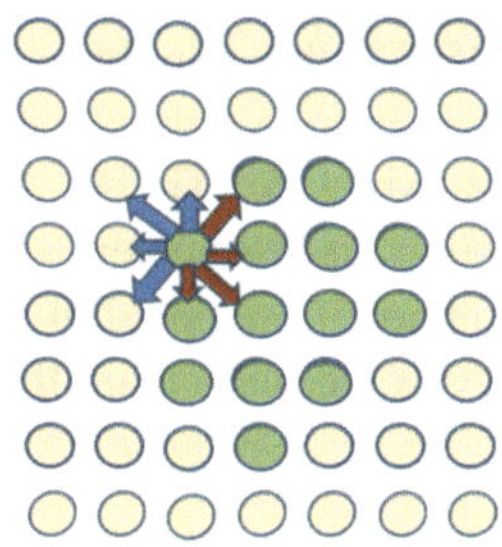

Figure 5.21. *Model principle: each cell interacts with its environment to die, proliferate, move, etc. For a color version of the figure, see www.iste.co.uk/andre/printing3.zip*

On this basis, there can be cellular proliferation, migration, so long as the nutrient and oxygen supply criteria are respected [MCC 14, SUN 17]. After numerous generations, the system must tend towards a dynamic balance, the cells present (if some exist) are those that respond to the sought "biological" programming. So long as the cells maintain their nature (no differentiation in this stage), it must be possible to demonstrate that a stationary state is effectively attained; moreover, it must be possible to take the potential attractiveness of the cells for scaffolding into consideration, thus favoring cellular migration towards the support. Insofar as the interaction mechanisms are known, it is foreseeable, but it is just a bit more complicated.

Reminder: a projected model must, using the knowledge of the present and past state of a system, allow its future behavior to be deduced. An obvious reduction principle must, however, be at work to try to seek the role of the chosen parameters in the system's behavior in a trend model.

Questions to ask:

– Invariance of scale: is the same phenomenon ruled by the same laws reproduced at every spatial scale?

– Existence of attractors that structure and organize the trajectory of the system in its dynamic?

– Existence of bifurcations?

In the systems to be conceived of the type "swarm on network", the cells interact randomly (but with probabilities linked to the assumed known laws of interaction) on a three-dimensional network and have no global representation of the architecture that they construct; they can only perceive the local configurations of matter around them, and maybe only a small number of these configurations will lead to an organized structure (see [THÉ 95]). Is this also the case for printing living tissues?

NOTE.– Other treatment models exist (see [TRI 11, BAL 13, BAT 13, MIS 11]). If the methods change, the question of parameters of interaction among cells and between cells and the environment remains [DES 06, BOI 10]. Ballet *et al.* [BAL 13] indicate several types of software that allow cell growth to be simulated (see Table 5.2).

Year	Software name	Reference
1992	ImmSim	www.immsim.org
1999	NetLogo	ccl.northwestern.edu/netlogo
2000	Cellular Automata Viewer	www.rennard.org
2001	MGS	Mgs.spatial-computing.org
2003	SimBioDyn	Virtulab.univ-brest.fr/SimBioDyn
2004	Smoldyn	www.smoldyn.org
2004	HCell	www.iri.fr/-pa/Hsim/index.html
2004	Flame & Flame GPU	www.flamepu.com
2007	NetBioDyn	Virtulab.univ-brest.fr/ NetBioDyn
2007	CompuCell3D	www.compucell3d.org
2011	Matrix Studio	Virtulab.univ-brest.fr/MatrixStudio

Table 5.2. *Cellular growth simulations software*

An example of work by Ballet *et al.* [BAL 13], who developed several of the pieces of software presented in Table 5.2, is presented in Figure 5.22.

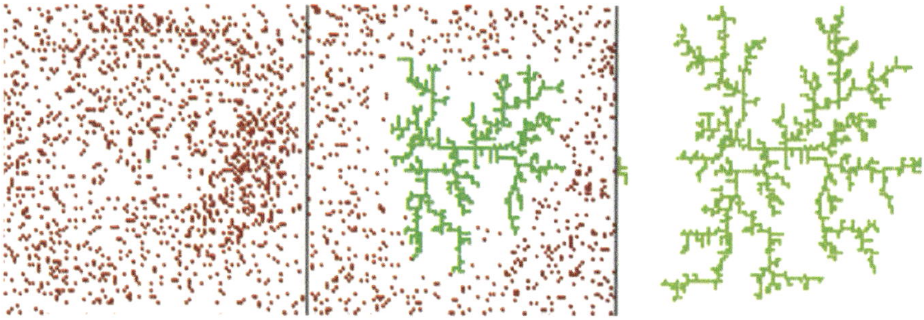

Figure 5.22. *Bacteria in the center feeding on glucose; left: beginning; center: intermediate situation; right: situation when all the glucose has been consumed (with permission from Pascal Ballet). For a color version of the figure, see www.iste.co.uk/andre/printing3.zip*

5.6.2. *Analogy with David Bohm's works?*

Peat [PEA 96], who drafted a biography of Bohn's works, reminds us that "when two negatively charged electrons are totally isolated, the interaction between them extends over a great distance. But in a plasma, such a gigantic number (of the order of a hundred million) of other charged particles recombine to protect this large-scale interaction. Each of the plasma's particles then only interacts with nearby particles over short distances. But the long-distance interactions have not yet disappeared. These are what allow the plasma to behave coherently". For Teodorani [TEO 11], there is a low potential but one with information thus giving shape to a strongly energetic structure (somewhat, all proportions maintained, in the way of Coriolis forces). Mathematically expressed, the concept thus requires the particles to be entirely determined by a higher-order description, that is, hidden variables [LOM 05]. Therefore, by this means, there is a possible introduction of an apparent form of determinism into the processes associated with a bio-printed object's formation. It is the framework of this apparent determinism that the PE must research.

In these conditions where the global and the local are associated, will it not be possible to examine how the conceptual approach stemming from plasma could be applied to the bio-printing issue? It is, incidentally, in this same direction, based more on biological competencies, that the work of Kupiec [KUP 08, KUP 13] is situated.

5.6.3. *Cellular differentiation*

Until now, we have tried to reflect on the spontaneous assembly of identical cells (without natural selection). It is possible (to discuss) that under the influence of the environment, cells A differentiate into cells B or C. Several occurrences can likely exist: proliferation of B relative to C (and vice-versa) or a coordinated proliferation (example of blood vessels that are built simultaneously with functional tissue). According to the types of interaction between cells B and C, Kupiec [KUP 08] has performed different simulations that lead to results that could be taken up by in-depth studies to judge their use for bio-printing (but this path seems promising, at least to start with). Figure 5.23 illustrates some results presented in his book and in [KUP 09]. Similar results were obtained in the same timeframe by Merks and Koolwijk [MER 09]. Indeed, a very important question is raised in a number of works on bio-printing, namely, that of transfer of matter and energy to printed cells [MER 09]. How, in a significant stage, to model angiogenesis? Even if refinements exist [BOU 15, LEJ 17, OKU 16, SUN 17], an in-depth study of this kind of simulation linked to the expertise of biologist colleagues may be likely to explain a potential for cell differentiation, leading to the appearance of micro-vessels; when the mechanism is understood (and experimentally validated), it will be necessary to research the conditions for the proper mastery of this evolution necessary for bio-printed tissue. This competence, still to be acquired, constitutes an essential element that should be found in the toolbox of every good bio-printer, even if this is not everything.

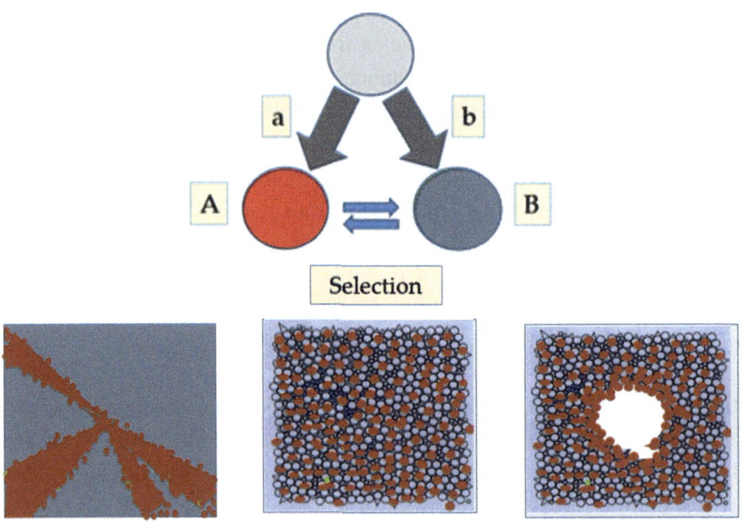

Different forms of selective balance according to the value of the influence parameters between A and B: separate areas; homogenous division; formation of lacunae; etc.

Figure 5.23. *Approximate drawings of simulations performed by Kupiec [KUP 08] and by Merks and Koolwijk [MER 09]. For a color version of the figure, see www.iste.co.uk/andre/printing3.zip*

Ontophylogenesis, a concept introduced by Kupiec, can clash with the usual way of thinking. In his framework, ontogenesis, instead of being the result of a deterministic mechanism controlled by the genes, is understood as an instrinsically probabilistic process at the level of intermolecular interactions, indeed, the proteins' lack of specificity consequently induces multiple combinatory possibilities in molecular interactions. Each of these combinations has a certain probability of occurring, but the ontogenetic process undergoes selection originating in the cellular (or multicellular) structure, which is itself selected by the organism's environment.

In the general theoretic framework, the random expression of genes, caused by the non-specificity of molecular interactions in the cells' nuclei, allows cells to change state without being directed by signals emanating from a genetic program. However, they are not given over to absolute probabilism. There is also a selective constraint that performs selection among the diversity of random cell states and steers embryogenesis towards the adult state. Each cell of an organism finds itself in a particular micro-environment that allows it to multiply and differentiate. This micro-environment determined by the embryo's multicellular structure is characterized by the concentrations of the metabolites to which the cell has access. The metabolism must be understood here in the broad sense; it is every reaction and every biochemical exchange, including molecules normally considered signals. As a function of this micro-environment's variations, the cells that express an adequate phenotype are selected or stabilized. This leads to cell differentiations at the origin of tissues constituting an adult being. This theory is based on a wealth of experimental data.

The proliferation of cells is controlled by activation or inhibition signals. With the Darwinian model, the process is completely different. There is no difference between the growing system and the stationary system that could be tied to the presence of specific signals to control proliferation. The cell population ceases to grow when it achieves a state of equilibrium and this state depends on the quantitative value of the model's parameters. In a real organism, the mutations that would survive in proteins involved in self-stabilization or interdependence would change the values of the parameters that govern these processes. For example, a mutation in a transcription factor involved in self-stabilization would modify its affinity for its target sequence in the DNA and, consequently, in the model, the parameter to self-stabilize the genetic expression would also be modified. The simulation for such an event shows that the equilibrium of the cellular bilayer is then broken and that proliferation begins again. When the value of the self-stabilization parameter is changed from a cellular biolayer having achieved its state of equilibrium, we witness a local recommencement of proliferation leading to the progressive appearance of cell masses evoking tumors. Likewise, a mutation could change the diffusion properties of a protein. In his computer model, this leads to changes in the value of the parameter regulating the molecules' diffusion speeds. The simulation shows that, in this case, this leads to an imbalance in the division of cells that also induce uncontrolled growth destroying the cellular bilayer.

Box 5.3. *Extracts from [KUP 09]*

5.6.4. *Scale change(s)*

From a computing standpoint, it may be possible to work with a few thousand cells [BOU 15], which represent a space of the order of 0.1 to 1 mm, which is far from representing a real tissue. Let us imagine that it is possible to define an average law of behavior for a set of cells that will constitute a voxel (non-vascularized and non-innervated at this stage). If this voxel corresponds to a representative element (with possibilities of threshold or random behavior), then it will be foreseeable to try to gain on spatial scales.

It may be possible to assume, for the needs of modeling, that the system studied can itself be composed of larger interacting parts. These parts can be physical or abstract entities. A system will be called self-organizing if it can, from its interactions with its environment, detect all the non-cooperative situations between its parts and try to delete them by reorganizing. A self-organizing system is typically capable of autonomously:

– creating, deleting, modifying parts within it;

– changing the organization (the inter-individual relations).

A system with variable connections can also be foreseen (but we are not there yet). However, on the basis of these first results, it will be foreseeable to examine whether, using biological knowledge, it is possible to bring out vascularization (always using Monte Carlo or equivalent modeling) or to examine whether manufacturing a bio-printed system, for example, a strongly porous one, allows this creation to be facilitated.

5.6.5. *Questions for realistic modeling*

Table 5.3 gathers a certain number of questions asked to prepare a basis for modeling.

Basic model scale < 0.1 mm	Questions
	Effects of the physical and biochemical environment on cell differentiation
	Laws of cell behavior among themselves; self-organization or hetero-organization?
	Laws of cell behavior with their scaffolding
	Laws of cell behavior vis-à-vis flows of matter and energy
	Effects of apoptosis

	Can an organization parameter like an order parameter be defined?
	Is there local "determinism"? Genetic algorithm?
Change of scale	Moving from a Monte Carlo system to something else; problem of networking
	Conception through a unique nucleus or an initial set of independent nuclei (sub-sets that respect the basic microscopic model form the start and whose evolutions will allow a more macroscopic, unique set with the desired functionality to be created)
	Cell differentiation
	Vascularization
	Heterogeneous bio-printing with the aid of several cell types (among those that will allow vascularization)?
	Conception of the bio-object with potential bio-resorbable micro-vessels [MAO 15]

Table 5.3. *Some questions posed for modeling*

5.6.6. *Provision of an operatory reference*

In most of the publications analyzed in Chapters 2 and 3, the authors show that their bio-tissue starts to develop, is capable of surviving x days or y months, differentiation appears, etc. In the spirit of what has been presented, the so-called R value or Shannon Redundancy Function, where R is defined by $R = 1 - H / H_{max}$ or H_{max}, corresponds to the quantity of information possessed by a diversity system in the absence of redundancy. If it were possible to define an R value (even on approximate, even semi-quantitative bases), it would be possible, with the help of this kind of "normalized" reference, to notice the positive (or not) evolution of a bio-printed tissue created in specific (and known) conditions.

This demand on the author's part finds its origin in its modest trust in global modeling contributing, in the short term, an ensured and robust vision of the bio-construct's development. It no longer dreams of resolving the inverse problem! In these conditions, instead of accumulating proofs of concept, certainly always interesting, it could be foreseen that, on the occasion of these often-original experiments, one capitalizes via an H factor (which is not that of Hirsch [HIR 05]!) on the quality of a direction aiming to create a bio-tissue with the desired shape, dimension, and functionality. It would only be, as with the thermometer in medicine, an integral indicator, but this would allow a comparison experiments between (as done by the Hirsch factor, but which, it must never be forgotten, requires a complementary analysis before judging, except if the gaps are notable).

Organogenesis is a dynamic, complex system that finds its origin in the cell stage. During embryogenesis and fetal development, the spatial–temporal organization of the cells is orchestrated by the biochemical environment (ion gradients and growth factors, etc.) and the biomechanical environment (solid and fluid mechanics) that will influence the genetic expression determinant for cell behavior (proliferation, migration, quiescence, differentiation, death) and the volumic architecture of tissues. Thus the parameter, which could be a self-organization parameter (depending, however, on an open environment), must inform (gray box) about the evolution of a biological system. This factor could be created on the bases of embryogenetic factors and should/could integrate a certain number of criteria, presented here to move forward with reflection. Some are already useless, others lacking:

– volume evolution factor;

– shape factor;

– influence of additives;

– influence of surrounding physical–chemical factors;

– vascularization;

– resorption of the initial scaffolding;

– creation of an internal structure;

– differentiation;

– orientation of the cells relative to the "instructions";

– mortality rate;

– autonomy.

It is well understood that this construction is certainly delicate to implement, if only because there are already, in this enumeration, interdependencies between sub-factors: hence the author's proposition of a demand, already expressed elsewhere, of an agreement on the part of specialists to come to terms on this important aspect to move forward in a slightly enlightened way.

5.6.7. *Organizational methodology*

Whatever we maintain, we do not think "past the end of our nose". Researchers and experts are often more familiar with previous disciplinary works than they are equipped to understand the events that they are meant to advance. The behavior of the committees is constituted, for different reasons, by certain forms of interests. The state of mind at the moment and fashionable subjects leads us to believe that they are lasting, that proofs of concept are enough to legitimize extraordinary promises. But, when nearly all the hypotheses are open, which is the case today, how do we decide? Or go about it? How to be sure that acting is better than abstaining, how to give oneself a chance to succeed when acting, how to deepen reflection and rule out facilities of weak or herd consensus? It should be feared that this is the relationship of forces, based on personal charisma and means of communication, that is gaining – and, with it, whatever economic ideology – or not – of the moment [AND 15].

According to Bouët and Duréault-Thoméré [BOU 11], the growing complexity of the issues that we must resolve imposes the idea that we cannot find the responses and solutions on our own. Thanks to its own creativity, each contributes, of course, to enriching this research. But, far from being the addition of each of these intelligences, methods of collective intelligence must stimulate the production of ideas, which should then be enriched exponentially, at least at the onset. They constitute the foundation of what is called "positive-sum economies".

The proposition to connect the concepts of creativity and collective intelligence poses the hypothesis of a possible link of causes and effects. The promoters of collective intelligence and its methods actually postulate that they are factors of creativity and innovation. Giving rise to a new form of intelligence, which, in principle, goes beyond people and their egos, methods of collective intelligence should encourage letting go in relation to the defense or paternity of ideas. They could offer the possibility to go where none have thought to go by significantly improving intellectual collaboration processes. These co-production processes could allow appropriation of the subject dealt with and recognition of the people who facilitate their implication.

They are, however, sensitive to the personalities involved in the process (batch or semi-batch), their common history and their dynamics: they echo a concrete and living reality. But, at the same time, subliminal forms of soft coercion (the mission) and sensitization (the group leader who, to advance the operation, may be tempted to make his/her contribution) intermix. Beyond a work on oneself that could have been conducted by each group member to take on appropriations, to make them his/her own, there is probably no means of going further in a given timeframe [GRO 15].

To draw closer to the learning organization, it is essential to develop "double-loop" learning, which involves not only learning to correct mistakes, but to modify one's way of thinking and posing the problem. Figure 5.24, inspired by works by Argyris and Schön [ARG 96], expresses the difference between these two learnings well. The challenge is to reach this second level of learning, so long as there is time and monitoring of the hierarchical system. This kind of learning, necessary in the case of bio-printing, will be difficult to achieve without collaborative work that allows ideas to be compared, confronted and mutualized to transform into operatory actions.

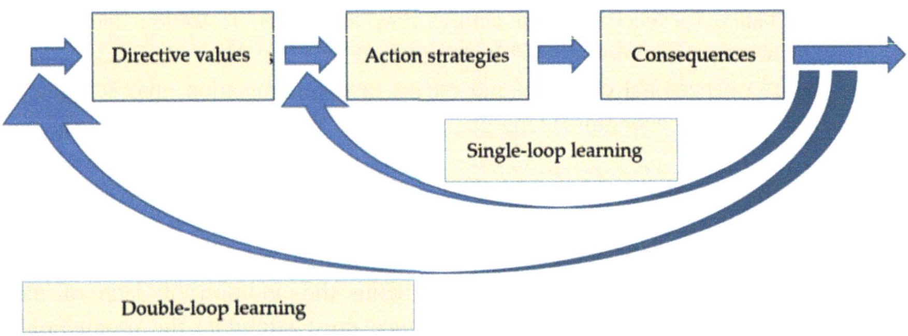

Figure 5.24. *Learning loops [ENA 16]*

There is a need for devices that allow each member to access existing knowledge in a scientific project on 3D bio-printing as quickly as possible (concepts, community of practice). Furthermore, it is best for the critical organization (to be put in place) to correct and adjust its knowledge base in a calm and trusting framework. The concept of socialization is thus seen as a meta-process. This is a global vision of interdisciplinary management focused on communication, interaction and cooperation (see Figure 5.25). It is likely another form of complexity to be dealt with.

Thus, a learning organization is a human organization that implements a set of practices and services to remain in line with its environment. In such a "learning" project, in principle, each member learns from the others (and vice-versa). This is by building on collective intelligence in order that the sustainable development of a project's organization is ensured to achieve its goal. The idea is: "I transform myself to transform our way of working". It is thus a matter of putting each partner in the center of reflection and encouraging environments, where divergence is creative rather than a harbinger of sterile confrontations. Table 5.4, taken from Lemyre and

Despatis [DES 06], reminds us what the expected qualities are on the parts' of the partners in such an operation.

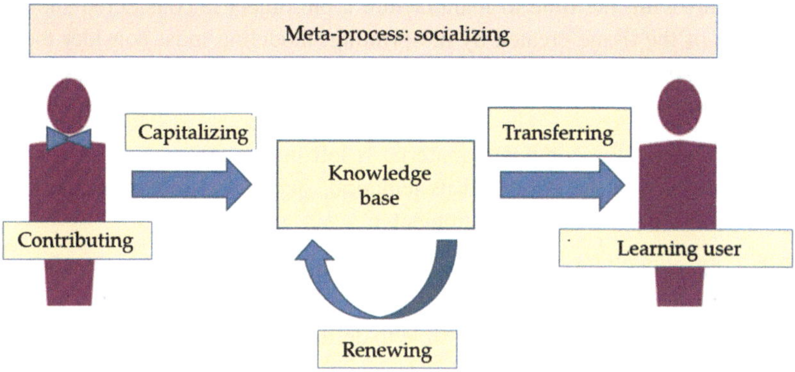

Figure 5.25. *Mutualization of knowledge [BAL 10]. For a color version of the figure, see www.iste.co.uk/andre/printing3.zip*

Domain	Sub-domain
Systemic thought	- System open to the internal and external environments - Work in networks - Concern with the results - Strategic monitoring
Mental models	- Flexibility - Innovation
Personal mastery	- Promotion of the ability to learn - Creative tension
Shared vision	- Shared decision-making - Oriented action - Collective aspiration
Collective learning	- Collective learning - Learning at all levels

Table 5.4. *Qualities expected from the different partners in the project*

In the approach desired for engagement in the project, several considerations can be recalled:

1) To start with, the time to manufacture a bio-object is considered short given the lifetime of the tissue created by bio-printing; modeling must consider the initial conditions (manufacturing the bio-object) to define (if possible) its evolution in terms of tissues.

2) In this stage, the composite material is left in the presence of environmental conditions defined at the onset and its temporal, spatial and functional evolution is followed (with all accessible measurements).

3) In a later stage, it will be interesting to examine how it is possible through the modification of environmental factors to know the "trajectory" (control-command) and potentially to examine if it is possible to define an optimal path.

4) From acquired experience, the question of the inverse problem can be dealt with considering the robustness of models, the nature of the biological systems and/or of tissues to be created, the quality of existing information, etc.

Over the unpredictable duration of such an engagement, whose interdisciplinary and heuristic aspect determines its interest and inter-relational difficulty, it does not seem foreseeable, to the author, to immediately deal with stages 3 and 4.

To advance on a possible project, in anticipation of a discussion in a group to be put together to reach an agreement on the means of proceeding, it may be proposed to operate at several moments (see [DUR 12]):

– Definition of the concept linked to the boundary-object of bio-printing while trying to make it operational, thus intelligible, in its own disciplinary practice.

– What does each member of the project expect, how can he/she contribute (significantly) to the operation?

– Is there a possibility of defining a common project? On what bases? What are the barriers to take into consideration?

– Use of an explanatory schema (e.g. flow-chart), for which each member is invited to express what he/she understands out loud, what seems intelligible to him/her (bring out the unspoken). The use of this "tool" allows biases to be grasped on one's own knowledge (recourse to "conceptual blackboxes") through interaction with "profane" scientists from other disciplines; each member can then not mobilize the common references from his/her discipline and must open blackboxes; the presences of other representatives in the interdisciplinary space imposes a demand for common understanding.

– Discussion of the importance of a particular discipline to specifically deal with the boundary-object and how it makes it potentially intelligible; disciplinary elements to consider?

– How do we integrate knowledge into the epistemological approach? What are the necessary elements to start this fusion of knowledge, what is known and what is obviously lacking? Are there acceptable compromises? Dynamic systems on multiple time scales? Once the objectives are defined, an exploration phase must help define the interaction domain, as well as the variables that can be acted on, the relations between elements, the circulation of the flows, and the critical points, the structural constraints, the regularities on which processes and singularities can be built justifying constant monitoring with great reactivity to avoid failure or seize opportunities; then will come the actual modeling phase to conduct simulations allowing the choice of a strategy that should ensure reproduction and development, taking into consideration the balance of flows and the reconstitution of both internal and external resources. The difficulty lies in visualizing what can be known about everything while underlining the risk and uncertainties.

– What kind of model to choose? Can it be tested with the help of the existing knowledge? If so, how to refine it? What is its sensitivity to the initial conditions (multi-stability)? What control is there of the process of regulating the living (difference between the living and cellular automatons, CPD (Cellular Particle Dynamics) from Kosztin, Vunjak-Novakovic, and Forgacs [KOS 12], LBM (Lattice Boltzman Method) from Cristea and Neagu [CRI 16])? Knowledge of self-organization mechanisms (direct, indirect, cooperative, extolled…)? What experiments should be planned to advance and on what kind of biological tissue? Will there be sufficient genericity to expand the application of a potential model?

– Is the epistemological problem predictable, temporarily unpredictable due to an absence of sufficient data or truly unpredictable? Are all the unpredictables under control (e.g. bifurcations)? The arborescence of new hierarchies demand a more precise definition of what constitutes them in organized wholes. It is not enough to say that there is more in the whole than the sum of the parts, it must be stated what sort of whole is being spoken of. What trust will be had of the prediction? How to avoid on, to the contrary, profit from possible bifurcations?

– How and according to what practical criteria should we distinguish between fate and complexity [LÉV 15]?

– Permanent feedback in debates on the member's practice, avoiding the by-passing of others to engage his/her own research activity (self-referral) and outside critique (going from "me" to "us" to achieve conceptual connivance).

– In exchanges, if each is led to express himself/herself in his/her scientific culture by attempting to avoid being in an "a-disciplinary no man's land", it is important to make others share their points of view (in an interdisciplinary way, in relation to one another, processes with an aim of defining the object of interdisciplinary research), and to accept that ignorance is not a flaw.

– To move forward, the comings-and-goings between different propositions and explanatory theories, on the one hand, and remarks and heuristic intuitions, on the other, must be accepted (returning to the complex problem from disciplinary reduction).

As indicated by Table 5.5, partly from Vinck [VIN 05], even if the partners have strong values for engagement, several tensions may be at work in an attempt to achieve the goal.

Internal success factors	External failure factors
Scientific method and questioning	Inertia
Trust and absence of prejudices	Beliefs; obscurantism
Relationship with the experiment	Persistence in error
Rationality and common sense	Non-openness to others
Rigor and traceability	Lack of rigor
Creativity	Herd mentality
A certain obstinancy	Lack of obstinancy

Table 5.5. *Difficulties and dissymetries in interdisciplinary investment*

This best integration of complexity into research is supported by the ISF (French Association of Engineers and Scientists) [IES 16] in terms of the initial training of engineers. This association writes that it is necessary "to engage pedagogical transformations to adapt the initial training of future engineers to the stakes of tomorrow, notably:

– Developing intelligence of complexity by teaching epistemological foundations and fundamental concepts of complex thought.

– Accompanying scientific and technical teachings with development of a true 'scientific spirit' and particularly critical thinking.

– Preparing for the digital universe, not only through mastery of technology, but also by taking a step back from the conception, uses, and stakes of data and algorithms.

– Reinforcing collaborative skills and opening students up through more multiform teaching and more transdisciplinary collaborative activities" (see also [HEF 16]).

In this investment of constructing a common, in-depth culture, it is best to limit as much as possible the recourse to operations that increase the risk of losing control of theorization processes. Yet, social communication, which lies at the heart of interactions, necessarily mobilizes common sense. Knowing that it is impossible to totally "purify" the rational dimension of a complex, high-risk project of its affective components and of personal, rarely expressed interest, it may be useful to limit and control the project through techniques allowing the loss of distance from the object during human interactions to be reduced. It is thus a matter of injunction, paradoxically, in the end, but it is perhaps the price to be paid for this enterprise.

5.7. Conclusion

The presentations or reflections presented in this seventh section are provisional; they characterize a state of reflection upon writing this Volume 3. They foresee a possible path that can only be engaged if our biologist colleagues bring to the start elements on cell behavior. On this learning base, it should be possible to model the growth of a free cell aggregate or one attached to a support, so long as the cells do not diversify. This stage can be considered a prerequisite for a change of scale allowing more realistic spatial domains to be reached.

Independent of this aspect, it could be useful for reflection to be made on the concept of the bio-printing operationalization indicator (H factor). If it is difficult to define well, because the volume of the bio-object is itself not defined, an estimation based on indicators considered pertinent could already allow a judgement of the experiments published (e.g. a question of interest: will the temporal evolution of R follow a monotonous curve or will it be more complex?). This global measure could thus allow a better grasp of possible organ construction via bio-printing methods using the results from the literature (insofar as there is access to the values of the indicators).

As has already been pointed out in this section on bio-printing, the scientific development of this exciting field poses considerable epistemological problems which do not seem as yet to have preoccupied many researchers and may not have concerned the hierarchy and the Agencies, interested in the most immediate results, at all. The path is thus still long before being able to make ourselves immortal and post-humanized (see Figures 5.21 and 5.22), as we see by taking a look at the somber facets of our anticipations.

From an operational standpoint, in agreement with André [AND 16]; Brynjolfsson and McAfee [BRY 14]; and Kalil [KAL 12], it must be possible to foresee the development of a "reflexive" HE operation (with strong PE support) in the field of bio-printing (see section 5.2). The principle is to stimulate, through divergent ways of thinking, the emergence of original ideas defining a project (association, application, modeling, technologists, between concerned disciplines). This operation depends on the principles defined in Figure 5.26 (inspired in part by Bila-Deroussy [BIL 15]). Figure 5.27 reminds us of the forms of operation necessary in a systemic approach.

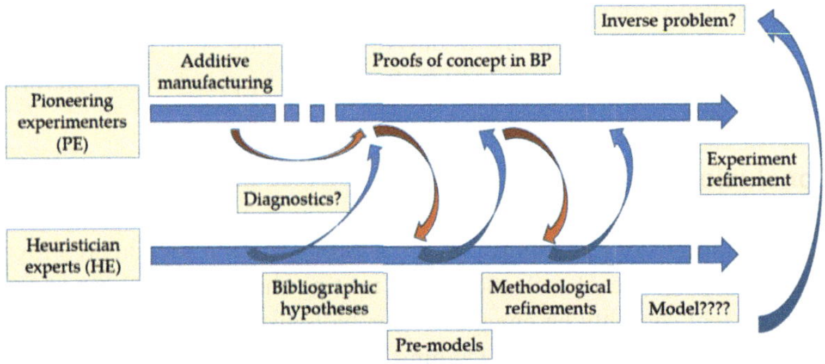

Figure 5.26. *Vision of action harmony between PE and HE. For a color version of the figure, see www.iste.co.uk/andre/printing3.zip*

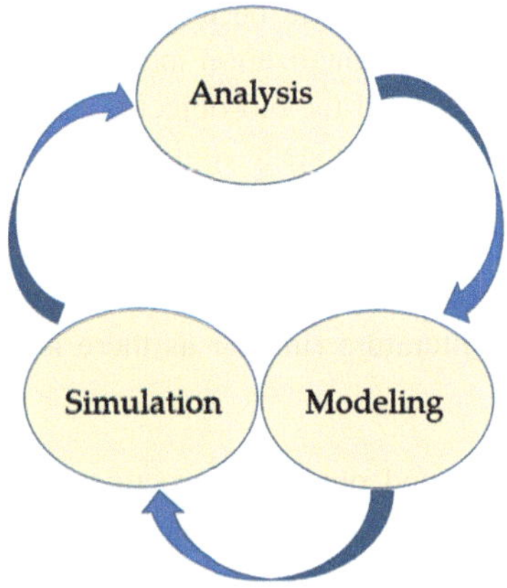

Figure 5.27. *The "virtuous" circle of the systemic approach*

Interdisciplinarity, which is indispensable in this type of operation (after defining the project's framework), is a process in which analysis and synthesis capacities are developed from perspectives from several disciplines. Its goal is to deal with an issue in its whole (principle of integration in organizational engineering), by identifying and integrating "all" the relationships between the different elements implied. It attempts to synthesize and link disciplinary knowledge and place it in a broader systemic framework, all while engaging each member from a discipline in the relationship to his/her own identity, all without forgetting the relationship between scientific practices and identity constructions.

This is, for academic research, perhaps a means of trying to "decipher", in light of the development of a theme as attractive as bio-printing, "mysteries" of an integral field of studies ensuring it, in case of success, real international legitimacy, rather than resolving "local" problems and seizing opportunities. It is necessary to go a bit further, no longer based on classic inertial foundations (because they have proven themselves ([MOR 12])) and by proposing that we leave the following conservative spiral: rather than producing knowledge adapted to situations as they are built, we adapt these to existing knowledge or as it is foreseen!

In doing this, those who do can naturally update technologies and/or methods that will respond in part to the stake. Would we then find, as in artificial intelligence relative to "natural" intelligence, an artificial bio-construction, and a new subject relative to what Nature knows how to do. This could open new fields of research, abilities to change the world, for the better and for the worse.

If the possible revolution, in which bio-printing submerges us, involves applying knowledge to knowledge itself, maybe it is best, passing the word of the near end to Jean Staune [STA 15], to conclude with this paradoxical and pessimistic sentence: "Why study problems that have no solutions or whose solution is too complex to be found?" Is it really possible to escape the exhaustion of concepts to be capable to predict the evolution of such a complex system from knowledge of its present state? Could there be incompleteness and consequently indecidability? How will we know if we do not try?

"In each discipline, there is, in fact, in relation to nearly every problem, an avant-garde: this is the group of researchers working in practice on the problem at hand. There is then the primary group: this is the official community. Finally, there are the more or less disorganized late-comers. [...] The avant-garde has no fixed position; overnight, it is in a different place. The primary group moves more slowly; it takes years, even decades, for it to change positions – often through pressure. Its path does not perfectly correspond to that of the avant-garde: the primary group certainly determines its path following the information provided by the avant-garde, but with a certain independence. [...] The paths must be transformed en route, the land leveled, etc., such that the landscape is significantly transformed until it becomes the parking place of the primary group". How many times will it be necessary for the trail, as it is foreseen by Fleck [FLE 08], to be blazed?

In this necessary quest for knowledge, there is particularly no place to oppose the PE and the HE but only, on the contrary, to bring them together. For the author, there is neither a debate nor a hierarchy between Epimetheus (who does and who reflects afterwards) and Prometheus (who reflects beforehand). The adventure needs every talent and it is not after having extolled interdisciplinary openness (as was presented in this reflection, in these three volumes) that the author wishes to engage a sterile quarrel between approaches that must interweave, mutualize to make gains. Otherwise, this would be mutilation (not reparable by bio-printing)!

> *"At the beginning of life, a sort of loop was created, a sort of natural machinery that returns on itself and produces ever more diverse elements that will create a complex being that will be living. The world produced itself in a very mysterious way. Knowledge must have instruments today, fundamental concepts that allow linking". [MOR 95]*

> *"A good model thus involved finding a structure of connections such that the calculation of the network's evolution from this structure shows it stabilizing itself in the correct states, effectively observed in reality. And it is then that I understood that it was possible to manufacture several different models, one as good as the other, without the available observations being able to decide in favor of one among them". [ATL 11a]*

5.8. Bibliography

[ABB 15] ABBOTT R.D., KAPLAN D.L., "Strategies for improving the physiological relevance of human engineered tissues", *Trends in Biotechnology*, vol. 33, pp. 401–407, 2015.

[ABE 04] ABELMALEK A.A., "Edgar Morin, sociologue et théoricien de la complexité: des cultures nationales à la civilisation européenne", *Sociétés*, vol. 86, pp. 99–117, 2004.

[AND 15a] ANDRÉ J.C., LARGER L., Approche épistémologique du bio-printing pour essayer de moins mutiler la recherche sur un objet-frontière, Report to CNRS, Paris, 2015.

[AND 15b] ANDRÉ J.C., "N'est pas Prométhée ou Mary Shelley qui veut: une petite histoire de la stéréo-photo-lithographie…", *Actualité Chimique*, vols 397–398, pp. 99–101, 2015.

[AND 16] ANDRÉ J.C., Peut-on modéliser de manière robuste le bio-printing ou l'impression 4D du vivant?, Report to CNRS, Paris, 2016.

[ARG 96] ARGYRIS C., SCHÖN D.A., *Organizational Learning II*, Addison-Wesley, Reading, 1996.

[ARM 05] ARMATTE M., "La notion de modèle dans les sciences sociales: anciennes et nouvelles significations", *Mathématiques et Sciences Humaines*, vol. 172, pp. 91–123, 2005.

[ASH 56] ASHBY W.R., *An Introduction to Cybernetics*, Chapman & Hall, London, 1956.

[ASH 60] ASHBY W.R., *Design for a Brain: The Origin of Adaptive Behavior*, Chapman & Hall, London, 1960.

[ASH 62] ASHBY W.R., "Principles of the self-organizing system", in VON FOERSTER H., ZOPF G. (eds), *Principles of Self-organization*, Pergamon Press, New York, 1962.

[ATL 06] ATLAN H., *L'organistion biologique et la théorie de l'information*, Le Seuil, Paris, 2006.

[ATL 11a] ATLAN H., *Le vivant post-génomique ou qu'est-ce que l'auto-organisation?*, Odile Jacob, Paris, 2011.

[ATL 11b] ATLAN H., *Qu'est-ce qu'un modèle?*, Manucius, Paris, 2011.

[BAL 10] BALLAY J.F., "Le transfert intergénérationnel des connaissances" available at: http://www.telescope.enap.ca/Telescope/docs/Index/Vol_16_no_1/Telv16n1_intergenerationnel.pdf, 2010.

[BAL 13] BALLET P., POTHET A., MISEVIC G. *et al.*, "Une approche multi-agent pour la simulation en biologie cellulaire", in GLADE N., STEPHANOV A. (eds), *Le vivant discret et continu*, Materiologiques, Paris, 2013.

[BAR 03] BAR-YAM Y., *Dynamics of Complex Systems*, Westview Press, Boulder, 2003.

[BAT 13] BATT G., Design, optimization and control in systems and synthetic biology, PhD Thesis, Paris Diderot University, Paris, available at: https://contraintes.inria.fr/~batt/publications/manuscript_hdr_batt.pdf, 2013.

[BEN 88] BENNETT C.H., "Logical depth and physical complexity", in HERKEN R. (ed.), *Turing Machines: A Half Century Survey*, Oxford University Press, Oxford, 1988.

[BEN 06] BEN AMOR S., LAVALLÉE I., BUI M., "Complexité, percolation et optimisation stochastique", *ROADEF'06*, available at: http://www.eurocontrol.int/eec/gallery/content/public/documents/EEC_INO_documents/INO_documents/BenamorLavalleeBui.pdf, 2006.

[BER 84] BERNIER R., "Potentialités morphogénétiques et auto-organisation", *Archives de Philosophie*, vol. 47, pp. 529–556, 1984.

[BIL 15] BILA-DEROUSSY P., Approche systémique de la créativité: outils et méthodes pour aborder la complexité en conception amont, Thesis, ENSAM, Paris, available at: https://pastel.archives-ouvertes.fr/tel-01315163/file/BILADEROUSSY.pdf, 2015.

[BLA 14] BLANCHOIN L., BOUJEMAA-PATERSKI R., SYKES C. *et al.*, "Actin dynamics, architecture, and mechanics in cell motility", *Physiological Reviews*, vol. 94, pp. 235–263, 2014.

[BLU 99] BLUMENBERG H., *La légitimité des temps modernes*, Gallimard, Paris, 1999.

[BOI 10] BOI L., "Méthodes mathématiques, processus biologiques et philosophie de la nature", *Eikasia. Revista de Filosofía, I,* vol. 35, pp. 267–297, 2010.

[BOU 11] BOUËT C., DURÉAULT-THOMÉRÉ M.-C., "Créativité et intelligence collective", available at: http://reseauculture21.fr/labo/2011/11/05/creativite-et-intelligence-collective/, 2011.

[BOU 15] BOUDON F., CHOPARD J., ALI O. *et al.*, "A computational framework for 3D mechanical modeling of plant morphogenesis with cellular resolution", *PLOS Computational Biology*, vol. 11, no. 1, 2015.

[BRY 14] BRYNJOFSSON E., MCAFEE A., *Le deuxième âge de la machine; travail et prospérité à l'heure de la révolution technologique*, Odile Jacob, Paris, 2014.

[BRY 15] BRYNJOFSSON E., MCAFEE A., "Will humans go the way of horses? Labor in the second machine age", *Foreign Affairs*, vol. 94, pp. 8–14, 2015.

[BUN 79] BUNGE M., *Treatise on Basic Philosophy II, Ontology – A World of Systems, volume 4*, Reidel, Dordrecht, 1979.

[CAI 16] CAIOZZO M., OKAWA Y., RANGA A. *et al.*, "Defined 3D micro-environments boost induction of pluripotency", *Nature Materials*, vol. 15, pp. 344–352, 2016.

[CAN 65] CANGUILHEM G., *La connaissance de la vie*, Vrin, Paris, 1965.

[CAN 16] CANCES E., "Simuler la matière à partir des lois fondamentales de la physique", *Science et Society*, Lorraine, France, available at: http://iecl.univ-lorraine.fr/Cycle-Conferences-Sciences-et-Societe/lanceur.php?action=confDetail&id=91, 2016.

[CAR 86] CARO P., *L'oiseau jardinier: la science est-elle un humanisme*?, Belfond, Paris, 1986.

[CER 07] CERTU, Une introduction à l'approche systémique; appréhender la complexité, Report, available at: http://lara.inist.fr/bitstream/handle/2332/1431/CERTU-RE_08-09.pdf?sequence=1, 2007.

[CHA 11] CHANG C.C., BOLAND E.D., WILLIAMS S.K., "Direct-write bio-printing three-dimensional bio-hybrid systems for future regenerative therapies", *Journal of Biomedical Materials Research Part B: Applied Biomaterials*, available at: https://www.ncbi.nlm.nih.gov/pmc/articles/PMC3772543/#R35, vol. 98, pp.160–170, 2011.

[CHA 12a] CHALINE J., "La relativité d'échelle dans la morphogenèse du vivant: fractal, déterminisme et hasard", *Annales de Chirurgie Plastique Esthétique*, vol. 5, pp. 517–526, 2012.

[CHA 12b] CHAPOUTHIER G., "Un modèle de complexité du vivant: Emergence en mosaïque et conséquences philosophiques", *Al-Mukhatabat*, available at: http://cmc.journals.yorku.ca/index.php/cmc/article/download/38014/35075, vol. 3, p. 18, 2012.

[CIN 02] CINQUIN O., DEMONGEOT J., "Positive and negative feedback: striking a balance between necessary antagonists", *Journal of Theoretical Biology*, vol. 216, pp. 229–241, 2002.

[CIN 05] CINQUIN O., DEMONGEOT J., "High-dimensional switches and the modelling of cellular differentiation", *Journal of Theoretical Biology*, vol. 233, pp. 391–411, 2005.

[COL 95] COLLECTIF S.M.I., "L'adéquation fonctionnelle comme limite des systèmes organisateurs", in THÉRAULAZ G. (ed.), *Evolution et organisation; hasard et contraintes dans la genèse des formes collectives*, Journées de Rochebrune and ENST, Paris, available at: http://perso.telecom-paristech.fr/~evolang/Rochebrune/Rochebrune%201995%20-%20Evolution%20et%20Organisation.pdf, pp. 58–67, 1995.

[COL 13] COLLINS D.H., HUZURBAZAR A.V., "Multi-state stochastic processes: a statistical flow-graph perspective", *International Statistical Review*, vol. 81, pp. 78–106, 2013.

[CON 11] CONCALVEZ DA SILVA E., "Introduction aux systèmes dynamiques et chaos", Lecture notes Institut Polytechnique de Grenoble, available at: https://cel.archives-ouvertes.fr/cel-00556972, 2011.

[CRI 16] CRISTEA A., NEAGU A., "Shape changes of bio-printed tissue constructs simulated by the Lattice Boltzmann method", *Computers in Biology and Medicine*, vol. 70, pp. 80–87, 2016.

[CUN 14] CUNCHILLOS C., *Les voies de l'émergence – Introduction à la théorie des unités de niveau d'intégration*, Belin, Paris, 2014.

[DAL 92] DALMEDICO A.D., CHABERT J.L., CHEMLA J.L., *Chaos et déterminisme*, Inédit Science, Paris, 1992.

[DAL 10] DALLON J.C., "Multiscale modeling of cellular systems in biology", *Current Opinion in Colloid & Interface Sciences*, vol. 15, pp. 24–31, 2010.

[DEL 15] DELIGNIÈRES D., "Systèmes dynamiques et morphogénèse", Lecture notes, available at: http://didier.delignieres.perso.sfr.fr/Supports-doc/systemes_dynamiques_et_morphogenese.pdf, 2015.

[DEM 14] DEMONGEOT J., HANSEN O., HAMIE A. *et al.*, "Dynalets: a new method for modelling and compressing biological signals. Applications to physiological and molecular signals", *Comptes Rendus Biologies*, vol. 337, pp. 609–624, 2014.

[DER 75] DE ROSNAY J., *Le Macroscope*, Le Seuil, Paris, 1975.

[DES 06] DESMEULLES G., Réification des interactions pour l'expérience in virtuo de systèmes biologiques multi-modèles, Thesis, University of Western Brittany, Brest, available at: https://hal.inria.fr/file/index/docid/142697/filename/these_gd.pdf, 2006.

[DHO 06] DHOMBRES J., KREMER-MARIETTI A., *L'épistémologie: état des lieux et positions*, Ellipses, Paris, 2006.

[DIM 06] DI MARZO SERUGENDO G., GLEIZES M.P., KARAGEORGOS A., "Self-organization and emergence in MAS: an overview", *Informatica*, vol. 30, pp. 45–54, 2006.

[DON 03] DONNADIEU G., DURAND D., NÉEL D. *et al.*, L'approche systémique; de quoi s'agit-il?, summary of the work of the AFSCET Group, available at: http://www.afscet.asso.fr/SystemicApproach.pdf, 2003.

[DOR 10] DORTIER J.F., "L'étonnant Monsieur Chapouthier et sa théorie de la mosaïque", available at: http://www.dortier.fr/letonnant-monsieur-chapouthier-et-sa-theorie-de-la-mosaique/, 2010.

[DRE 10] DREW E., *Symposium on Opportunities and Challenges in the Emerging Field of Synthetic Biology*, OECD and Royal Soc. Ed., available at: https://www.oecd.org/sti/biotech/45144066.pdf, 2010.

[DUF 11] DUFOURMENTELLE A., *Eloge du risque*, Payot, Paris, 2011.

[DUR 12] DURRIVE B., FAURY M., HENRY J., "Réflexivité et dialogue interdisciplinaire: un retour sur soi selon l'autre", *Symposium Proceedings, Réflexivité en contexte de diversité: un carrefour des sciences humaines*, Limoges, France, December 1–2 2012, available at: https://halshs.archives-ouvertes.fr/hal-00671217/document, pp. 1–13, 2012.

[DUR 13] DURAND D., *La systémique*, PUF, Que sais-je, Paris, 2013.

[ENA 16] ENAP, Intelligence collective: comment s'assurer que les apprentissages servent aux organisations, available at: http://servicesauxorganisations.enap.ca/fr/lectureNouvelle.aspx?idNews=27550, 2016.

[FER 02] FERRARI J., *Aleph zero*, Babel, Paris, 2002.

[FLE 08] FLECK L., *Génèse et développement d'un fait scientifique*, Champs-Sciences, Paris, 2008.

[FLE 12] FLENNER E., JANOSI L., BARZ B. *et al.*, "Kinetic Monte Carlo and cellular particle dynamics simulations of multicellular systems", *Physical Review*, vol. 85, 031907, 2012.

[FOU 16] FOUREL L., VALAT A., FAUROBERT E. *et al.*, "β3 integrin–mediated spreading induced by matrix-bound BMP-2 controls Smad signaling in a stiffness-independent manner", *The Journal of Cell Biology*, vol. 212, pp. 693–706, 2016. doi: 10.1083/jcb.201508018 and CNRS: available at: http://www.cnrs.fr/inc/communication/direct_labos/picart.htm.

[FOX 09] FOX-KELLER E., "Organisms, machines, and thunderstorms, a history of self-organization", *Historical Studies in the Natural Sciences*, vol. 39, pp. 38–39, 2009.

[FRA 97] FRANK A., GOUYON P.H., "Information et complexité: questions sans réponses", *La Recherche*, vol. 296, pp. 106–109, 1997.

[FRO 04] FROMM J., *The Emergence of Complexity*, Kassel University Press, Kassel, 2004.

[GAL 01] GALLOPIN G.C., FUNTOWICZ S., O'CONNOR M. *et al.*, "La science pour le XXIème siècle: du contrat social aux fondements scientifiques", *Revue Internationale des Sciences Sociales*, vol. 168, pp. 239–250, 2001.

[GAS 10] GASCUEL O., Modélisation des systèmes biologiques, bio-informatique, Section 43 CNRS, available at: http://www.cnrs.fr/comitenational/doc/rapport/2010/43_conj_2010.pdf, 2010.

[GIU 11] GIULIANI A., LICATA I., MODONESI C.M. *et al.*, "What is artificial about life?", *The Scientific World Journal*, vol. 11, pp. 651–654, 2011. doi: 10.1100/tsw.2011.73.

[GLE 08] GLEICK J., *La théorie du chaos*, Champ Sciences, Paris, 2008.

[GOD 04] GODIN B., "L'organisation innovante: D'un système d'indicateurs appropriés", CSIIC, available at: http://collections.banq.qc.ca/ark:/52327/bs64427, 2004.

[GOD 11] GODIN B., Innovation and Politics: The Controversy on Republicanism in Seventeenth-Century England, Working Paper No. 10, INRS, Montreal, 2011.

[GOD 12] GODIN B., Social Innovation: Utopias of Innovation from c. 1830 to the Present, Working Paper No. 11, INRS, Montreal, 2012.

[GOD 15] GODIN B., Innovation: a conceptual history of an anonymous concept, working paper no. 21, INRS, Montreal, available at: http://www.csiic.ca/PDF/WorkingPaper21.pdf, 2015.

[GRE 17] GREEN S., BATTERMAN R., "Biology meets physics: reductionism and multi-scale modeling of morphogenesis", *Studies in History and Philosophy of Science Part C: Studies in History and Philosophy of Biological and Biomedical Sciences*, vol. 61, pp. 20–34, 2017.

[GRO 15] GROJEAN O., "Sensibilisation et coercition; à propos de quelques questions relatives au façonnage institutionnel des individus", in TAÏNI C. (ed.), *Emotions et expertises; les modes de coordination des actions collectives*, Presses Universitaires de Rennes, Rennes, pp. 73–92, 2015.

[GUI 10] GUILLEMOT F., MIRONOV V., NAKAMURA M., "Bioprinting is coming of age: report from the International Conference on Bioprinting and Biofabrication in Bordeaux", *Biofabrication*, vol. 2, 010201, 2010.

[GUI 15] GUILLIN V., "Scientisme et politique", in PRUD'HOMME J., DORAY P., BOUCHARD F. (eds), *Sciences, technologies et sociétés de A à Z*, Presses de l'Université de Montréal, Canada, 2015.

[HEA 13] HEARD E., *Epigénétique et mémoire cellulaire*, Fayard, Paris, 2013.

[HEF 16] HEFCE, Higher Education Funding Council for England, Interdisciplinary Research Policy and Practice Report, available at: http://www.hefce.ac.uk/media/HEFCE,2014/Content/Events/2016/Interdisciplinary,research,Policy,and,practice/IDR_conference_report.pdf, 2016.

[HEI 98] HEISENBERG W., *Philosophie: le livre de 1942*, Le Seuil, Paris, 1942/1998.

[HEU 98] HEUDIN J.C., *L'évolution au bord du chaos*, Hermès, Paris, 1998.

[HIR 05] HIRSCH J.E., "An index to quantify an individual's scientific research output", *Proceedings of the National Academy of Sciences*, vol. 102, pp. 16569–16572, 2005.

[IBD 16] IBDM, Approches physiques de la dynamique cellulaire et de la morphogénèse des tissus, Report, available at: http://www.ibdm.univ-mrs.fr/fr/equipe/approches-physiques-de-la-dynamique-cellulaire/, 2016.

[IES 16] IESE, Relever les défis d'une économie prospère et responsable, available at: http://home.iesf.fr/752_p_43968/parution-des-propositions-des-iesf.html, 2016.

[IZA 04] IZAGUIRRE J.A., CHATURVED R., HUANG C. *et al.*, "Compucell, a multi-model framework for simulation of morphogenesis", *Bioinformatics*, vol. 20, pp. 1129–1137, 2004.

[JAC 70] JACOB F., *La logique du vivant*, Gallimard, Paris, 1970.

[JUR 13] JURA J., BILA J., "Model of cognitive functions for description of the creative design process with computer support: improving of the interpretation method for the computer conceptual re-design", *ISCS 2013: Interdisciplinary Symposium on Complex Systems*, vol. 8, pp. 163–171, 2013.

[KAC 71] KACHALSKI A., "Biological flow-structures and their relation to chemico-diffusional coupling", *Neurosciences Research Program Bulletin*, vol. 9, pp. 397–413, 1971.

[KAE 05] KAERN M., ELSTON T.C., BLAKE W.J. *et al.*, "Stochasticity in gene expression: from theory to pheno-types", *Nature Reviews Genetics*, vol. 6, pp. 451–464, 2005.

[KAL 12] KALIL T., Implementation of Federal price authority: Progress report, US Office of Science and Technology Policy, available at: https://www.nasa.gov/sites/default/files/competes_report_on_prizes_final.pdf, 2012.

[KAN 85] KANT E., *Prolégomènes à toute métaphysique future*, Gallimard, Paris, 1985.

[KIR 13] KIRSCHNER C.M., ANSETH K.S., "Hydrogels in healthcare: from static to dynamic material micro-environments", *Acta Materalia*, vol. 61, pp. 931–944, 2013.

[KOS 12] KOSZTIN I., VUNJAK-NOVAKOVIC G., FORGACS G., "Colloquium: modeling the dynamics of multicellular systems: application to tissue engineering", *Reviews of Modern Physics*, vol. 84, pp. 1791–1805, 2012.

[KUH 62] KUHN T., *The Structure of Scientific Revolutions*, University Chicago Press, Chicago, 1962.

[KUN 86] KUNDERA M., *L'art du roman*, Gallimard, Paris, 1986.

[KUP 98] KUPIEC J.J., "Un mécanisme de hasard-sélection pourrait expliquer la différenciation de plusieurs types cellulaires", *Médecine/Sciences*, vol. 14, pp. 115–121, 1998.

[KUP 08] KUPIEC J.J., *L'origine des individus*, Fayard, Paris, 2008.

[KUP 09] KUPIEC J.J., "Une approche darwinienne de l'ontogenèse", in HEAMS T., HUNEMAN P., LECOINTRE G. *et al.* (eds), *Les Mondes Darwiniens*, 2009.

[KUP 13] KUPIEC J.J. (ed.), *La vie, et alors? – Débats passionnés d'hier et d'aujourd'hui*, Belin, Paris, 2013.

[LAP 16] LAPOINTE J., L'approche systémique et la technologie de l'éducation, Unpublished paper, Yves Noblet, available at: http://www.sites.fse.ulaval.ca/reveduc/html/vol1/no1/apsyst.html, 2016.

[LAR 17] LARRÈRE C., LARRÈRE R., *Bulles Technologiques*, Wildproject, Paris, 2017.

[LAU 13] LAURENT M., *Systèmes biologiques à dynamique non linéaire*, Ellipses, Paris, 2013.

[LEG 02] LEGRAND P., "De quelle discipline relève l'étude des objets environnementaux?", in KOURILSKY F. (ed.), *Ingénierie de l'interdisciplinarité*, L'Harmattan, Paris, 2002.

[LEG 08] LE GUYADER H., "Morphogenèse, structures physiques et évolution biologique", *Comptes Rendus Chimie*, vol. 11, pp. 186–191, 2008.

[LEJ 17] LEJEUNE E., LINDER C., "Quantifying the relationship between cell division angle and morphogenesis through computational modeling", *Journal of Theoretical Biology*, vol. 418, pp. 1–7, 2017.

[LEM 99] LE MOIGNE J.L., *La modélisation des systèmes complexes*, Dunod, Paris, 1999.

[LEM 04] LE MOIGNE J.L., "Quoi de plus naturel que les sciences de l'artificiel?", in SIMON H.A. (ed.), *Les sciences de l'artificiel*, Gallimard, Paris, 2004.

[LEM 06] LE MOIGNE J.L., *La théorie du système général*, Presses Universitaires de France, 2006.

[LEM 06] LEMYRE D., DESPATIS F., L'organisation apprenante démystifiée, available at: http://slideplayer.fr/slide/1775809/, 2006.

[LEN 17] LENTINI R., MARTÍN N.Y., FORLIN M. *et al.*, "Two-way chemical communication between artificial and natural cells", *ACS Central Science*, vol. 3, pp. 17–123, 2017.

[LES 82] LESOURNE J., *La notion de systèmes dans les sciences contemporaines*, Librairie de l'Université, Aix en Provence, 1982.

[LEV 10] LEVIN J., *Un fou rêve de machine de Turing*, Markus Haller, Geneva, 2010.

[LEV 15] LÉVY J.C.S. (ed.), *Complexité et désordre; éléments de réflexion*, EDP Sciences, Paris, 2015.

[LIC 08] LICATA I., "Vision as adaptive epistemology", *Proceedings of the 5th National Conference on System Science*, Marche Polytechnic University, Italy, 14–16 October, available at: https://arxiv.org/ftp/arxiv/papers/0812/0812.0115.pdf, 2008.

[LIC 10] LICATA I., "Living with radical uncertainty: the exemplary case of folding protein", in LICATA I., SAKAJI A. (eds), *Crossing in Complexity*, Nova Science, New York, 2010.

[LIC 16] LICATA I., "Les règles du jeu sont simples, mais les règles ne sont pas le jeu", in LÉVY-LEBLOND J.M. (ed.), *Lettres à Alan Turing*, T. Marchaise, Vincennes, 2016.

[LOM 05] LOMBRY T., *Un siècle de physique*, Aegus, Paris, 2005.

[MAN 73] MANDELBROT B., *Les objets fractals: forme, hasard et dimensions*, Flammarion, Paris, 1973.

[MAN 08] MANCION T., Modélisation de la complexité et de la dynamique des simulations multi-agents 2005 application pour l'analyse des phénomènes émergents, Thesis, University of Evry Val d'Essonne, Evry, 2008.

[MAO 15] MAO A.S., MOONEY D.J., "Regenerative medicine: current therapies and future directions", *Proceedings of the National Academy of Sciences*, vol. 112, pp. 14452–14459, 2015.

[MAR 10] MARTINS M.L., FERREIRA S.C., VILELA M.J., "Multiscale models for biological systems", *Current Opinion in Colloid & Interface Sciences*, vol. 15, pp. 8–23, 2010.

[MAT 70] MATURANA H., Biology of cognition, Report, Biological Computer laboratory, Urbana, 1970.

[MAT 14] MATHIEU N., SCHMID A.F., "Reconsidérer le lien entre modélisation et interdisciplinarité", in MATHIEU N., SCHMID A.F. (eds), *Modélisation et interdisciplinarité; six disciplines en quête d'épistémologie*, Quae, Paris, 2014.

[MCC 65] MCCULLOCH W.S., *Embodiments of Minds*, MIT Press, Cambridge, 1965.

[MCC 14] MCCUNE M., SHAFIEE A., FORGACS G. *et al.*, "Predictive modeling of post bioprinting structure formation", *Soft Matter*, vol. 10, pp. 1790–1800, 2014.

[MER 09] MERKS R.M.H., KOOLWIJK P., "Modeling morphogenesis in silico and in vitro: towards quantitative, predictive, cell-based modeling", *Mathematical Modelling of Natural Phenomena*, vol. 4, pp. 149–171, 2009.

[MIS 11] MISBAH C., *Dynamiques complexes et morphogénèse; introduction aux sciences non linéaires*, Springer-Verlag, Paris, 2011.

[MOL 12] MOLIER P., La biologie de synthèse face à la complexité du vivant, Report, INRA, available at: http://www.inra.fr/Chercheurs-etudiants/Economie-et-sciences-sociales/Tous-les-dossiers/biologie-de-synthese-et-sciences-sociales/La-biologie-de-synthese-face-a-la-complexite-du-vivant/(key)/1, 2012.

[MOM 05] MOMBACH C.M., ROBERT D., GRANER F. *et al.*, "Rounding of aggregates of biological cells: experiments and simulations", *Physica A*, vol. 352, pp. 525–534, 2005.

[MON 08] MONCION T., Modélisation de la complexité et de la dynamique des simulations multi-agents: application pour l'analyse des phénomènes émergents, PhD Thesis, University of Evry Val d'Essonne, Evry, 2008.

[MON 14] MONOD J., *Le hasard et la nécessité: Essai sur la philosophie naturelle de la biologie moderne*, Points Essais, Paris, 2014.

[MOR 74] MORIN E. "L'émergence d'un nouveau mode de pensée", *Revue Hermès*, pp. 86–92, 1974.

[MOR 95] MORIN E., *La Méthode – Les Idées. Leur habitat, leur vie, leurs moeurs, leur organization*, Le Seuil, Paris, 1995.

[MOR 04] MORENO A., "Auto-organisation, autonomie et identité", *Revue internationale de philosophie*, vol. 2, no. 228, pp. 135–150, 2004.

[MOR 12] MORANGE M., *Les secrets du vivant: contre la pensée unique en biologie*, La Découverte, Paris, 2012.

[MOR 13] MORANGE M., "Comment la biologie est-elle devenue moléculaire?", in KUPIEC J.J. (ed.), *La vie, et alors? – Débats passionnés d'hier et d'aujourd'hui*, Belin, Paris, 2013.

[MUG 08] MUGNIER L., "Introduction aux problèmes inverses: application à l'imagerie optique à haute resolution en astronomie", in LÉNA P. (ed.), *Des données à l'objet: le problème inverse*, EDP Sciences/CNRS, Paris, 2008.

[MUN 15] MUNJAI A., PHILIPPE J.M., MUNRO E. *et al.*, "A self organized biomechanical network drives shape changes during tissue morphogenesis", *Nature*, vol. 524, pp. 351–355, 2015.

[MYO 11] MYON L., FERRI J., CHAI F. *et al.*, "Ingénierie du tissu osseux oro-maxillofacial par combinaison de biomatériaux, cellules souches, thérapie génique", *Revue de Stomatologie et de Chirurgie Maxillo-faciale*, vol. 112, pp. 201–211, 2011.

[NAD 99] NADEAU R., *Vocabulaire technique et analytique de l'épistémologie*, PUF, Paris, 1999.

[NAD 16] NADEAU R. (ed.), *Philosophies de la connaissance*, Presses Universitaires de Montréal, Montreal, 2016.

[NEA 06] NEAGU A., KOSZTIN I., JAKAB K. *et al.*, "Computational modeling of tissue self-assembly", *Modern Physics Letters B*, vol. 20, pp. 1217–1231, 2006.

[NIC 77] NICOLIS G., PRIGOGINE I., *Self-organization in Non-equilibrium Systems; from Dissipative Structures to Order through Fluctuations*, John Wiley & Sons, New York, 1977.

[NIC 02] NICOLESCU B., "Niveaux de réalité", in D'ESPAGNAT B. (ed.), *Implications philosophiques de la science contemporaine*, PUF, Paris, 2002.

[NIE 07] NIETZSCHE F., *Le gai savoir*, Livre de Poche, Paris, 2007.

[NON 00] NONGA HONLA J., LE MOIGNE J.-L., La modélisation des systèmes complexes, Lecture notes, CNAM, available at: http://mip-ms.cnam.fr/servlet/com.univ.collaboratif.utils.LectureFichiergw?ID_FICHIER=1295877018078, 2000.

[NOT 08] NOTTALE L., "Scale relativity, non-differentiability and fractal time-space", in ABRY P., GONÇALVÈS J., LÉVY VEHEL M. (eds), *Scaling, Fractals and Wavelets*, ISTE, London and John Wiley & Sons, New York, 2008.

[OKU 16] OKUDA S., INOUE Y., EIRAKU M. *et al.*, "Modeling cell apoptosis for simulating three-dimensional multicellular morphogenesis based on a reversible network reconnection framework", *Biomechanics and Modeling in Mechanobiology*, vol. 15, pp. 805–816, 2016.

[OLL 16] OLLE-VILA A., DURAN-NEBREDA S., CONDE-PUEYO N. *et al.*, "A morphospace for synthetic organs and organoids: the possible and the actual", *Integrative Biology*, 2016.

[OTT 04] OTTINO J.M., "Complex systems", *AIChE Journal*, vol. 49, pp. 292–299, 2004.

[PAR 65] PARETO V., *Les systèmes socialistes*, Droz, Geneva, 1965.

[PAV 06] PAVÉ A., "By way of introduction: modelling living systems, their diversity and their complexity: some methodological and theoretical problems", *Comptes Rendus Biologies*, vol. 329, pp. 3–12, 2006.

[PAV 12] PAVÉ A., *Modeling Living Systems*, ISTE, London and John Wiley & Sons, New York, 2012.

[PEA 96] PEAT F.D., *Infinite potential – the life and times of David Bohm*, Helix Books, New York, 1996.

[PER 03] PERRIER P., "Modélisation et complexité", in D'ESPAGNAT B. (ed.), *Implications philosophiques de la science contemporaine*, PUF, 2003.

[PIC 15] PICTET R., "Complexité, ordre et hasard en biologie: le cas de l'évolution", in LÉVY J.C. (ed.), *Complexité et désordre; éléments de réflexion*, EDP Sciences, Paris, 2015.

[PRI 93] PRIGOGINE I., *Les lois du chaos*, Flammarion, Paris, 1993.

[PRI 09] PRIGOGINE I., STENGERS I., *Le temps et l'éternité*, Champs Sciences, Paris, 2009.

[PRO 03] PROULX S., "Heinz Von Foerster (1911–2002) – Le père de la seconde cybernétique", *Hermès*, vol. 27, pp. 253–260, 2003.

[PRO 12] PROCHIANTZ A., *Qu'est-ce que le vivant?*, Le Seuil, Paris, 2012.

[RAM 88] RAMOS-MACDONADO M., Vers un pilotage auto-organisant de systèmes flexibles de fabrication, PhD Thesis, Nancy-Université, France, 1988.

[RAM 08] RAMASUBRAMANIAN A., TABER L.A., "Computational modeling of morphogenesis regulated by mechanical feedback", *Biomechanics and Modeling in Mechanobiology*, vol. 7, pp. 77–91, 2008.

[RAS 54] RASHEVSKY N., "Topology and life: in search of general mathematical principles in biology and sociology", *The Bulletin of Mathematical Biophysics*, vol. 16, pp. 317–348, 1954.

[REV 12] REVAULT D'ALLONNES M., *La crise sans fin – essai sur l'expérience moderne du temps*, Le Seuil, Paris, 2012.

[RIC 03] RICARD J., "Complexité, émergence, information et causalité dans les systèmes biologiques", in D'ESPAGNAT B. (ed.), *Implications philosophiques de la science contemporaine*, PUF, Paris, 2003.

[RIC 16] RICHARD A., BOULLU L., HERBACH U. *et al.*, "Single-cell-based analysis highlights a surge in cell-to-cell molecular variability preceding irreversible commitment in a differentiation process", *PLoS Biology*, vol. 14, e1002585, 2016. doi: 10.1371/journal.pbio.1002585, available at: https://www.ncbi.nlm.nih.gov/pmc/articles/PMC5191835/.

[RIE 88] RIEDL R., "Les conséquences de la pensée causale", in WATZLAWICK P. (ed.), *L'invention de la réalité – Contributions au constructivisme*, Le Seuil, Paris, 1988.

[ROS 58] ROSEN R., "The representation of biological systems from the standpoint of the theory of categories", *The Bulletin of Mathematical Biophysics*, vol. 20, p. 25, 1958.

[ROS 85] ROSENBERG A., *The Structure of Biological Science*, Cambridge University Press, Cambridge, 1985.

[ROS 14] ROSEN D.W., "Research supporting principles for design for additive manufacturing", *Virtual & Physical Prototyping*, vol. 9, pp. 225–232, 2014.

[ROT 62] ROTHSTEIN J., "Information and organization as the language of the operational viewpoint", *The Philosophy of Science*, vol. 29, pp. 406–411, 1962.

[SAA 96] SAATDJIAN E., MIDOUX N., GASTOU-CHASSAING M.I. *et al.*, "Chaotic mixing and heat transfer between confocal ellipses: experimental and numerical results", *Physics of Fluids*, vol. 8, pp. 677–691, 1996.

[SCH 10] SCHMID A.F., "Epistémologie et éthique technologique", Conférence au Centre Franco-russe de Moscou, 20 September, 2010.

[SEV 05] SÈVE L., "De quelle culture logico-philosophique la pensée du non-linéaire a-t-elle besoin?", in GUESPIN-MICHEL J. (ed.), *Emergence, complexité et dialectique*, Odile Jacob, Paris, 2005.

[SHE 03] SHELDRAKE R., *Une nouvelle science de la vie*, Editions du Rocher, Monaco, 2003.

[SPE 08] SPERBER F., PARIS R., "Qu'est-ce que le chaos déterministe?", available at: http://www.matierevolution.fr/spip.php?article474, 2008.

[STA 15] STAUNE J., *Les clés du futur; réinventer ensemble la société, l'économie et la science*, Plon, Paris, 2015.

[SUN 13] SUN Y., WANG Q., "Modeling and simulations of multicellular aggregate self-assembly in bio-fabrication using kinetic Monte Carlo methods", *Soft Matter*, vol. 9, pp. 2172–2186, 2013.

[SUN 17] SUN M., ZAMAN M.H., "Advanced review modeling, signaling and cytoskeleton dynamics: integrated modeling-experimental frameworks in cell migration", *WIREs Systems Biology and Medicine*, vol. 9, e1365, 2017.

[TEO 11] TEODORANI M., *David Bohm – la physique de l'infini*, Macro, Cesena, 2011.

[THE 95] THÉRAULAZ G., BONABEAU E., "Essaims sur réseaux et modélisation des processus de construction collective chez les insectes sociaux", in THÉRAULAZ G. (ed.), *Evolution et organisation; hasard et contraintes dans la genèse des formes collectives*, Journées de Rochebrune and ENST, Paris, 1995.

[THO 72] THOM R., *Stabilité structurelle et morphogenèse*, Inter Editions, Paris, 1972.

[THO 83] THOM R., *Paraboles et catastrophes*, Flammarion, Paris, 1983.

[THO 89] THOM R., *Structural Stability and Morphogenesis: an Outline of a General Theory of Models*, Addison-Wesley, Reading, 1989.

[THO 04] THOMAS R., BASIOSA V., EISWIRTH M. *et al.*, "Hyperchaos of arbitrary order generated by a single feedback circuit, and the emergence of chaotic walks", *Chaos: An Interdisciplinary Journal of Nonlinear Science*, vol. 14, p. 699, 2004.

[THO 14] THOMAS G.L., MIRONOV V., NAGY-MEHEZ A. *et al.*, "Dynamics of cell aggregates fusion: experiments and simulations", *Physica A: Statistical Mechanics and its Applications*, vol. 395, pp. 247–254, 2014.

[TIN 94] TINLAND F., "Le site de la technique: éclairages théoriques et enjeux pratiques", in TINLAND F. (ed.), *Ordre Biologique; Ordre Technologique*, Champ-Vallon, Paris, 1994.

[TOM 06] TOMASI DI LAMPEDUSA G., *Le guépard*, Points, Paris, 2006.

[TRI 11] TRIPOLI S., Etude de l'auto-organisation des cellules basées sur le modèle de Potts cellulaire, Thesis, Université de Bretagne Occidentale, Brest, available at: https://www.google.fr/url?sa=t&rct=j&q=&esrc=s&source=web&cd=2&cad=rja&uact=8&ved=0ahUKEwissdLu_JHOAhXIvBoKHYT6AUwQFggiMAE&url=http%3A%2F%2Flabsticc.univ-brest.fr%2F~rodin%2FFTP%2FRecherche%2FTheses_BiologieVirtuelle%2FtheseSebastienTripodi.pdf&usg=AFQjCNELo9sQqVHn12bc7yshywyfs2EJhw, 2011.

[TUR 52] TURING A., "The chemical basis of morphogenesis", *Philosophical Transactions of the Royal Society of London*, vol. B237, pp. 37–72, 1952.

[VAN 12] VANDEGINSTE P., "Modéliser le vivant", La Recherche, vol. 464, pp. 47–49, 2012.

[VAR 07] VARENNE F., *Du modèle à la simulation*, Vrin, Paris, 2007.

[VAR 13] VARENNE F., "Théorie mathématique des catégories en biologie et notion d'équivalence naturelle chez Robert Rosen", *Revue Histoire des Sciences*, vol. 66, pp. 167–197, 2013.

[VIN 00] VINCK D., *Pratiques de l'interdisciplinarité: mutations des sciences, de l'industrie et de l'enseignement*, Presses Universitaires de Grenoble, Grenoble, 2000.

[VIN 05] VINCK D., Regard sociologique sur les pratiques interdisciplinaires, available at: http://slideplayer.fr/slide/1656600/, 2005.

[VON 60] VON FOERSTER H., "Sur les systèmes auto-organisateurs et leurs environnements", in YOVITS M.C., CAMERON S. (eds), *Self-organizing Systems*, Pergamon Press, London, 1960.

[VON 12] VON BERTALANFY L., *Théorie générale des systèmes*, Dunod, Paris, 2012.

[WAT 88] WATZLAWICK P., "Avec quoi construit-on des réalités idéologiques?", in WATZLAWICK P. (ed.), *L'invention de la réalité – Contributions au constructivisme*, Le Seuil, Paris, 1988.

[WAT 91] WATZLAWICK P., *Les cheveux du baron de Münchhausen*; *psychothérapie et 'réalité'*, Le Seuil, Paris, 1991.

[WOL 84] WOLFRAM S., "Cellular automata as models of complexity", *Nature*, vol. 311, pp. 419–424, 1984.

[YAT 12] YATCHINOVSKY A., *L'approche systémique; pour gérer l'incertitude et la complexité*, ESF, Issy-les-Moulineaux, 2012.

[ZIN 03] ZIN J., "La complexité et son idéologie", available at: https://jeanzin.fr/ecorevo/sciences/complexi.htm, 2003.

Conclusion

Interdisciplinarity in science, like additive manufacturing in society, is "in", so rather than rejecting or embracing it *a priori*, let us question its pertinence, its potential and its cross-fertilizations, for beyond being a trend, interdisciplinarity is above all else a reality faced by a large number of researchers working on boundary-objects (like 3D printing) at some point or another in their in-depth scientific work. Furthermore, beyond the simple use of results from different disciplines (toolbox syndrome), where knowledge is effectively mutualized through interdisciplinarity (questions of convergence, mutualization, systemic approaches), Volumes 1–3 have attempted to engage the debate applied to the field of additive manufacturing on forms of dialogue that may (must) be established, concretely between the 3D boundary-object and the different disciplines that compete in its development.

Interdisciplinarity often results in a display or strong constraint, as shown in this book in the particular case of bio-printing, and for a whole range of difficulties that have been presented, considering the time, we do the best possible with the scientific and technological skill of the researchers involved in this action. Yet, at the initiation of "historic" processes in 1984, it was (still) possible to function in the way of a craftsman, with forms of knowledgeable DIY (Do-It-Yourself) in a simply "bottom-up" logic. The initial proof of concept in stereolithography was performed in a few days, the extension of a demonstrator was a bit longer, approximately a year, but with researchers and students excited by the magical aspect of the process, it was not necessary to seek outside support. It is roughly the same for the six other "basic" technologies of 3D printing that managed to emerge with the applicational results presented in detail in Volume 1 and incrementally numerous innovations in Volume 2. With a consequent market (in the neighborhood of 10 billion €/year) and a progression of the order of 20%/year, the dynamic is largely in favor of the development of 3D technologies with more and more consequent application markets.

What has been tempted is a will to question one's own scientific practice by exemplifying the use of interdisciplinarity and the associated methodological choices, the need for dialogue between those concerned by the field of 3D printing, particularly in emerging domains. There are perhaps as many philosophers as people who attempt to think, which stems from the principles of "unique thought", inertias with different origins, and involvement in the consensus induced by paradigms (whose interest in terms of efficiency must absolutely not be rejected). If specialization has played a major role in the progress of science, does it not, in the field concerned by these three volumes, lead to a certain fragility? Then, returning to the basics, perhaps a significant cultural advantage to participate responsibly in the robust construction of our futures, near and far, in a world led at all levels to quantitative reductionism, particularly when the social and economic demands are clearly high.

To achieve these ambitious goals, as may be the case of bio-printing, it is necessary to change methods, accept exchange, failure and questioning, because the interdisciplinarity called for is profound, complex and thus difficult. So long as it does not voluntarily seek to confront "recalcitrant" problems, the field will remain alive, so long as it keeps "acceptable" contact with the anticipation of society's needs (by different means), which poses the delicate question of voluntarily breaking away from habits perpetuated by the "system", its social organization, its ends (a bit like in the cartoons by Tex Avery, where the hero continues to run over the chasm before falling). The question of unanticipated *post-mortem* survival towards the opening of new fields of research is, however, posed in a world that too strongly supports *a posteriori* risk-taking, creativity and breakthroughs, but which recovers them with ease.

Yet, what is expected in research is risk-taking in disruptive terms and not a presentation of comfortably seated professionals who satisfy all of financers' criteria for incremental approaches (absence of risk-taking in particular, it is enough to be "good"). Beyond crippling incoherences (e.g. a common thesis with researchers from different cultures located several thousand kilometers from one another), finally satisfying the administrative criteria, it is still possible to share a bit of money with "milestones" regulated to the millimeter. Everyone is happy, except those in creative research.

For some years now, this situation has become rather critical, however, because in a Malthusian context, the population of researchers varies exponentially while the resource varies linearly (or is even stable). To come out unscathed in a system that creates starvation, we either perform rather costly research on the computer or we accept the rules of the game. However, we are increasing the number of experts with considerable linear losses to the extent that we can sometimes ask ourselves the question of the effective expertise of said evaluators with lower and lower success rates at the end of the day, associated with a considerable loss of time.

With a system that finances "from the top down", there is no longer a real transfer of responsibility in research and instead there is an inability to propose modifications for human activity allowing defuse debate, to avoid by reinventing our relational modes of forms of resignation. Retaking the anticipatory role (but nothing more) of research could likely prevent reductive discourse on the impotence of national society to escape crisis, because we are stuck in the same habits of not thinking. The present situation illustrated in these three volumes should force us to challenge ourselves, to revisit our points of view.

Just as stagecoach drivers did not invent the airplane, future projection (in three dimensions!), which is a real barrier, cannot take place using extrapolations from the past nor simple planning for the future. One solution implies being able to distance ourselves to propose links between science and society based on reinforcement paths. The link between policies should then concern the flexible framework of the movement in which society must be integrated to reorient growth and/or employment, not the optimal form of a responsible research activity at all levels.

Additive manufacturing corresponds in an unarguable way to a true market boost for research and R&D environments. The offered domain can serve as a laboratory for experimenting on original methods of operation, so long as research, abundant considering the economic market, can be profitable to the field; but for this, we must accept to operate differently, or skill will be elsewhere.

For some scientists, the complexity of the world may be an insurmountable obstacle through simplifications of reputedly complex situations or reduction of these to the analysis of their elementary components (disciplinary approach). However, the number of researchers knowing how to bring them together and capable of criticizing them and then drawing legitimate conclusions from them through logical reasoning based on all previous experiments, personal or not, is probably modest. How then can disciplinary scientists manage heterogeneous networks – both internal and external – that cross the boundaries between science, Society, economy and policy (in the broad sense) and that can have the property of being highly complex, even if they are harbingers of the future?

That is a (good) question!

We will soon be able to revisit this question and see whether society has woken up… We will get there; additive manufacturing deserves it!

Postface

Tomorrow and the Future?

"Faced with the future, men have a choice between four attitudes: the passive ostrich who undergoes change, the reactive firefighter who expects someone to yell fire so it can be fought, the pre-active insurance salesman who is prepared for foreseeable changes because he knows that repairs are more expensive than prevention, and finally, the proactive conspirator who acts to provoke the desired changes". [GOD 01]

"The moment capitalist societies undergo the dominating tropism of profit and socialist societies remain tributes to bureaucracy, we could sense that the French – allergic to all options – might acquire both defects. That is indeed what they are doing". [ELG 66]

"By daring to imagine the future, humanity is balancing between pleasures and shuddering. As Science and Technologies take significant steps, Man immediately builds barriers out of a fear of being surpassed". [DE 05]

"What events, what innovations will remain free of consequence, what others are likely to affect the global regime, to irreversibly determine the choice of an evolutionary path, what are the areas of stability?". [PRI 90]

> *"Scenarios allow the mental horizons to be broadened and a greater openness of the mind towards new knowledge to be developed. They are multidimensional [...]. They constitute an interdisciplinary and multicultural exercise". [BAR 00]*

> *"We live on ideas, moralities, sociologies, philosophies, and a psychology that belong in the 19th century. We are our own great-grandparents". [PAU 72]*

> *"The threats that attest to the fragility of personal or collective identity are not illusory: it is remarkable that the ideologies of power are beginning, with worrying success, to manipulate these fragile identities through symbolic mediations of action". [RIC 04]*

> *"The individual is no longer under the strong dependence of traditions and ideologies that apparently impose themselves upon him/her. In our societies with an individualistic nature, the actor tends to determine more and more often the values on which he/she intends to base his/her existence on his/her own". [LE 97]*

> *"Saint Augustine had included curiosity in the 'catalogue of vices' and heresies on the grounds that Man's interest for the world must be limited to its use in views of safety. Pleasure must only be connected to the relationship he maintains with God". [BLU 99]*

> *"Understanding the world, like Cervantes, as ambiguity, having to face a wealth of relative truths that contradict themselves rather than a single absolute truth [...], thus having only the certainty that the wisdom of uncertainty demands no less great a force". [KUN 86]*

> *"First of all, you will learn that thinking too quickly could indeed be a very bad idea. Infobesity attacks us from all sides and it is a serious blow to our creativity". [LEW 16]*

Micheau [MIC 09] writes, "A legislator, supreme engineer, or senior technocrat, a central committee or administrative council of engineers or technocrats defines social needs, functions to respond to these needs, and the principles of dividing populations meant to occupy these functions". In these conditions, we find writings by Bertrand Russel (Scientific Outlook of the World) or Aldous Huxley (Brave New World) describing a scientifically organized society, but the idea of generalized meritocracy stumbles over difficulties linked to the complexity of research systems, boundary objects, the absence of a good prospective, particularly in fields where initiative may be great, and finally, on the rarity of talents (and the difficulty of bringing them out) in a profoundly massified world of research.

In this décor, there may be several versions of social utility:

– either it is a matter of benefitting from all the innovations, inventions and progress – particularly technological progress – skills from an elite of enlighteners who can, in other respects, be considered atypical, even dissidents relative to the ordinary classifications applied in the world of academic research;

– or it is necessary for a majority of the scientific environment to get involved in large action programs to benefit the same society. In this latter case, forms of innovation result from a more top-down approach and leave far less space for individual initiative.

In reality, there is probably no scientific creation without the cooperation from the principle of profusion and freedom, imagination and disorder relative to the "given rules" and rigorous principles, methods, regulated tidying up, criticism and robust validation. The two methods seem necessary and naturally must "work" together, which may limit the efficiency of an overly closed strategy and induce paradoxical injunctions of "harmonious" operations of research.

But what is left to ensure our viability if not inventing more new technologies to resolve problems created by our ancestors as they arise? The principle of creating freedom in all common institutions has been replaced by the direct influence of technologies, new habits concerning interactions that we have with ourselves and Nature. This is indeed what has been the object of these three volumes concerned with additive manufacturing and its environment.

Since the first patent of stereolithography in 1984, the world has profoundly changed and strong trends threaten the (partial) social agreement in which the French lived at the end of the "Trente Glorieuses". If this is the case for a society confronted with numerous interdependent difficulties, the world of innovation will likely not be able to continue to function as in the past for much longer, just with insidious transformations and no frank breakthroughs because they will be incremental. Moreover, it is now known that France can no longer cover every field in terms of scientific research, which imposes choices, certainly difficult ones, to maintain its excellence, pertinence and role as an active interface between science, technology and society and thus to satisfy the need for novelty necessary to society through science. 3D printing will not escape this general rule unless efforts are made.

Every research and innovation system only has a true meaning if it is thought of and challenged at regular intervals, so as to maintain its level of excellence relative to a certain technoscientific vision of the future, which constitutes the soul of additive manufacturing. What is observed is that more and more refined control mechanisms and procedures are implemented each and every day, and their operations continue to make things work as they should ("business as usual"). Yet, the world changes with strong trends that must be considered with alternative visions of the future that can/must escape tradition.

Strong trends

According to a recent report moderated by the author [AND 15] in the framework of a prospective study, the following elements considered to be strong trends should be taken into consideration.

– internationalization of production processes (and its dematerialization);

– the public's permanent demand for the new despite the exhaustion of reserves; health and well-being are favored;

– social demand necessitating more and more "zero" risk and NIMBY ("Not In My Back Yard"); individualization and disintermediation;

– marginalization of material production activities (and, in any case, with notable evolutions); shift from production to services;

– omnipresence of "digital" technology, communicating objects (with its consequences in terms of electric consumption);

– weakness of the public perception of the importance (and effects) of processes for transforming matter and energy;

– deterioration of the standard employment relationship, a new look at training and the empowerment of humanity;

– obligated decrease… (the word "decrease" is never quoted in international strategic documents), with the possibility of defining other growth criteria based on other foundations;

– globalization of scientific research and the attractiveness of relationships within the economy;

– ambivalent national relationships between industrial innovation and "academic" research, between conceptual training and practice;

– other strong trends: access to energy, water, food, global warming, demand for a better and healthy life, material well-being, exhaustion of reserves, ethics, powers delegated to companies and financing, shorter and shorter timeframes, weakness of policies, etc.

Research and socioeconomic relationships

In this report, the standpoints of representatives of the French industry allowed situational relationships between the socioeconomic actors and those in research to be analyzed and possible evolutions of the social system linked to heavy trends to be brought up (particularly energy, as the French have been exceeding the limit of rewewal that the planet can allow thanks to solar energy since 1976, as is happening in most developed countries).

Thus, three axes were brought up:

– the financial framework of companies implies more and more flexibility in production processes and the rapid use of incremental innovations, which is opposed to significant investment on their part in the long term (if only associated with the increasing higher cost of R&D): this constraint imposes a complex (but lucrative) approach like research of scientific "nuggets" (thus relevant to the notion of excellence or reference in a given domain);

– for some, the industrial world has abandoned too many of its own responsibilities in terms of research. The investments from the public sector are strengths with an assumption that the private sector will profit. In the perspective of productive development, this is a shortsighted policy because it risks hampering the possibility of radically innovative useful discoveries in the medium term. This conception risks trapping scientific knowledge in an instrumental subordinate role to the detriment of its intellectual and cultural values. Moreover, we can ask the question of knowing if the private sector gives the public sector the benefits that it reaps from the relationship it creates with academic laboratories;

– on a more social register, an "economy of contribution" is starting to emerge in which everyone becomes both a producer and a consumer – and in the end neither of the two. Reciprocal exchanges, gratuitous as they are of service, are generalized, sometimes by the bias of local currencies and in other places, of different forms of mutualization of consumption or investment (see maker movement). The third sector is assuming growing importance through cooperative systems or solidarity financing. Imperceptibly, it is actually a completely other social operation that is being put in place under the influence of the context and new communication technologies.

In these descriptions summarized above, there are also aspects of ambivalence between continuity and breakthroughs, between centralization and local action, between industrial production and a certain emerging revisit to craftsman forms. It is diversity that is being built which may need to be analyzed more closely in its dynamics, at the same time as that of the phenomena linked to opposition between democracies and certain forms of integrism.

Extreme scenarios

In the prospective process, it is indeed a matter of objectifying an idea that is in (partial) rupture with what is acquired to introduce new dimensions, and that must bring new, unexpected effects. It is on this basis that innovations emerge, innovations that, in our context, induce a cycle that cannot be easily broken. According to Van Der Leeuw [VAN 94], this leads "to a loss of control that Man had in his relationship with his environment. The more Man transforms his surroundings, the less he understands its consequences". In an attempt to reduce this effect, Wismann [WIS 04] recalls that "interrogation raises consequences of the action towards its premises, lending value to the need to be ensured, even on the normative plan, of the soundness of the affirmations that it comes from". One of the obstacles for the scientist in this approach is that he/she is often concerned with his/her goal to gain in-depth but rather monodisciplinary knowledge and through his/her lacking consideration of non-technical recommendations not subject to "measurable" aspects relative to expected gains [BIE 06]. It is thus a matter of breaking away from the principle that the future will not exactly be the image of the present (prevision of the identical proposed by Bronner [BOR 03], with breakthroughs, see for example NIC [NIC 17], Viossat [VIO 16], Du Granut [DU 16], Portnoff [POR 16], EU [EU 17]). Yet, De Gaulejac, Bonetti and Fraisse [DE 95] recall, "in the public domain, the dynamics of supply and demand are often inversed: it is supply that structures demand to the extent that suppliers seek to bring out a dynamic of demand… […] The problem is all the more complex as:

– "the absence of demand does not mean that there is no need;

– "the presence of a supply does not mean that it corresponds to a need;

– "the inadequacy between supply and demand is not sanctioned by the market".

In any case, it is a matter of defining paths of progress by considering what is possible and of excellence on pre-chosen matters limiting risk-taking by half. On the basis of these works, it is possible to be a significant element of dynamic science by aiding the state in decision-making. "The Man who thinks cannot be content with enlightening leaders; he must also convince them that the best ideas are not those that are given, but those that are brought about: appropriation is indispensable to move from anticipation to action. [...] It is necessary to know how to integrate reason and passion to succeed in the action" [GOD 01a]. It is then not impossible to approach the prospective through a possible form of manipulation!

Scenarios

Today, the absence of effectively driven technical choices is noticed with the difficulties for the politician to pose conditions of a sustainable dialog on the options (see failure of the debates on nanotechnologies, on the nuclear industry, on synthesis biology, etc., but social agreement for renewable energies) and the "bitter" polarization of some decision-makers who know how to make themselves understood. It is more and more the market's activity (encounter between the solvable needs of consumers and the supplying capacities of producers) to guide economic change and make society evolve. Politicians find themselves in an uncomfortable situation of adapting public policies to the socioeconomic reality, which is much more globalized, essentially targeting employment and GDP [ACC 17].

Faced with this notion, several options present themselves to us today: continuing to count on the technosciences to maintain (*a minima*) well-being, even to enhance human performances, continuing to consume, or thinking of human progress as a work on oneself (humanist, even spiritual option) with a change in the social context (downturn). Extreme, more or less democratic scenarios are presented below:

Scenario 1, "inertia"

In the end, it is a matter of being included in the continuity, in "Business As Usual". Problems with reserves, that of access to energy and water, etc. lead to the evolution of the economic system towards an overly "cosmetic" mobilization of citizens and companies: circular economy, recycling, renewable energies, but ongoing maintenance of the demand for an enhanced quality of life, all this while trying to do more by spending less, following the example of smart mobility, energy efficiency in buildings, etc. However, these strategies of incremental technological adaptations will only have limited effects in the future for human society in the long-term for different reasons: impossibility of managing both the increase in the human population and the development of technological progress, with aspects to be mastered: food and water resources, energy, integrisms, precarity and urban chaos, etc.

Scenario 2, "isolated society"

"We will witness an increase in tensions between nations and within countries, with the increase of interdependencies without symmetric progress of regulation instances and processes on a planetary scale" [DE 17]. The inability to deal with interfaces between populations, cultures, ways of life and standards of living in a democratic and peaceful way can be translated by development of the exclusion of those lacking links with the dominant power. Beyond losses of freedom for all citizens, demands for the development of sophisticated means of control, surveillance, protection and defense of citizens will be posed in connection to the dominant power from an economic and political standpoint. "Protectionism has spread across the globe and walls are being rasied on every state border. Global commerce is facing a net reduction and economic growth is low" [NIC 17]. The city is becoming the space in which most citizens are meant to live, the problem of food, energy and material good autonomy will need to be dealt with as such, with the difficulties linked to certain forms of segregation. The target would then move from the search for well-being as is taking place more or less today to other forms targeting the survival of the "fittest" with support for ethically arguable operations (enhanced man, for example, synthetic biology, RFID microchips, etc.) made inevitable by premises of civil conflicts.

Scenario 3, "sober society"

"Sustainable development is that which responds to the needs of the present without compromising the ability of future generations to satisfy theirs"… Perhaps it will be necessary to expect a generational change to achieve (possibly) "greener" consumerism, observing its limitations with citizen initiatives and anti-globalist movements. Will simplicity be considered a need or controlled choice with a social or integrist vision of energy, climate transition and local experiments"?

Challenging classical models of development attempts to reposition humanity at the center of every policy instead of a statistical entity defined by an economic GDP/inhabitant ratio. Development will henceforth be revendicated as a successful articulation of the global and local, whose success depends, among other conditions, on support with cultural dynamics and local economic and social organizations (with questioning of the risks to which the most promising technologies today like gene therapies or, in another field, nanotechnologies could expose us tomorrow).

This possible transition concerning the uses and real frugality could lead to another relationship with technological objects, but should be part of the rediscovery of proximity and coexistence with a reconquest of public spaces and quality of urban services. A certain relocalization of (more artisanal?) productions and new economic models like the food-producing city, the transition towards "the age of knowledge" and the economy of the link, etc. would then be expected.

Scenario 4, "'integrist ecologist' society"

This is an ecologist philosophy that is characterized by its defense of the intrinsic value of living beings, i.e. a value independent of their utility for humans. While classical ecology poses the satisfaction of human needs as an end (anthropocentrism) and attributes the status of "resources" to all other living things, in-depth ecology brings human ends back into a broader perspective, that of the living (bio-centrism) in order to consider the needs of the whole biosphere, notably species with which the human line has co-evolved for millions of years. Ecology does not escape partisan thinking (certain schools of ecological thought sink into ideology, that is, into forms of fascistic dogmatism at the level of the planet by basing themselves, particularly with arguments, on the finiteness of resources, the calamitous management of natural resources, and our lack of respect for the environment). In the name of "anti-productivism", there is a distrust of all forms of technical progress, contributing to a new point of view that, in order to preserve the planet, wants to return to possible medieval lifestyles.

Note: blockage factors; probable scenario(s)

In a society that has seen the benefits of technological adventure for more than a century with a headlong rush to provide new consumer goods, ever more numerous, and benefits of habits associated with technology, forms of independence and individual freedom, the changes to be proposed would be based on strong constraints, even radical breakthroughs. These are changes that would be accepted not only by French citizens, but more broadly by the planet, that the different powers in place may not be interested in supporting and even less in promoting (democratic aspects, choices of society, social stability, decentralization of decisions, etc.). Among the blockage factors, let us mention the following:

– the French citizen's desire to continue being a consumer with an attraction maintained by new technologies;

– the difficulty escaping a culture of multisecular technical progress against a transition led to grapple with possible abusive simplifications and failures; the publication of a robust and credible critique of technology is a matter of civic courage;

– the relationship between (centralized) employment and the maintenance of its purchasing power;

– pollution among others and NIMBY ("Not In My Back Yard");

– the uncertainties and controversies that make decisions, whatever they may be, difficult to make. This is a general problem of legibility added to a lack or complexity of available information;

– confusion on timeframes that expels *a priori* a whole set of possibilities, largely determining the margins of available moves;

– an insufficient perception of the benefits (?) linked to "post-coal" policies whether they are local or collective, in terms of jobs, economy, attractiveness, innovation, risk management or reducing the oil bill. These benefits are still too vaguely measured, when they are measured at all; whereas the potential costs and risks are made more explicit by existing actors (e.g. automobile jobs today);

– an overly great "closing in" of climate and energy questions in technical and sectoral debates, with the consequences of reduced citizen involvement, but particularly an overly strong cut-off between technical and social innovation, at the expense of the latter (whereas they are often questions essential to daily life);

– the lack of adaptation of local governance structures to question that involve strong coordination at the scale of population catchment areas, leadership problems, mutualization of means, articulation of sectoral actions (property, infrastructure, etc.), and intercommunal cohesion;

– the difficulty of defining transition paths that would not affect social groups that are already the most precarious and vulnerable. In other words, it would be a matter of moving towards simplicity not applied to those who are already the most constrained.

On these bases, even if this is a vision of a “sustainable” world that is drawn in the scenario of a “sober society”, it is not certain that it will be explored as long as dramatic situations inducing major tensions and obligations to act do not appear. In the first two scenarios (1 and 2), technology and the sciences that compete for its development will obviously maintain an eminent role. Furthermore, could it possibly be too late, even if we could change something in the culture of citizens and their way of consuming? However…

Questions asked for additive manufacturing

Managing research, operation on projects, overly centralized evaluation of scientific productions, weak management of interdisciplinary projects, the position of academic research vis-à-vis companies, the aptitude to improvable responsibility and ethical notion, etc. are questions asked in these three volumes, at the same time that the notion of innovation, a key element in the development of promising emerging technology was tackled.

“By encouraging the emergence of these new modes, digital technologies then no longer constitute only a new brick allowing us to move further in matters of task automation and production process optimization: they radically challenge the operation and organization of certain sectors and branches and allow the emergence of new ones, which raises regulation stakes” [DGE 16].

Reminders on creative innovation

If the true markets of industrial enterprises are globalized in all domains, including that of 3D printing, there is no longer real training needed in terms of employment because services do not fill this void. Then, how should we act so as to make different logics agree and work in national innovation, i.e. the entrepreneurial marketing logic, scientific and technological logic, the relationship with knowledge and the citizen logic of the democratic process, so as to achieve innovation where the different levels enrich one another? It may be a matter of immersive innovation, with the expertise to ask the right questions, very quickly produce prototypes and multiply tests (the precise contribution of additive manufacturing). Yet, these good practices (immersion, questioning, creativity, prototypes, tests) that contemporary innovators use regularly are pillars of “design thinking” and are not practiced often enough in France [OJA 16]. This method, which is settling down in Europe today, can provide the ideal catalyst for all contributors of projects involved in 3D printing determined to move the lines in his/her scientific activity open to society. Seven keys open the doors of this "responsible innovation" targeting the realization of original demonstrators, which are as follows:

1) *Serendipity*: the most promising advances are connected to marginal behaviors and the creative exploitation of the unexpected, behaviors linked to this research or technological invention. A certain disinterest for creative original talent is a risk for leaving companies and academic research with no solutions for a new competitive technology with social utilities like 3D, 4D, bio-, etc. printing technologies. It is still necessary to know how to find this catalyst. Conservatism today privileges a uniform approach and represents a waste of alterity and diversity. Creative innovation has trouble making its way (see [SNR 15]).

2) *Disorder, imagination*: innovation is born from disorder. The excessive incremental planning of R&D (research and development) passes by opportunities to create. The rigidity of financing research methods and R&D too often prevents reinvention, the proposal of visions, risk-taking, choosing original and astute personnel for research, the perception of possible games/stakes of getting involved (for a time) in interdisciplinary projects, etc. In our Western societies and particularly in France, we avoid taking risks as much as possible.

3) *Cartography*: the cartographic method allows a set of experiences to be brought together on a single plane of existence and new paths to be discovered among ideas that, for some, had not been connected until then. This is a precious tool for innovating in the complex society that we know manages a multitude of standpoints.

4) *Weak signals, prospective vision and scientific monitoring*: the prospective vision of weak signals calls our attention to partial information, fragmentary indices or early warning signs. It is a matter of reimagining the future outside marked paths; in other words, of developing a "vision", certainly a risky one (and that is always difficult to support).

5) *The deliberate management of uncertainty*: the unstable environment of the economy requires a system of innovation that is not only built on the known proven solutions or opportunists, but a more flexible and adaptive system, capable of facing uncertainty, of deciding in the presence of risks, and of turbulence (see scenario "isolated society").

6) *Cooperation and learning organizations*: "Living-Labs", FabLabs are innovation devices that are currently seeing clear success. They are devices both of open innovation (they are multi-partner) and ascending innovation. Thanks to these open devices, they can constitute a performing means to coproduce new solutions.

7) *The long-term*: It is often utilitarian or economic reasons that justify the need for research (see [SNR 15]). There cannot be research that some call "applied" without fundamental research, without going more in-depth. Innovation, linked to goal-oriented basic research, is thus a long-term affair and normally depends on the scenario or choices made. In some countries like France, it remains too much the victim of the ideology of the result that stops it from freely producing often-unsuspected effects (see [SNR 15] with a strong display of support for incremental innovation).

In today's competitive economy, it is important to undertake "risky" investments, allowing long-term reductions of timeframes between conception and production on fast-growing subjects like those mentioned in these three volumes. The new order of things is perhaps to integrate aspects linked to the limits of reserves and energy austerity. Whatever scenario takes place, there is no place to systematically oppose support for the search to satisfy disposable caprices, the fear of others and the protection of one's own life and goods and the major decrease in "simple society". It is a question of better considering the strong trends in 3D printing research processes, if only to anticipate an inexorable future. In the future, it would also be necessary to base ourselves on the ability of systems to evolve so as to maintain their operations in a world that will be more and more disturbed (complexity, resilience, degraded modes of operation).

Yet, in the research laboratory, whoever does not contribute directly to performance (defined by the only quantifiable production) through measurable time implication in the process of scientific innovation risks finding himself/herself rejected by the work-collective research unit. Moreover, the time characterizing the dynamics of scientific and technological changes will become the new norm, excluding from the race the performance of competitors that are slower to react, at least for a while. In this sense, a reductive global process targeting evaluation is contagious, affecting those who, internally, only imperfectly participate and are potentially considered individually inept, and/or externally, all those who did not consider the necessary anticipation early enough. Those defeated by the ordeals of quantitative performance must attribute their failures to themselves and nothing else. In other words, the more the conditions of performance reign, the more complex the participation in the research collective and thus the very standing of the research unit. This reduction is translated by forms of collective ineptitude affecting the research personnel involved more and more in individualism and disciplinary "silo" operation (because it is evaluated by peers according to known methods). These different reduction processes can destroy, if care is not taken, the notion of knowledge's incommensurability to achieve the implementation of cold quantification common to the research system, reaching bases of irrefutability because they are impartial.

These inertial characteristics are certainly one of the concerning factors in some research activities functioning too much within "disciplinary silos" and recognized themes. Arguments on potential risks, unanalyzed or unperceived by researchers involved in a new action, can naturally encourage the positions of rejecting scientific and technological breakthroughs that can only be risky. How to extract those who will take the initial risks allowing the elaboration of new associations that are difficult to imitate in the eyes of competitors from the work collective? How to support these non-conformist factors of scientific innovation to involve the unit in the new and excellent? How can a capital event take into consideration the originality, creativity and ability to found a school of thought (at least for a time)?

In a general way, innovators and the creatives only show themselves in situations where uncertainty is sufficient to exercise scientific and/or technical competence that is not "regulated" with them (this is possible in academic research). But to be the object of an active appropriation, these spaces of intellectual play imply being inhabited by actors with a level of scientific and/or technical competence that is greater and/or different from that for which they were "placed" in this position. Furthermore, beyond the original idea, will there be the necessary funding (and particularly men/women) to take the risk of a new approach that may, moreover, be interdisciplinary? Without this uncertainty the normal situation in research cannot be transformed into a creative space. These aspects that cannot flout more contractual methods of financing may limit creativity through ill-adapted frameworks of financial support and/or the absence of support for risk-taking. Yet, in the long term, it is indeed on this early risk-taking individual in its birth and then more collectively in a research team that the research unit's pertinence will be judged. This statement means that the organization of the laboratory must have reflexive and solidary partners who anticipate change, who are not only involved in the "values of protection" (satisfying passive precepts of quantitative evaluation, for example) and in hierarchical logics of an old regime and ordinary careers.

Furthermore, in order to explore the complexity inherent in new 3D processes, interdisciplinarity is a process in which an ability to analyze and synthesize is developed from perspectives from several disciplines. Its goal is to solve an issue in its whole (principle of integration) by identifying and integrating all of the relationships between the different elements involved. It attempts to synthesize and connect disciplinary knowledge and place it back into a broader systemic framework. The governance of such an operation has very significant power from a standpoint of optimizing staff and means with an end in sight on subjects like 4D printing and bio-printing. It would be best to reflect on the support of interdisciplinarity, with thinking globally, collective intelligence, encouraging convergences, and getting involved in exemplary creations/actions on the systemic approach with a view to create new processes (scientific and/or technological). Interdisciplinarity must be the creator of values. For this, it is necessary to be familiar with the whole value change and think of new structures. It is necessary to be able to break existing structures, especially those that are inhibiting. There are not enough experiments allowing interdisciplinarity to be confronted in order to develop openness, curiosity, welcoming, and finally, a habit of interdisciplinarity.

Implementing norms, references or indicators corresponds to representations of an organized system, relatively long-lasting, of material production or not, more or less standardized (with the classical risk of reductive simplification not allowing interdependencies and distributive effects of scientific activities to be taken into consideration: employment, balance of payments, standing and attractiveness, etc.). This situation may mean that in France, and more broadly in the developed West, there is a dominant form of "legitimate" thought that could have a tendency to impose itself as a revealed and uncontestable truth. There is then a risk of confining the interrogations and scrupules of on-site actors who could find themselves rejected because they are considered more or less heretical. On these modeling bases, necessarily reductive, the management criteria are always simpler than the system itself in its complexity, particularly if a man/woman is associated to it (and how, in scientific research, can we do otherwise?), and the unsteadiness linked to new scientific evolutions, to a competitive world that is more open than before. For additive manufacturing, the difficulty of managing changes in research structures leads to concerns over growing tension between an adaptation to constraints (in-depth pertinence, creativity, opposition) and coherence (activities of incremental in-depth research).

These comments thus imply being able to, with confidence, make flexible decisions of targets, organization, time mastery, etc. all while escaping the old "romantic" myth of the solitary creator. If it is the job of public and private research structures to know how to detect talent (what talent?), it is important for the activity to not be translated by a formatting factor of ensured production and quality, but to conform. "The true birthplace is that where one first took an intelligent look at oneself" [YOU 77]. Let us then try to reach this. It is indeed a matter of reflecting on how to organize divergent thought, openness, even a certain scientific nomadism to value skills and make society profit from them in its dynamics and technological legitimacy. Will we then know how to reach a certain coherence of scientific and/or technological innovation by:

– considering the socio-cultural universe that it covers?

– respecting the heterogeneity, even the disjunction, of the knowledge assembled, inaccessible to the disciplines alone?

– the heterogeneity of the concerned audiences?

– reinforcing the desire to create and undertake together?

The worksite is certainly important, but it rather concerns organizational and responsibility aspects (at every level). The comments presented above force us to find places where the social links weave together other than through the simple juxtaposition of opportunities or simple mechanical operations. If there must be a cultural breakthrough, it must be the object of an in-depth reflection. Indeed, it is necessary to rediscover Norbert Wiener's reasoning, "that of the engineer whose look does not stop at material facts, but spreads to human facts".

It is thus a question of desire.

Scenarios–additive manufacturing relationships

If we pay no heed to these considerations by posing the hypothesis that we will know how to escape this situation limiting creativity in the broad sense, and that this is necessary for the development of additive manufacturing processes relative to the subject, this section examines the foreseeable evolutions of additive manufacturing in light of the four scenarios presented. First of all, within the current paradigm, Table 1 gathers the author's vision as it was presented in these three volumes.

If additive manufacturing is often presented as likely to radically transform the way in which objects are conceived and created and the organization of the productive system, its large-scale spread still remains largely dependent on a significant number of technical advances whose speed is difficult to anticipate. "At this time, machines and processes do not allow us to respond to all of the constraints of industrial production, and progress should notably be made to accelerate the rhythm of production" [COE 17].

	"Classical" 3D printing	4D printing; programmable matter; Bio-Printing (BP)
Current market	5-10 billion €/year	Less than 1 billion €/year (cosmetics, toxicology)
Market possible in 10 years	20-40 billion €/year (growth rate > 20%/year)	In case of effective success of BP > 1000 billion €/year (risk-taking to be overcome)
Innovation	Process engineering Material sciences (and interactions with processes) Software Hygiene, safety, and environment Human and social sciences: organization, social perception, economy, etc.	See "Classical" 3D printing + Biology + Truly interdisciplinary actions
Creativity	New processes (i.e. elimination of layers, etc.)	Creativity and systemic approach training Exploration of complexity; epistemology; new "smart" materials; etc.
Comments on desired EU support	Modest because most processes stem from companies outside the EU, except on the exploration of application niches and creativity (taking back the industrial future in the EU)	Involvement in risky projects to rediscover a pioneering spirit linked to a considerable economic market Changes in ways of thinking and acting and better relationships between training and creative research Escaping herd mentality: fewer than 2000 scientific publications in the world with less than 1-2% coming from EU countries… Need for real European integration on an original theme with social interest

Table 1. *"Roadmap" proposed to the EU to avoid being a "shining second" in additive manufacturing.*

This synthesis [AND 17] attempts to show the need for the European Union to be able to take back over in fields that have escaped it notably in emerging areas [PRE 14]. It will be understood that this comes from the author's limited power as a simple citizen, but it seems essential to him to continue supporting a branch that he saw the birth of and for which he maintains a strong interest. In particular, to ensure an ability to innovate on the national and/or European level and to facilitate the adoption of new uses, governments must adopt mechanisms for responding to disruption. "The goal is to give visibility to entrepreneurs, to on-site actors, and to citizens by bringing a pro-innovation policy to every level of decision on public action" [CHA 17].

Independent of these conclusive comments linked to a continuity of the global socio-economic schema, Table 2 is concerned with the evolutions foreseen for additive manufacturing within each scenario that could take place in the near future.

Field	Scenario 1	Scenario 2	Scenario 3	Scenario 4
3D printing				
Innovation (new processes, creation time, variable-size voxels, creation without layers, multi-materials, etc.)				?
Materials				
Recycling				
Energy savings				
Application niches				
Mechanics (metal, ceramics, multi-materials)				
Repair				
Space		?	?	?
Military and confined systems				
Robotics				?
Nano- and micro-applications				?
Electronics				?
3D medical				
Breakthroughs in additive manufacturing				
4D printing and programmable matter				
Bio-printing				

Table 2. *Developing fields in additive manufacturing according to the scenario (from light: insignificant to dark: significant)*

Obviously, scenario 1, to pursue a consumer society, supports the pursuit of developing the emergence of additive manufacturing (which is only the beginning in terms of revenue, whether we speak of "classical" 3D manufacturing or foreseen breakthroughs). This situation is certainly less clear for other scenarios, but it remains quite positive, even for the least favorable scenario. Indeed, by bringing localized production means, smart methods of repair, methods of maintaining health with fewer exogenous substances, etc., there is room for a whole set of processes that can continue to help humans in their everyday lives. However, the framework of technologies may be different even if the methodological bases remain more or less the same.

The field of 3D printing can, whatever the scenario may be, fall into the same traps of promising hazardous solutions being instrumentalized to justify all sorts of political decisions, or on the contrary to establish major results independent of the public powers. Nothing too surprising. The same fate is reserved for the results of other scientific disciplines that oscillate between promises and letdowns in a strongly growing universe.

In 1959, on the occasion of the annual gathering of the American Society of Physics, Richard Feynman wrote the following famous words: "There's plenty of room at the bottom" (i.e. towards atomic manipulations). We could surely paraphrase this as "There's plenty of room in 3D printing".

> *"The future does not bring us anything, does not give us anything; it is we who, in order to build it, must give it everything, give it our life itself. But to give, it is necessary to own, and we own no other life, no other lifeblood, than the treasures inherited from the past and directed, assimilated, recreated by us. Of all the needs of the human soul, there is none more vital than the past." [WEI 90]*

> *"Another major factor is that, for years, organizational management has been developing methods for increasing productivity and minimizing risks and errors that tend to stifle creative experimentation. The predominant approach to management that evolved during the industrial era, known as scientific management, broke jobs down into specific, sequential tasks, which could be allocated appropriate times for completion in order to optimize efficiency." [SIM 11]*

"Technology can allow survival through the sophistication of models or systems that are not optimal. In other words, it could prevent latent epistemological revolutions in the future. I will call this vice of the technician vision of the world, 'Ptolemy's computer syndrome'." [MAL 11]

Bibliography

[ACC 17] ACCENTURE, "Accenture technology vision 2017 forecasts a future of technology for people, by people", available at: https://newsroom.accenture.com/news/accenture-technology-vision-2017-forecasts-a-future-of-technology-for-people-by-people.htm, 2017.

[AND 15] ANDRÉ J.C., DOMBRE E., AMRA C. *et al.*, Quelles ruptures pour INSIS-CNRS? Des propositions de la cellule prospective de l'Institut, Internal report, p. 35, Paris, 2015.

[AND 17] ANDRÉ J.C., Some comments after the meeting held in Brussels, workshop-tackling the skills gap in additive manufacturing, Brussels, Belgium, 13 March 2017.

[BAR 00] BARBIERI-MASINI, *Penser le futur*, Dunod, Paris, 2000.

[BIE 06] BIEDER C., *Les facteurs humains dans la gestion des risques*, Hermès-Lavoisier, Paris, 2006.

[BLU 99] BLUMENBERG H., *La légitimité des temps modernes*, Gallimard, Paris, 1999.

[BRO 03] BRONNER G., *L'empire des croyances*, PUF, Paris, 2003.

[CHA 17] CHARRIÉ J., JANIN L., "Répondre à l'innovation disruptive", available at: http://www.strategie.gouv.fr/sites/strategie.gouv.fr/files/atoms/files/2017-2027-actions-critiques-innovation-disruptive-ok.pdf, 2017.

[COE 17] COE (CONSEIL D'ORIENTATION POUR L'EMPLOI), "Automatisation, numérisation et emploi – Tome 1 – Les impacts sur le volume, la structure et la localisation de l'emploi", available at: http://www.coe.gouv.fr/IMG/pdf/COE_170110_Rapport_Automatisation_numerisation_et_emploi_Tome_1.pdf, 2017.

[DE 95] DE GAULEJAC V., BONETTI M., FRAISSE J., *L'ingénierie sociale*, Syros, Paris, 1995.

[DE 05] DE RIEDMATTEN E., *XXIe siècle: les innovations qui vont changer notre vie*, L'Archipel, Paris, 2005.

[DE 17] DE JOUVENEL H., "Penser l'impensable", available at: https://www.lesechos.fr/idees-debats/sciences-prospective/0211875118553-penser-limpensable-2072073.php, 2017.

[DGE 16] DGE (DIRECTION GÉNÉRALE DES ENTREPRISES), "Technologies Clés 2020, Préparer l'industrie du futur", available at: http://www.entreprises.gouv.fr/politique-et-enjeux/technologies-cles-2020, 2016.

[DU 16] DU GRANUT C., "La croissance économique en voie de disparition", *Futuribles*, vol. 415, pp. 39–50, 2016.

[ELG 66] ELGOZY G., *Le paradoxe des technocrates*, Denoël, Paris, 1966.

[EU 17] EU, "Livre blanc sur l'avenir de l'Europe: cinq scénarios", available at: https://ec.europa.eu/commission/white-paper-future-europe-five-scenarios_fr, 2017.

[FEY 60] FEYNMAN R., "There's plenty of room at the bottom", *Caltech Engineering and Science*, vol. 23, pp. 22–36, 1960.

[GID 72] GIDE A., *Les faux monnayeurs*, Poche, Paris, 1972.

[GOD 01a] GODET M., "De la rigueur pour une indiscipline intellectuelle", in LESOURNE J., STOFAÈS S. (eds), *Prospective stratégique d'entreprise*, Dunod, Paris, 2001.

[GOD 01b] GODET M. cited by COHENDET P., "Décision, prospective et auto-organisation", *Futuribles*, vol. 263, pp. 29–35, 2001.

[KUN 86] KUNDERA M., *L'art du roman*, Gallimard, Paris, 1986.

[LE 97] LE BRETON D., "Jeux symboliques avec la mort", *Religiologiques*, vol. 16, pp. 55–65, 1997.

[LEW 16] LEWIS C., *Too Fast to Think: How to Reclaim Your Creativity in a Hyper-connected Work Culture*, Kogan Page, Paris, 2016.

[MAL 11] MALRIEU J.P., *La science gouvernée*, Rue des gestes, Toulouse, 2011.

[MIC 09] MICHEAU Y., *Les habits neufs du mérite*, Bourin, Paris, 2009.

[NIC 17] NIC (National Intelligence Council), "Global trends – paradox of progress", available at: https://www.dni.gov/files/images/globalTrends/documents/GT-Full-Report.pdf, 2017.

[OJA 16] OJASALO J., KAUPPINEN H., "Collaborative innovation with external actors: an empirical study on open innovation platforms in smart cities", *Technology Innovation Management Review*, vol. 6, no. 12, pp. 49–60, 2016.

[PAU 72] PAUWELS L., BERGIER J., *Le matin des magiciens*, Folio, Paris, 1972.

[POR 16] PORTNOFF A.Y., "Sept défis pour les entreprises et le travail", *Futuribles*, vol. 415, pp. 51–76, 2016.

[PRE 14] PRETZ K., "Ten technologies that could change the world by 2022", available at: http://theinstitute.ieee.org/ieee-roundup/members/achievements/10-technologies-that-could-change-the-world-by-2022, 2014.

[PRI 90] PRIGOGINE I., *Loi, histoire et désertion; la querelle du déterminisme*, Collection le débat, Gallimard, Paris, 1990.

[RIC 04] RICŒUR P., *Parcours de la reconnaissance*, Stock, Paris, 2004.

[SIM 11] SIMS P., *How Breakthrough Ideas Emerge from Small Discoveries*, Business Book, London, 2011.

[SNR 15] SNR, "Stratégie Nationale de Recherche", available at: http://www.gouvernement.fr/sites/default/files/contenu/piece-jointe/2015/11/rapport_strategie_nationale_de_recherche_snr.pdf, 2015.

[VAN 94] VAN DER LEEUW J., *Pour penser la technique*, vol. 21, La Découverte, Paris, 1994.

[VIO 16] VIOSSAT L.C., "La santé à l'horizon 2030-2050", *Futuribles*, vol. 415, pp. 93–123, 2016.

[WEI 90] WEIL S., *L'enracinement*, Folio-Essais, Paris, 1990.

[WIS 04] WISMAN H., "Le savant et la Cité – Pour une éthique de l'expertise scientifique", *Le Banquet*, vol. 19, pp. 1–5, 2004.

[YOU 77] YOURCENAR M., *Les mémoires d'Hadrien*, Folio, Paris, 1977.

Figure. *Poietis' bio-printing machine (2017)*

Index

H, I, L, M

O, P, R

S, T

Other titles from

ISTE

in

Systems and Industrial Engineering – Robotics

2017

ANDRÉ Jean-Claude
From Additive Manufacturing to 3D/4D Printing 1: From Concepts to Achievements
From Additive Manufacturing to 3D/4D Printing 2: Current Techniques, Improvements and their Limitations

ARCHIMÈDE Bernard, VALLESPIR Bruno
Enterprise Interoperability: INTEROP-PGSO Vision

CAMMAN Christelle, FIORE Claude, LIVOLSI Laurent, QUERRO Pascal
Supply Chain Management and Business Performance: The VASC Model

FEYEL Philippe
Robust Control, Optimization with Metaheuristics

MARÉ Jean-Charles
Aerospace Actuators 2: Signal-by-Wire and Power-by-Wire

POPESCU Dumitru, AMIRA Gharbi, STEFANOIU Dan, BORNE Pierre
Process Control Design for Industrial Applications

RÉVEILLAC Jean-Michel
Modeling and Simulation of Logistics Flows 1: Theory and Fundamentals
Modeling and Simulation of Logistics Flows 2: Dashboards, Traffic Planning and Management
Modeling and Simulation of Logistics Flows 3: Discrete and Continuous Flows in 2D/3D

2016

ANDRÉ Michel, SAMARAS Zissis
Energy and Environment
(Research for Innovative Transports Set - Volume 1)

AUBRY Jean-François, BRINZEI Nicolae, MAZOUNI Mohammed-Habib
Systems Dependability Assessment: Benefits of Petri Net Models (Systems Dependability Assessment Set - Volume 1)

BLANQUART Corinne, CLAUSEN Uwe, JACOB Bernard
Towards Innovative Freight and Logistics (Research for Innovative Transports Set - Volume 2)

COHEN Simon, YANNIS George
Traffic Management (Research for Innovative Transports Set - Volume 3)

MARÉ Jean-Charles
Aerospace Actuators 1: Needs, Reliability and Hydraulic Power Solutions

REZG Nidhal, HAJEJ Zied, BOSCHIAN-CAMPANER Valerio
Production and Maintenance Optimization Problems: Logistic Constraints and Leasing Warranty Services

TORRENTI Jean-Michel, LA TORRE Francesca
Materials and Infrastructures 1 (Research for Innovative Transports Set - Volume 5A)
Materials and Infrastructures 2 (Research for Innovative Transports Set - Volume 5B)

WEBER Philippe, SIMON Christophe
Benefits of Bayesian Network Models
(Systems Dependability Assessment Set – Volume 2)

YANNIS George, COHEN Simon
Traffic Safety (Research for Innovative Transports Set - Volume 4)

2015

AUBRY Jean-François, BRINZEI Nicolae
Systems Dependability Assessment: Modeling with Graphs and Finite State Automata

BOULANGER Jean-Louis
CENELEC 50128 and IEC 62279 Standards

BRIFFAUT Jean-Pierre
E-Enabled Operations Management

MISSIKOFF Michele, CANDUCCI Massimo, MAIDEN Neil
Enterprise Innovation

2014

CHETTO Maryline
Real-time Systems Scheduling
Volume 1 – Fundamentals
Volume 2 – Focuses

DAVIM J. Paulo
Machinability of Advanced Materials

ESTAMPE Dominique
Supply Chain Performance and Evaluation Models

FAVRE Bernard
Introduction to Sustainable Transports

GAUTHIER Michaël, ANDREFF Nicolas, DOMBRE Etienne
Intracorporeal Robotics: From Milliscale to Nanoscale

MICOUIN Patrice
Model Based Systems Engineering: Fundamentals and Methods

MILLOT Patrick
Designing Human–Machine Cooperation Systems

NI Zhenjiang, PACORET Céline, BENOSMAN Ryad, RÉGNIER Stéphane
Haptic Feedback Teleoperation of Optical Tweezers

OUSTALOUP Alain
Diversity and Non-integer Differentiation for System Dynamics

REZG Nidhal, DELLAGI Sofien, KHATAD Abdelhakim
Joint Optimization of Maintenance and Production Policies

STEFANOIU Dan, BORNE Pierre, POPESCU Dumitru, FILIP Florin Gh., EL KAMEL Abdelkader
Optimization in Engineering Sciences: Metaheuristics, Stochastic Methods and Decision Support

2013

ALAZARD Daniel
Reverse Engineering in Control Design

ARIOUI Hichem, NEHAOUA Lamri
Driving Simulation

CHADLI Mohammed, COPPIER Hervé
Command-control for Real-time Systems

DAAFOUZ Jamal, TARBOURIECH Sophie, SIGALOTTI Mario
Hybrid Systems with Constraints

FEYEL Philippe
Loop-shaping Robust Control

FLAUS Jean-Marie
Risk Analysis: Socio-technical and Industrial Systems

FRIBOURG Laurent, SOULAT Romain
Control of Switching Systems by Invariance Analysis: Application to Power Electronics

GROSSARD Mathieu, REGNIER Stéphane, CHAILLET Nicolas
Flexible Robotics: Applications to Multiscale Manipulations

GRUNN Emmanuel, PHAM Anh Tuan
Modeling of Complex Systems: Application to Aeronautical Dynamics

HABIB Maki K., DAVIM J. Paulo
Interdisciplinary Mechatronics: Engineering Science and Research Development

HAMMADI Slim, KSOURI Mekki
Multimodal Transport Systems

JARBOUI Bassem, SIARRY Patrick, TEGHEM Jacques
Metaheuristics for Production Scheduling

KIRILLOV Oleg N., PELINOVSKY Dmitry E.
Nonlinear Physical Systems

LE Vu Tuan Hieu, STOICA Cristina, ALAMO Teodoro, CAMACHO Eduardo F., DUMUR Didier
Zonotopes: From Guaranteed State-estimation to Control

MACHADO Carolina, DAVIM J. Paulo
Management and Engineering Innovation

MORANA Joëlle
Sustainable Supply Chain Management

SANDOU Guillaume
Metaheuristic Optimization for the Design of Automatic Control Laws

STOICAN Florin, OLARU Sorin
Set-theoretic Fault Detection in Multisensor Systems

2012

AÏT-KADI Daoud, CHOUINARD Marc, MARCOTTE Suzanne, RIOPEL Diane
Sustainable Reverse Logistics Network: Engineering and Management

BORNE Pierre, POPESCU Dumitru, FILIP Florin G., STEFANOIU Dan
Optimization in Engineering Sciences: Exact Methods

CHADLI Mohammed, BORNE Pierre
Multiple Models Approach in Automation: Takagi-Sugeno Fuzzy Systems

DAVIM J. Paulo
Lasers in Manufacturing

DECLERCK Philippe
Discrete Event Systems in Dioid Algebra and Conventional Algebra

DOUMIATI Moustapha, CHARARA Ali, VICTORINO Alessandro, LECHNER Daniel
Vehicle Dynamics Estimation using Kalman Filtering: Experimental Validation

GUERRERO José A, LOZANO Rogelio
Flight Formation Control

HAMMADI Slim, KSOURI Mekki
Advanced Mobility and Transport Engineering

MAILLARD Pierre
Competitive Quality Strategies

MATTA Nada, VANDENBOOMGAERDE Yves, ARLAT Jean
Supervision and Safety of Complex Systems

POLER Raul *et al.*
Intelligent Non-hierarchical Manufacturing Networks

TROCCAZ Jocelyne
Medical Robotics

YALAOUI Alice, CHEHADE Hicham, YALAOUI Farouk, AMODEO Lionel
Optimization of Logistics

ZELM Martin *et al.*
Enterprise Interoperability –I-EASA12 Proceedings

2011

CANTOT Pascal, LUZEAUX Dominique
Simulation and Modeling of Systems of Systems

DAVIM J. Paulo
Mechatronics

DAVIM J. Paulo
Wood Machining

GROUS Ammar
Applied Metrology for Manufacturing Engineering

KOLSKI Christophe
Human–Computer Interactions in Transport

LUZEAUX Dominique, RUAULT Jean-René, WIPPLER Jean-Luc
Complex Systems and Systems of Systems Engineering

ZELM Martin, *et al.*
Enterprise Interoperability: IWEI2011 Proceedings

2010

BOTTA-GENOULAZ Valérie, CAMPAGNE Jean-Pierre, LLERENA Daniel, PELLEGRIN Claude
Supply Chain Performance / Collaboration, Alignement and Coordination

BOURLÈS Henri, GODFREY K.C. Kwan
Linear Systems

BOURRIERES Jean-Paul
Proceedings of CEISIE'09

CHAILLET Nicolas, REGNIER Stéphane
Microrobotics for Micromanipulation

DAVIM J. Paulo
Sustainable Manufacturing

GIORDANO Max, MATHIEU Luc, VILLENEUVE François
Product Life-Cycle Management / Geometric Variations

LOZANO Rogelio
Unmanned Aerial Vehicles / Embedded Control

LUZEAUX Dominique, RUAULT Jean-René
Systems of Systems

VILLENEUVE François, MATHIEU Luc
Geometric Tolerancing of Products

2009

DIAZ Michel
Petri Nets / Fundamental Models, Verification and Applications

OZEL Tugrul, DAVIM J. Paulo
Intelligent Machining

PITRAT Jacques
Artificial Beings

2008

ARTIGUES Christian, DEMASSEY Sophie, NERON Emmanuel
Resources–Constrained Project Scheduling

BILLAUT Jean-Charles, MOUKRIM Aziz, SANLAVILLE Eric
Flexibility and Robustness in Scheduling

DOCHAIN Denis
Bioprocess Control

LOPEZ Pierre, ROUBELLAT François
Production Scheduling

THIERRY Caroline, THOMAS André, BEL Gérard
Supply Chain Simulation and Management

2007

DE LARMINAT Philippe
Analysis and Control of Linear Systems

DOMBRE Etienne, KHALIL Wisama
Robot Manipulators

LAMNABHI Françoise *et al.*
Taming Heterogeneity and Complexity of Embedded Control

LIMNIOS Nikolaos
Fault Trees

2006

FRENCH COLLEGE OF METROLOGY
Metrology in Industry

NAJIM Kaddour
Control of Continuous Linear Systems

Printed and bound by CPI Group (UK) Ltd, Croydon, CR0 4YY

15/09/2026

14962346-0001